Newnes
Electronic Testing and Inspection
Pocket Book

Other books by Keith Brindley

Electronics Assembly Handbook
Electronics Assembly Pocket Book
Electronics Engineers Pocket Book
Soldering in Electronics Assembly (co-written with Mike Judd)
Modern Electronic Test Equipment
Sensors and Transducers
QuarkXpress for Windows
QuarkXPress for the Macintosh

Newnes

Electronic Testing and Inspection Pocket Book

Edited by Keith Brindley

NEWNES

Newnes
An imprint of Butterworth-Heinemann Ltd
Linacre House, Jordan Hill, Oxford OX2 8DP

A member of the Reed Elsevier group

OXFORD LONDON BOSTON
MUNICH NEW DELHI SINGAPORE SYDNEY
TOKYO TORONTO WELLINGTON

First published 1994

British Library Cataloguing in Publication Data
Brindley, Keith
Newnes electronic testing and inspection pocket book
I. Title II. Series
621

ISBN 0 7506 1753 5

Produced by Co-publications, Loughborough
Graphics by Sylvester Publications, Loughborough
Printed in Great Britain by Clays Ltd, St Ives plc

Contents

Preface vii

Inspection 1
Standards and specifications 3
Basic tools 7
Meters 11
Practical use of meters 72
Signal generators 91
Cathode-ray oscilloscopes 106
Spectrum analysers 147
Time domain reflectometers 150
Logic analysers 154
Inspection procedures 157
Inspection techniques 163
Final inspection 172
Component identification 175
Inspection of printed circuit boards 187
Inspection of component mounting 190
Inspection of soldered joints 198
Inspection of wrapped joints 207
Inspection of crimped joints 214
Inspection of wires and cableforms 218
Inspection of connectors 226
Inspection of equipment assembly 233
Test instructions and drawings 243
Connecting equipment 250
Conventions used on circuit diagrams 257
Special test jigs 263
Measurands in test and inspection 271
Glossary 282

Index 286

Preface

This book is another example of collaboration between the Engineering Training Authority (ETA) — formerly known as the Engineering Industry Training Board (EITB) — Butterworth-Heinemann the publishers, Sylvester Publications the typesetting and layout, and myself as editor.

The book is essentially a new edition of previous work published under the ETA auspices. Going back over the years, the ETA's output contains a wealth of information for the engineering industry. In total, therefore, this book represents considerable work over many years; researching, compiling, writing, editing and publishing work relevant to all aspects of electronic instrumentation and testing. Indeed, in this form the book is merely the latest variation of earlier presentations, repackaged for a different market. Despite some of the information available being a little long in the tooth, however, this edition ensures it is all still perfectly relevant in today's high-tech electronics laboratory.

Newnes pocketbooks are intended as concise guides to designated subject areas. This book — Newnes Electronic Testing and Inspection Pocketbook — is by far the most concise such work covering electronics instrumentation and testing procedures and practices we feel has ever been published in book form. My aim in editing it is to present as practical a book as possible, within the limits of the technicalities of the subject. Holding just a little short of 600 line drawings it is also highly illustrated.

Newnes Electronics Testing and Inspection Pocketbook was produced and typeset electronically in 8 point Times, using Apple Macintosh computers. Original text was input using Word 5, while pages were laid out using QuarkXPress version 3. Camera-ready copy was presented as film from imagesetter.

Inspection

Each product has a specification which describes its qualities, characteristics and capabilities. The purpose of inspection is to ensure that the product is made accurately to the design and at the required level of quality to meet its specification. Deviations from the required standard of work are called defects.

The product is inspected at a series of predetermined inspection points. At each point all defects are located and corrected.

Quality control

Quality Control works to improve the level of quality by:

- improving the standard of work produced
- improving inspection techniques
- analysing information on defects.

Inspection aids Quality Control by:

- providing information on defects
- reporting recurrent defects
- using the new techniques provided.

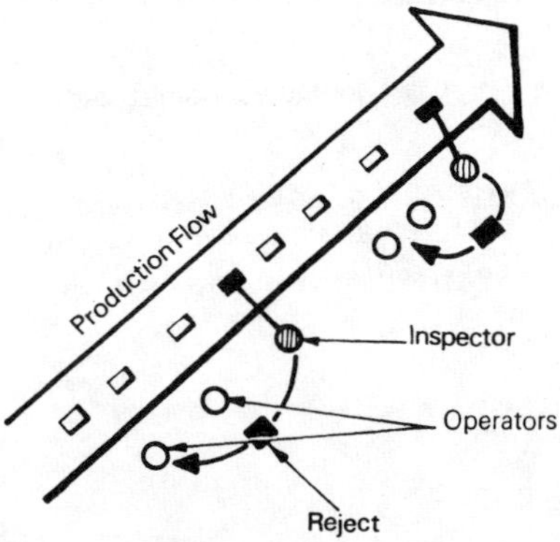

Inspection schedules

Inspection schedules detail the features to be inspected at particular inspection points.

Drawings, production sequences and completed assemblies are used to prepare inspection schedules.

They usually follow the sequence of assembly. They detail the basic aspects of the inspection of each type of component, or sub-assembly, and then note the special aspects of the inspection of the assembly in question.

Always use the inspection schedules provided.

EQUIPMENT ABC

Inspection sched. No 23

Component Board 12345

Use this ins. sched in conjunction with DWG 14762

1/ CHECK ITEM 1 (BOARD ASSY) FOR DAMAGE

2/ ENSURE TERMINALS SECURELY SOLDERED ON REAR FACE

3/ CHECK ITEMS 7–13 (9 RESISTORS) ENSURE THEY ARE CORRECT TYPE & VALUE & ARE MOUNTED CENTRALLY ABOUT THEIR RESPECTIVE FIXING HOLES

4/ CHECK ITEMS 17–24 (7 CAPACITORS) ENSURE THEY ARE CORRECT TYPE & VALUE

5/ CHECK RESISTORS AROUND LINE ABC FOR DAMAGE

6/ CHECK FOR DAMAGE ON PRINTED CIRCUIT BOARD DUE TO EXCESS HEAT

7/ ENSURE THAT CABLEFORM ON RIGHT HAND UPPER FRAME OF XYX UNIT IS LAID FLAT SO AS NOT TO FOUL THE UNIT ABOVE

Standards and specifications

Standards and specifications are documents which give detailed descriptions of qualities and requirements of a product or process, or parts of a product or process.

This product or process may be:

- a component used in an assembly
- the assembly itself, or a part of it
- any process used in making a product.

A description of the specific process recommended to achieve the required characteristics of the product may be included in the standard or specification.

Standards and specifications can be at one of four levels:

- international level — followed by manufacturers throughout the world
- regional level — followed by manufacturers within a region, say, Europe
- national level — followed by manufacturers within a country
- company level — followed by an individual manufacturer.

International standards

Internationally, general standards are organised by the International Standards Organisation (ISO), while technical standards are organised by the International Electrotechnical Commission (IEC).

Regional standards

The European regional standards organisation for general standards is the Comité Européen Normalisation (CEN). The electrotechnical counterpart of CEN is CENELEC. Together CEN and CENELEC form the Joint European Standards Institution.

National standards

In the UK, the national standards organisation is the British Standards Insitution (BSI). Generally, each country has its own standards body, perhaps coordinating the work of other smaller related institutions and bodies.

Company standards

Company standards are generally specifications of products and working standards are generally specifications of products and working practices defined by a particular manufacturer to give detailed descriptions of working procedures and manufacturing processes. Generally, company standards should mirror national, regional and international standards in what they specify, although particular company regulations will be perhaps emphasised.

Other standards-making bodies

Often in the past, and still occasionally, standards and specifications are produced by other bodies:

- industrial associations — such as the Electrical Engineering Association
- governmental departments — often specifying particular conditions to be met for government contracts.

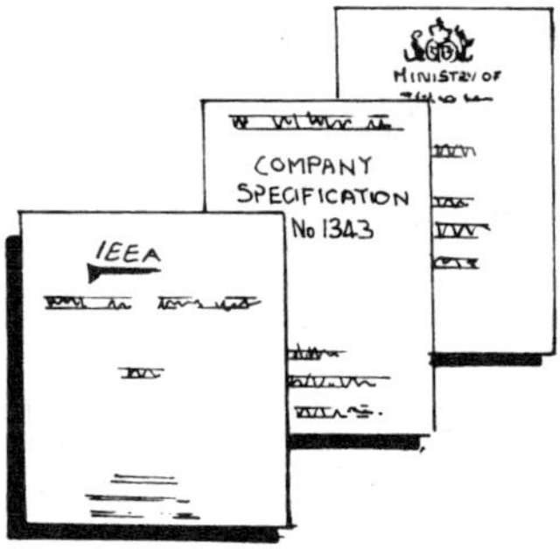

However, there is a worldwide trend away from such standards and specifications, instead making national, regional, and international standards bodies responsible for all standards.

Aims of standards and specifications

Standards and specifications exist simply to ensure products are built, and processes are maintained, to sufficiently high qualities. By defining the standard any product or process should meet, and monitoring the products and processes to ensure they do meet the standard, products and processes are known to be of sufficient quality for the purpose they are intended.

British Standards

The British Standards Institute has produced specifications for a wide range of products and processes. A list of available standards is given annually in the British Standards Catalogue.

British Standards are used widely thoughout industry.

They describe standard ranges of products and so provide a product identification which is easily understood by:

- the designer — who is saved the task of drawing up specifications for the materials needed
- the purchaser — who can order to a standard code
- the supplier — who knows the standard of product to be supplied.

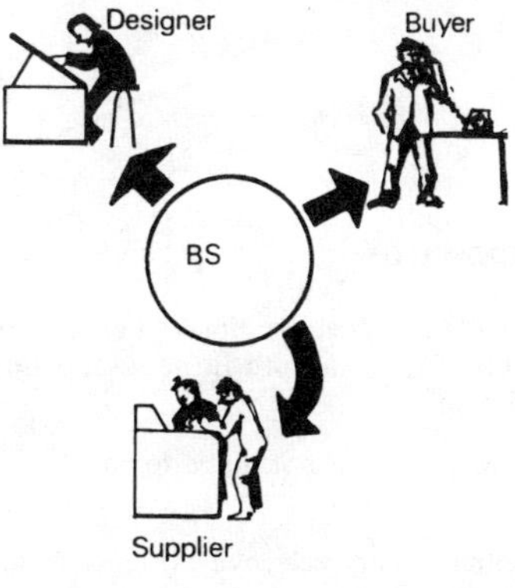

Standards in the electronics field

There is a wide range of British Standards covering electronic assembly components, many covering processes and a growing number covering specific electronic component types.

Commercial equipment

Commercial equipment is usually built to British Standards where these are available, and to individual company standards otherwise.

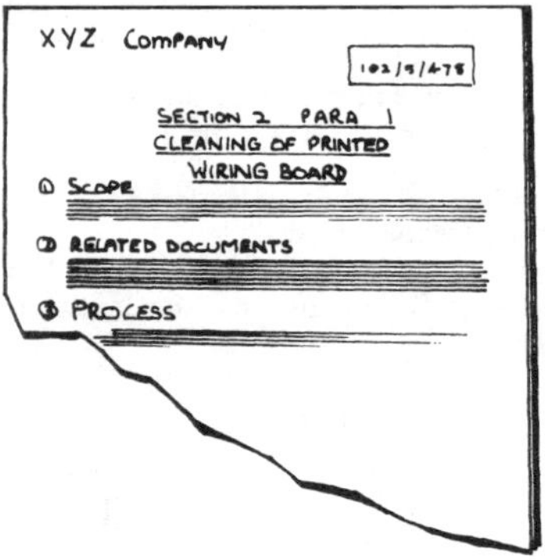

Use of standards

Read standards and specifications carefully. Pick out those sections relevant to the job in hand and make sure you understand them.

Obtain and read any other standard referred to on the point in question.

If contradictions occur, seek advice from your supervisor.

Ensure you have the correct issue number of each standard you use. Check this with your company's standards department (often the drawing department). Remember the issue number quoted may be out of date.

Many Government standards and specifications are marked with date of issue rather than having an issue number.

Basic Tools

The lens-light

A lens-light is usually available to all inspectors. It provides:

- excellent illumination of the work
- magnification of about x 3.

Eye-piece lens

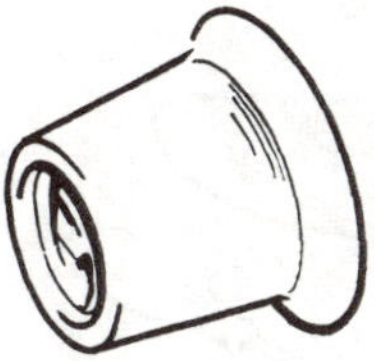

The eye-piece lens also gives magnification of about x 3 but its focal point is nearer, allowing closer scrutiny of the work.

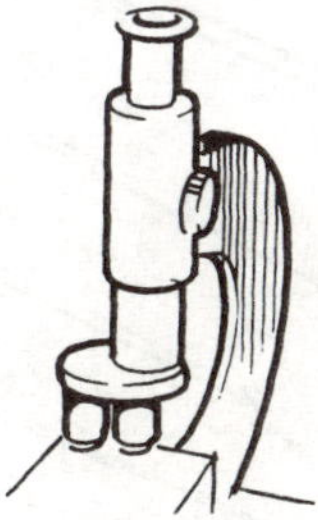

Microscope

A microscope is usually available for the scrutiny of miniaturised work, e.g. the reflow soldering of *flat-packs* to printed circuit boards.

Lamp probes and mirrors

Kits of probes, fitted with adjustable mirrors and small lamps, are usually available on the shop floor. They are used to inspect in dark or blind areas of the equipment.

Dental mirrors may be used if such a kit is not available.

Inspection stages are usually planned so as to avoid the need to use these instruments.

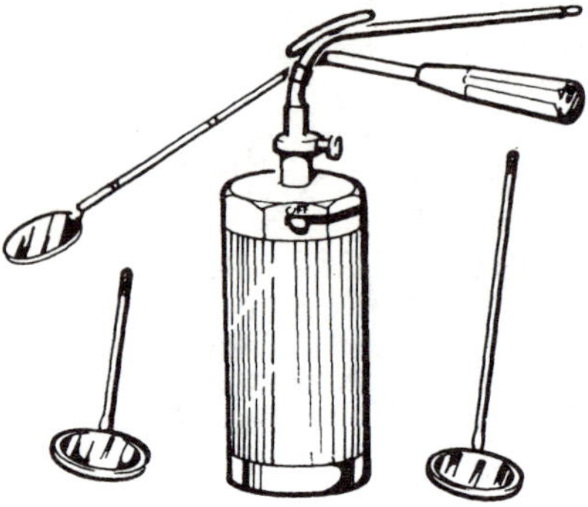

Probes

Probes are used to arrange wiring and cableforms so as to gain access to work underneath the wiring.

Avoid prodding at joints and components as damage may occur.

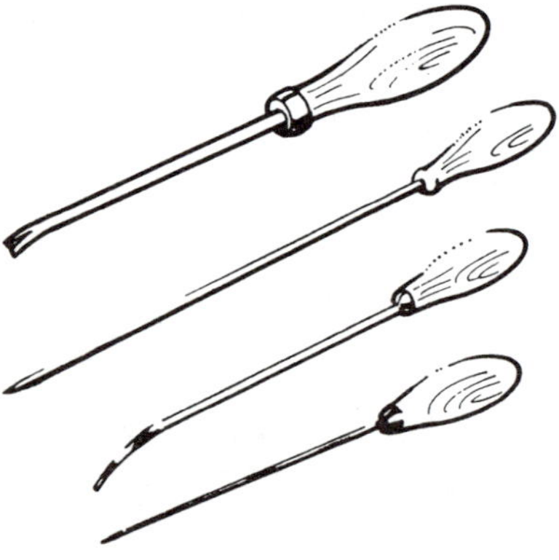

Mechanical strength test instruments

- spring press gauges are used to test the pressure which can be borne by, for instance, connector pins

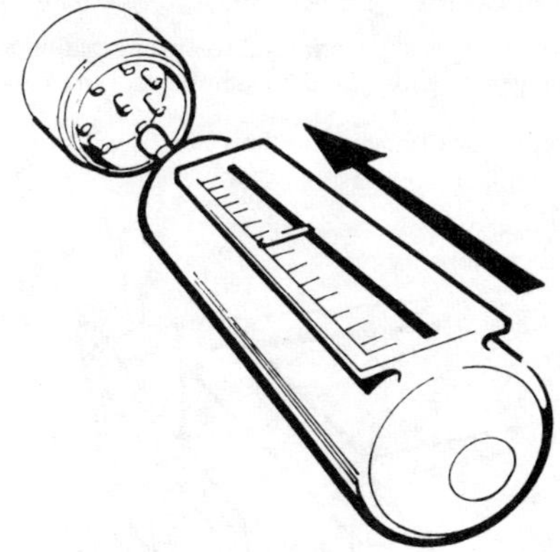

● tension test gauges are used to test the strength of wrapped and crimped joints, and of the retaining clips in crimped connectors.

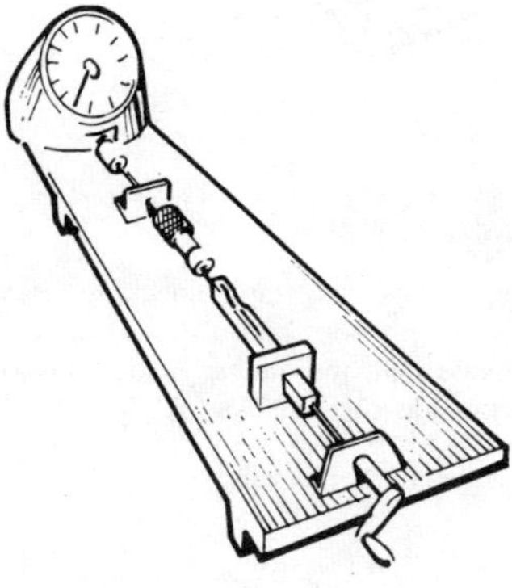

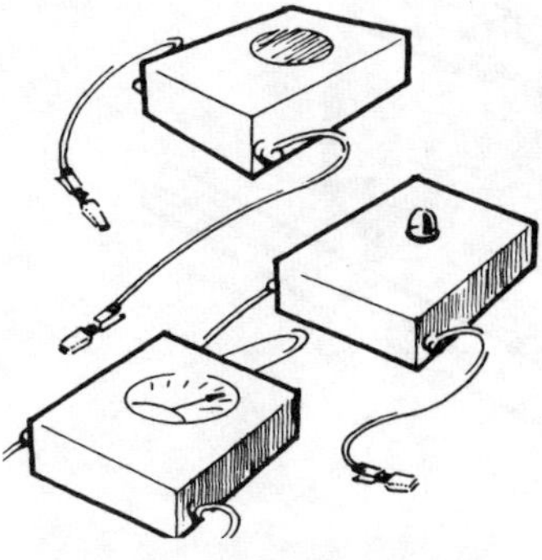

Electrical test instruments
Buzzers, continuity testers and megohmmeters are used to carry out continuity and insulation tests.

Interchangeability gauges

Interchangeability gauges are used to ensure that an assembly is interchangeable with all other assemblies of the same type.

If the assembly will mate with the gauge it will mate with all equipment of that type.

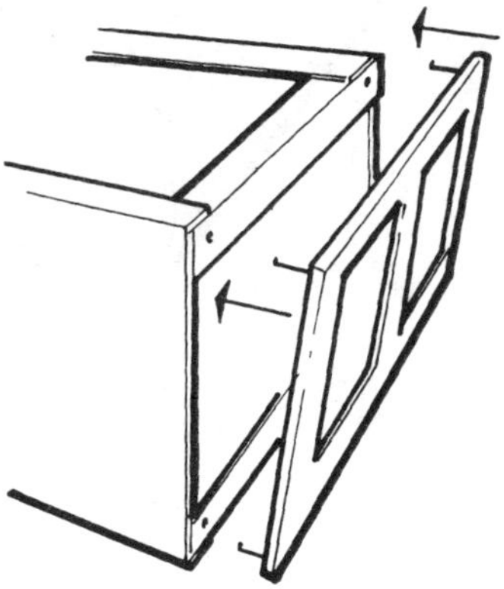

Go no-go gauges

Go no-go gauges are used for rapid checking of tolerances. If the dimensions to be checked are within tolerance the assembly will mate with the gauge. If they do not mate, the dimensions are outside tolerance.

Check gauges at specified intervals.

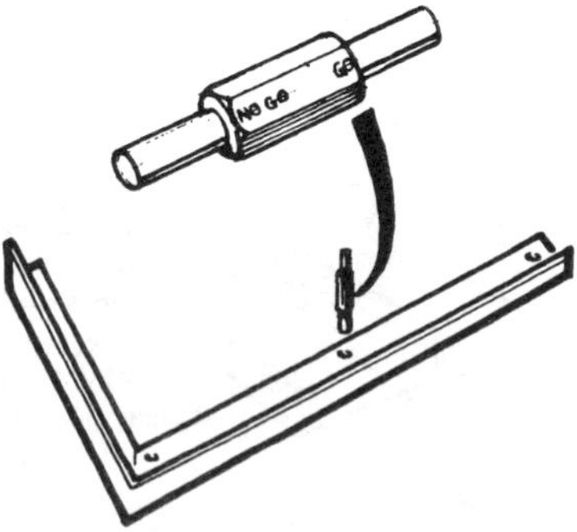

Meters

Care and handling of meters

Handle meters carefully. They are delicate and expensive.

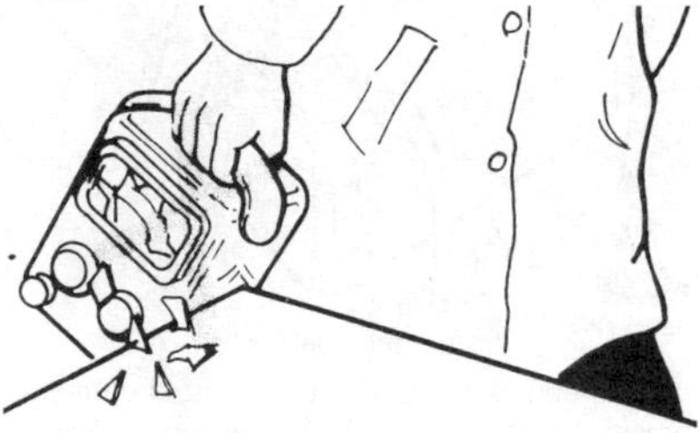

Place meters in a safe position where they cannot fall.

Avoid trailing test leads over the edge of the bench.

Report any damage. Damage may cause inaccurate measurements.

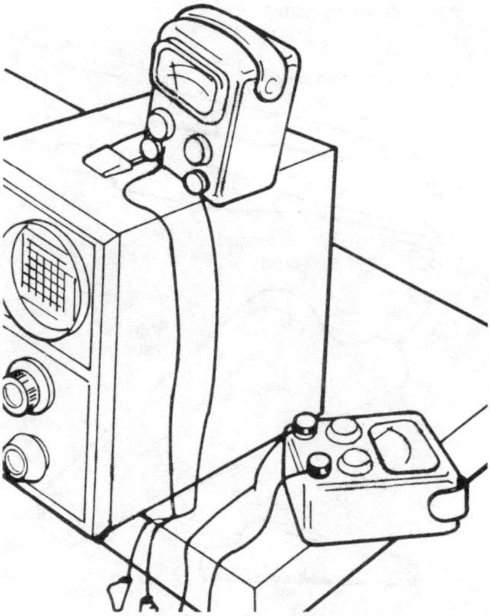

Carry the meter by the handle or strap provided.
Avoid trailing the test leads.
Avoid applying pressure to the glass. Do not tap the glass.

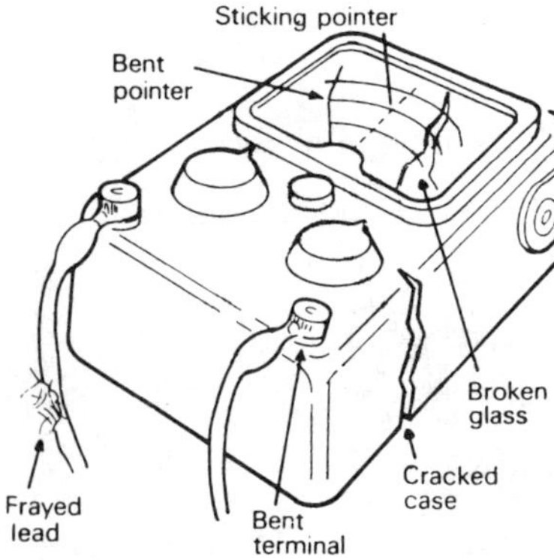

Use the recommended test leads and terminal connectors. Various types of test prods and clips are available — use the recommended type.

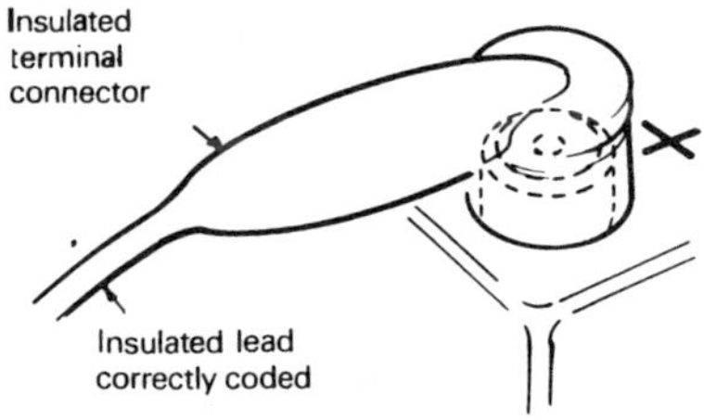

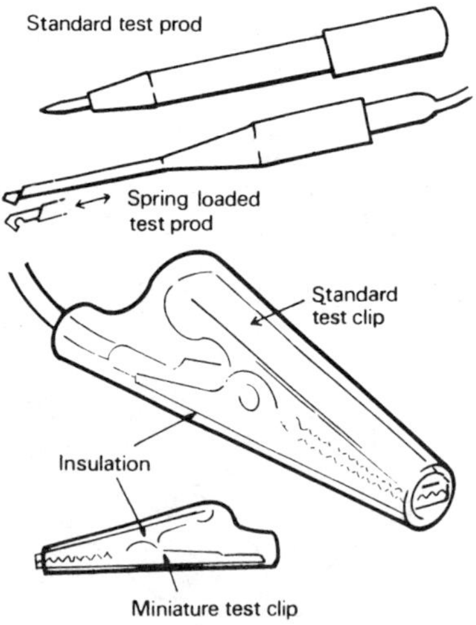

Examine the insulation of test leads, clips and prods frequently. Check that they are fitted correctly.

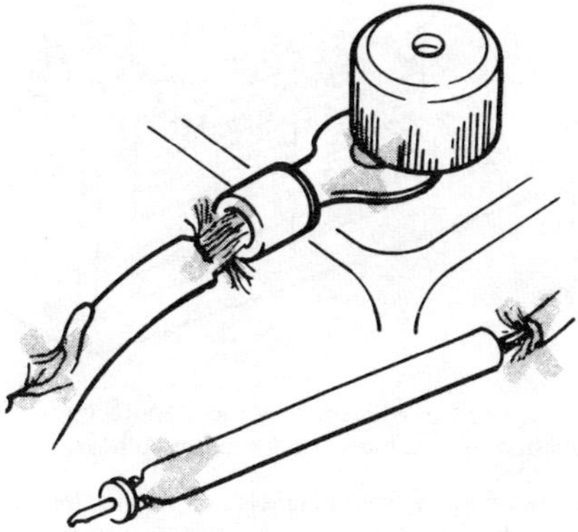

Make sure that you know how to use the meter correctly.

Read the manufacturer's instruction book.

Calibration should be checked frequently.

The date of calibration is sometimes indicated on a label attached to the instrument.

If in doubt → ask your supervisor.

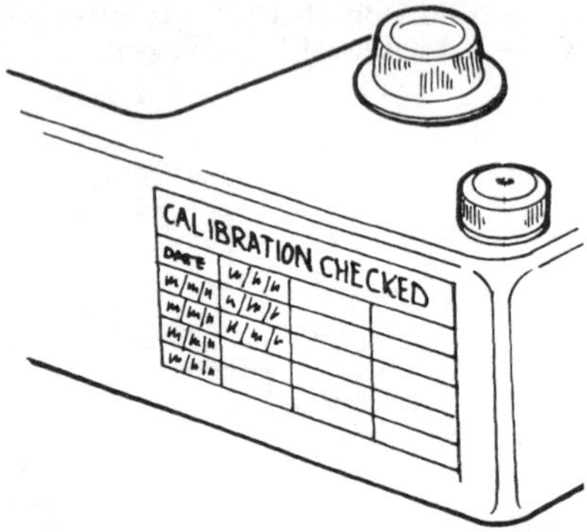

Position the meter as specified by the manufacturer. This will be the position in which it has been calibrated.

Keep the meter away from magnets and ferrous materials, e.g. do not place the meter on the metal case of another equipment.

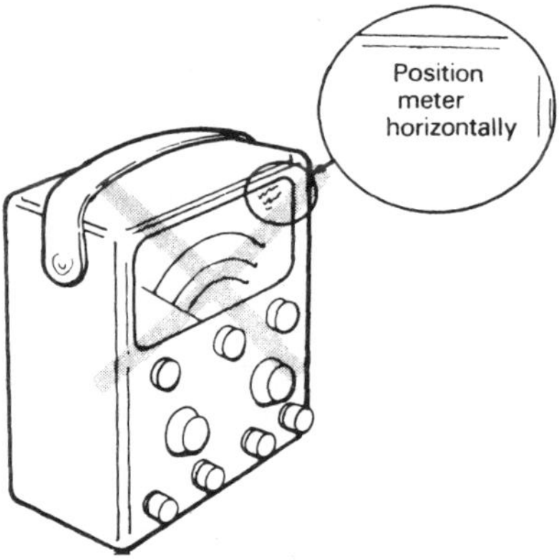

Before using the meter, check that the pointer reads zero.

If necessary adjust it to zero by means of the pointer adjusting screw.

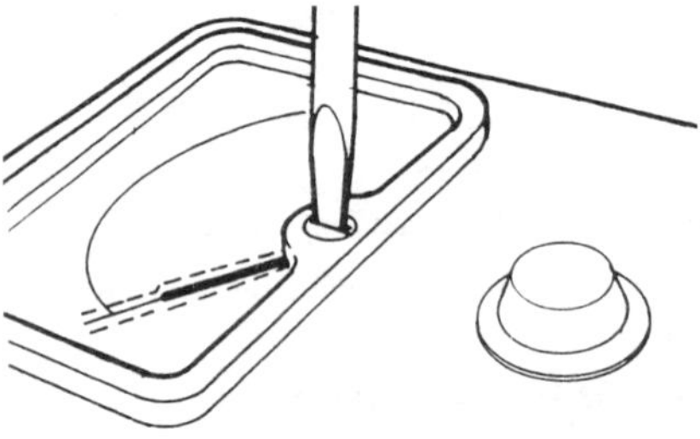

Ohmmeters and multimeters contain batteries. The responsibility for changing them varies → **ask your supervisor**.

Do not use steel-cased leakproof batteries.

Observe correct polarity when fitting batteries.

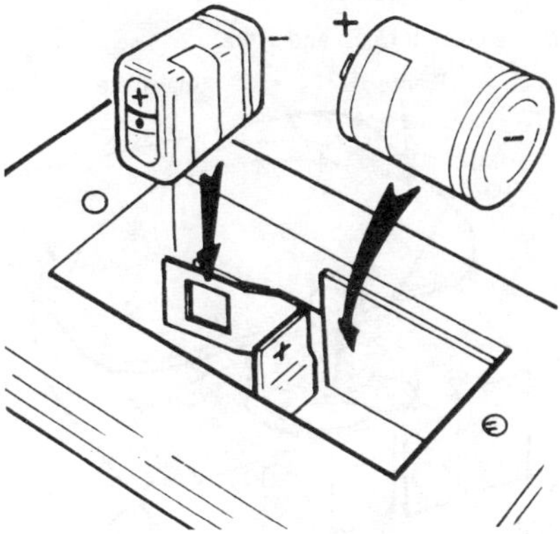

Before connecting a multimeter to a circuit check that:

- the function switch is set correctly — to a.c. in the example illustrated
- the meter is switched to its highest range for the initial reading.

Switch to a lower range only when you are sure that the meter will not be overloaded.

Never switch the meter to a current or resistance range when the meter is connected to measure voltage.

After use, switch the meter to its highest a.c. range.

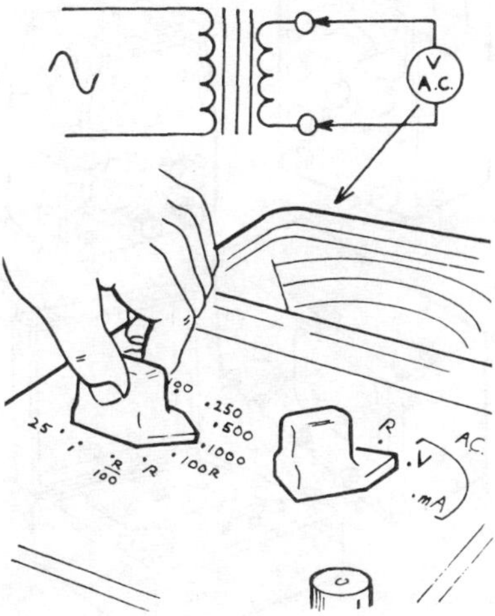

Moving-coil meters

A moving-coil meter is used to measure direct current (d.c.). The meter can be adapted to measure:

- d.c. voltage
- alternating current (a.c.) and voltage.

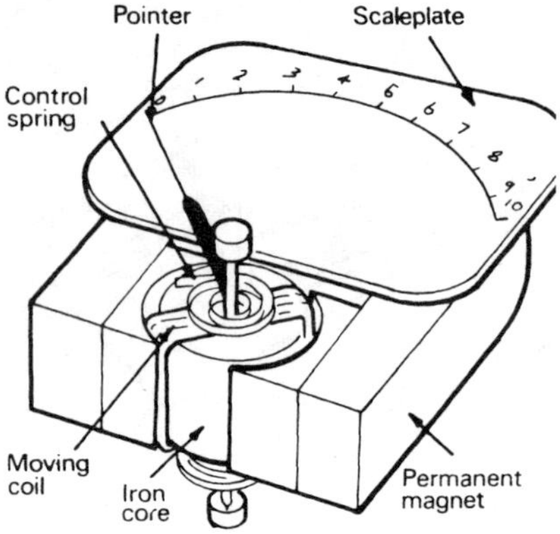

Many measuring instruments incorporate a moving-coil meter.

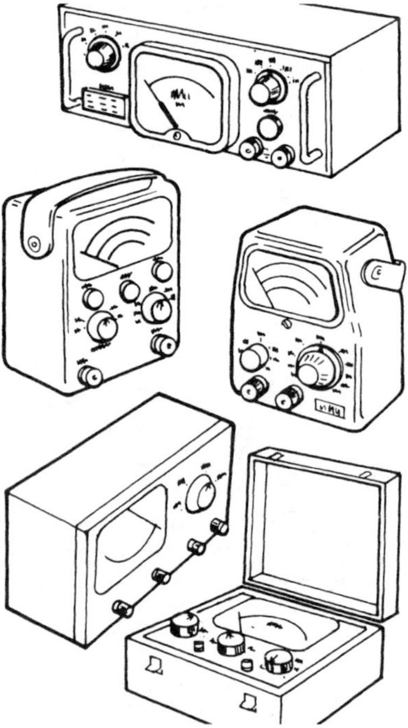

Meter sensitivity

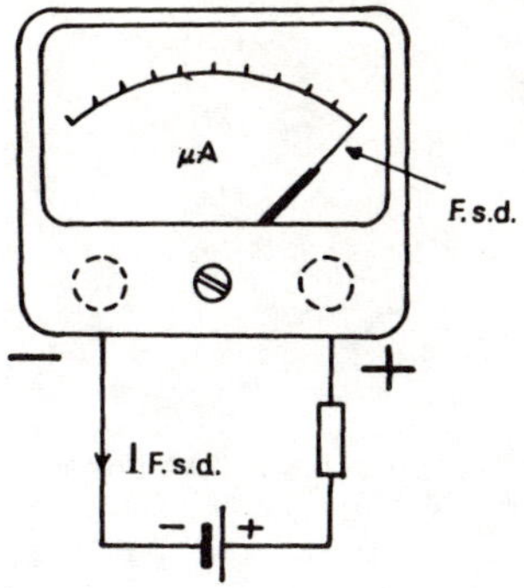

This is defined as that value of current which causes full-scale deflection (f.s.d.). It is commonly denoted as:

$$I_{f.s.d.}$$

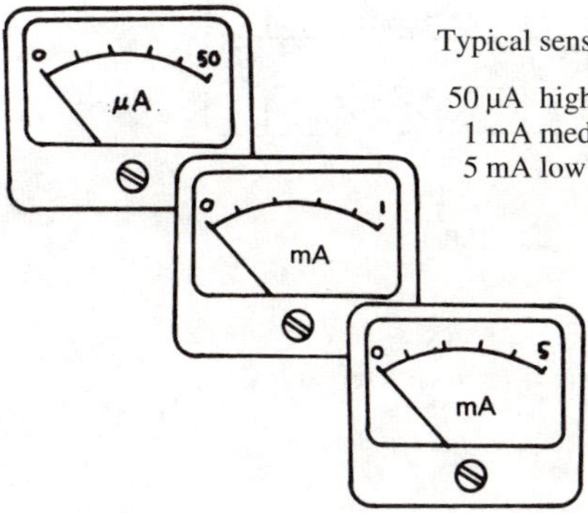

Typical sensitivities are:

50 μA high sensitivity
1 mA medium sensitivity
5 mA low sensitivity.

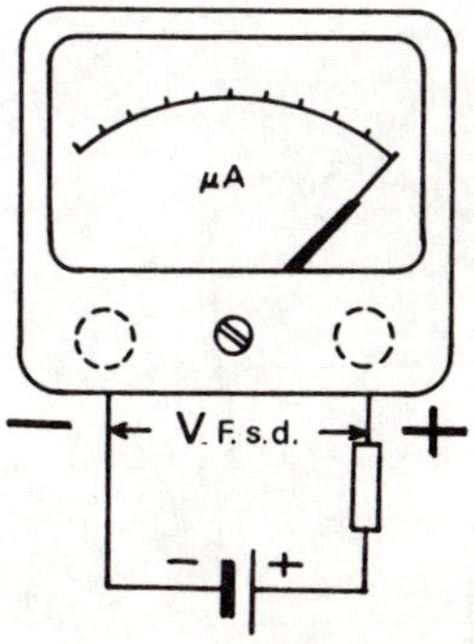

Movement sensitivity is sometimes defined as the voltage across the meter which causes full-scale deflection. It is commonly denoted as:

$$V_{f.s.d.}$$

Meter resistance

This is the resistance of the moving coil. Meter resistance is commonly denoted as: R_m

Typical meter resistances are:

2,000 Ω	high sensitivity
100 Ω	medium sensitivity
10 Ω	low sensitivity.

Note that from Ohm's law:

$$\frac{V_{f.s.d.}}{I_{f.s.d.} \times R_m} = 1$$

Therefore if any two of the above are known, the third can be calculated.

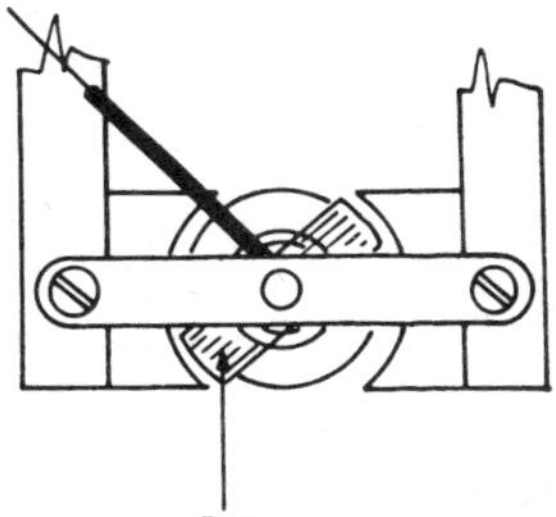

Example

If $I_{f.s.d.} = 50\mu A = 50 \times 10^{-6}$ A

and $R_m = 2{,}000\,\Omega$

then $\dfrac{V_{f.s.d.}}{50 \times 10^{-6} \times 2{,}000} = 1$

Therefore

$$V_{f.s.d.} = 50 \times 10^{-6} \times 2{,}000$$
$$= 0.1\text{V}$$
$$= 100\,\text{mV}$$

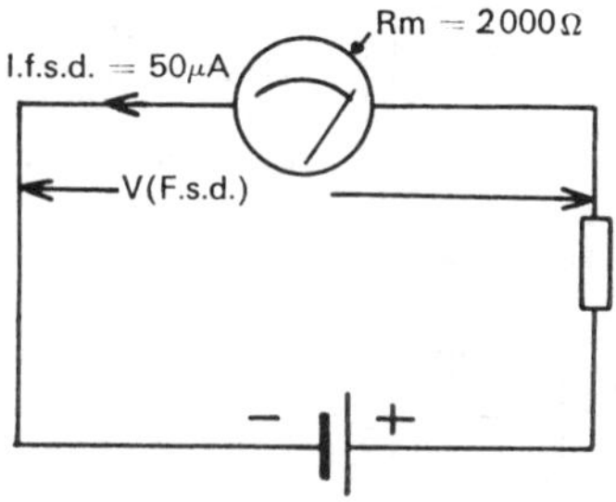

Measuring meter sensitivity and resistance

It is sometimes necessary to measure the sensitivity ($I_{f.s.d.}$ or

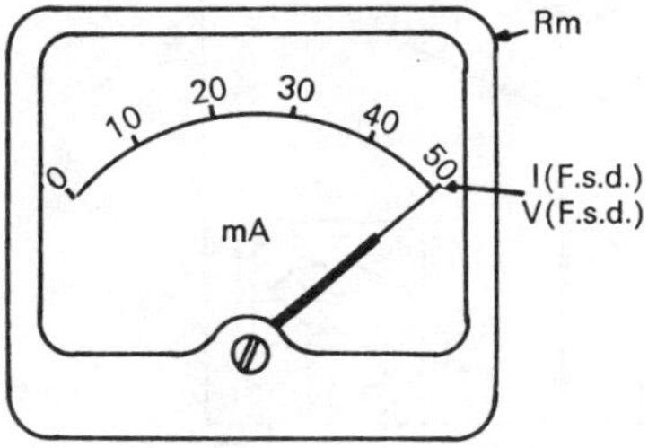

$V_{f.s.d.}$) and resistance (R_m) of a milliammeter or microammeter.

To measure $V_{f.s.d.}$

Use the equipment and test circuit shown overleaf. M1 is the meter under test.

- before connecting up the equipment set the decade resistor to maximum resistance and ensure that meter pointers are zeroed
- connect up the equipment
- set the standard voltmeter to a suitable voltage range (d.c.)
- adjust the decade resistor to obtain f.s.d. on M1
- note the reading of the standard voltmeter. This reading is the $V_{f.s.d.}$ of meter M1 to the accuracy of the standard voltmeter.

To measure $I_{f.s.d.}$

Use the equipment and test circuit shown overleaf, M1 is the meter under test.

- before connecting up the equipment set the decade resistor to maximum resistance and ensure that meter pointers are zeroed
- connect up the equipment
- set the standard milliammeter to a suitable current range (d.c.)
- adjust the decade resistor to obtain f.s.d. on M1
- note the reading of the standard milliammeter. This reading is the $I_{f.s.d.}$ of meter M1 to the accuracy of the standard milliammeter.

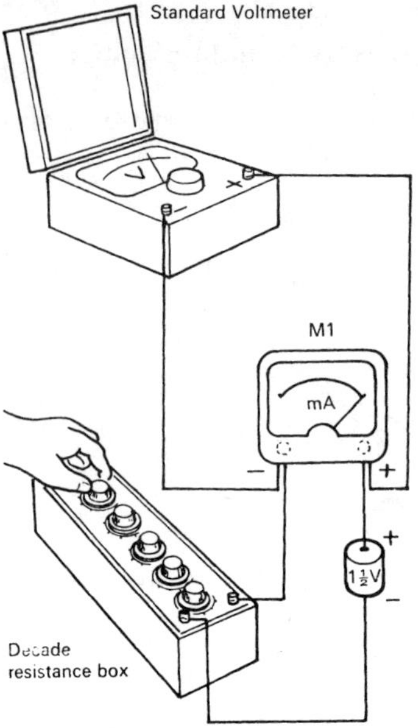
Standard Voltmeter
V
+
−
M1
mA
−
+
+
1½V
−
Decade
resistance box

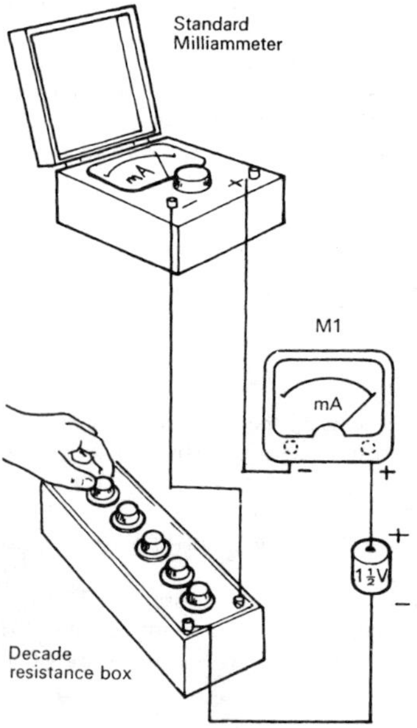
Standard
Milliammeter
mA
−
+
M1
mA
−
+
+
1½V
−
Decade
resistance box

D.C. ammeters

Use a d.c. ammeter to measure direct current (d.c.) in amperes (A).

Where the current to be measured is less than 1 A:

Use a milliammeter or microammeter.

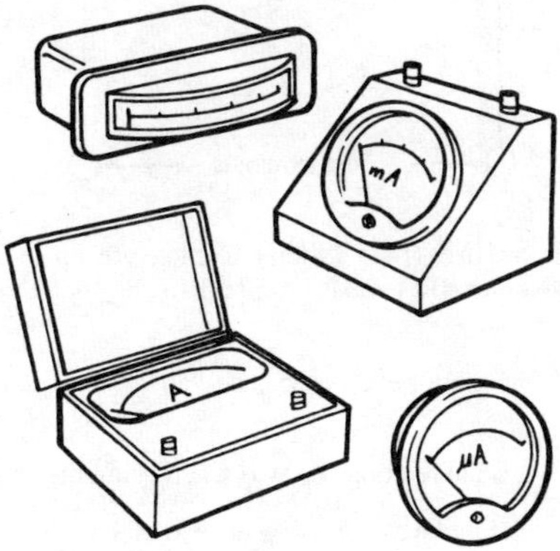

The simplest milliammeter or microammeter consists of a moving coil movement only.

The meter is connected in series with the circuit load.

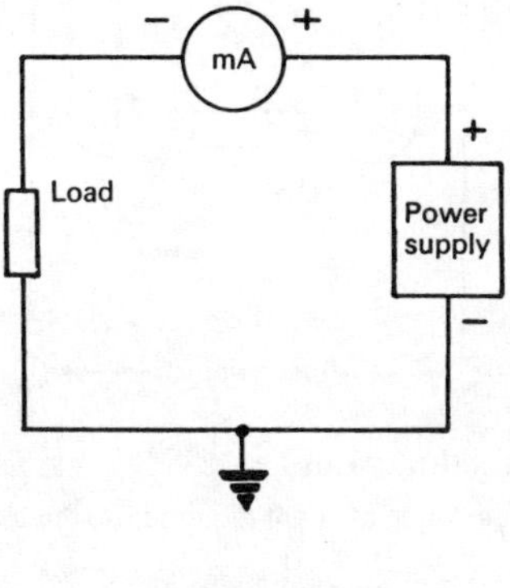

Extending the measuring range

Use a shunt resistor to extend the measuring range of the basic meter.

Connect the shunt in parallel with the meter.

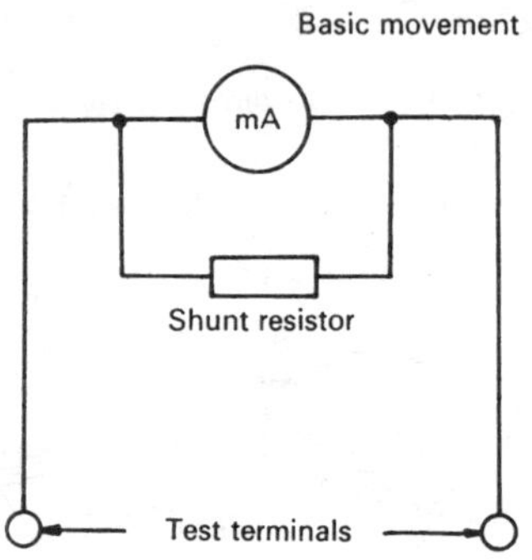

To enable a 1 mA (f.s.d.) meter to measure up to 2 mA the shunt resistance (R_s) must be equal to the meter resistance (R_m).

The meter and shunt each carry one half of the circuit current (2 mA).

Multiply the scale readings by two when using the shunt.

Full scale reading becomes 2 mA instead of 1 mA.

Half scale reading becomes 1 mA instead of 0.5 mA.

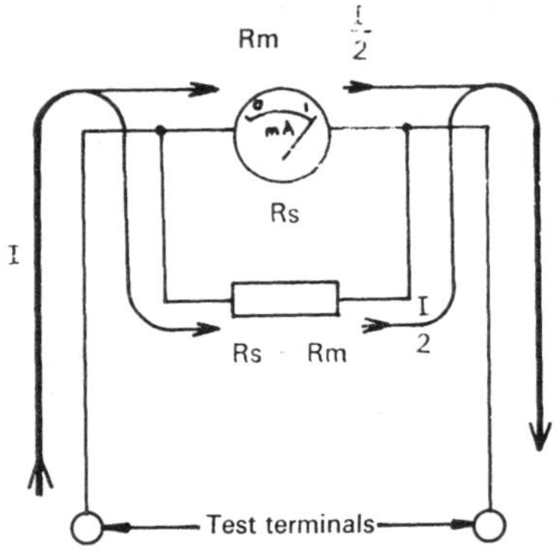

Calculating shunt values

To calculate the value of shunt required use the formula:

$$R_S = \frac{I_{f.s.d.}}{I_S} \times R_m, \text{ where}$$

R_s is the resistance of the shunt.
I_m is the f.s.d. of the basic meter.
I_s is the current required to flow through the shunt.
R_m is the resistance of the meter.

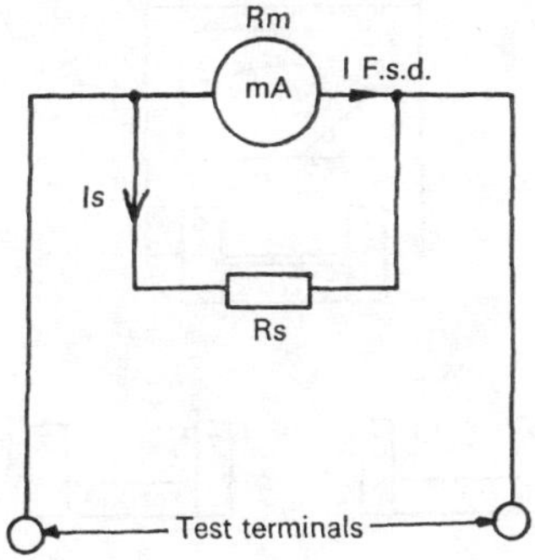

Example

A meter with $R_m = 1{,}000\,\Omega$ and $I_{f.s.d.} = 1\,mA$ is required to measure up to 10 mA.

$$R_s = \frac{I_{f.s.d.}}{I_s} \times R_m$$

$$\text{where } I_s = 10\,mA - 1\,mA = 9\,mA,$$

$$\text{therefore } R_s = \frac{1 \times 10^{-3}}{9 \times 10^{-3}} \times 1{,}000 = \frac{1{,}000}{9}$$

$$= 111.11\Omega$$

As the shunt increases the range of the meter by ten times the f.s.d. of the basic movement — multiply the scale readings by 10.

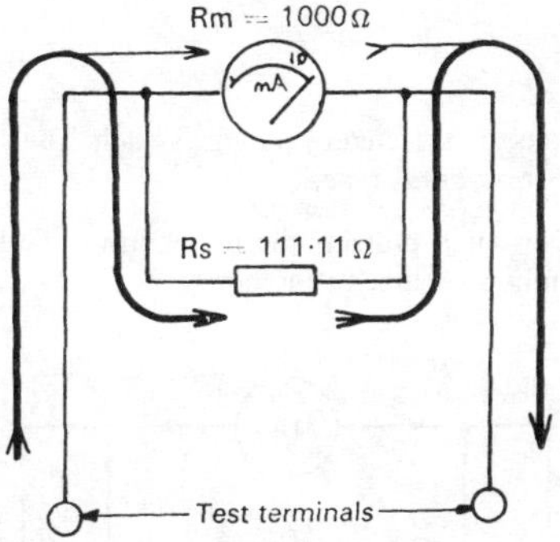

Several current ranges can be provided by using different values of shunts, one for each range.

The shunt is connected to the meter either externally or internally.

External shunts are marked with a multiplying factor.

If the basic meter has a f.s.d. of 1 mA and the shunt multiplying factor is 10, multiply all scale readings by 10.

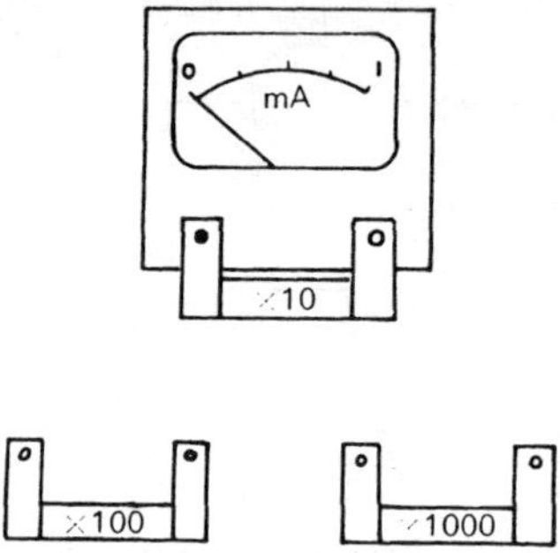

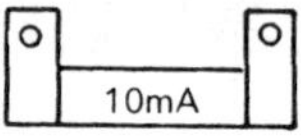

Alternatively, the shunts may be marked with the new f.s.d. reading.

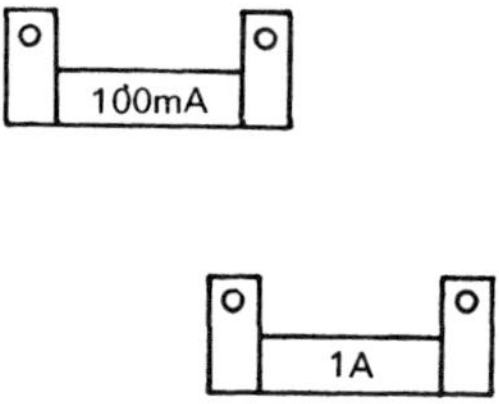

Internal shunts are selected by a range switch. The switch must be a *make-before-break* type.

This type of switch protects the meter from overload whilst switching from one range to another.

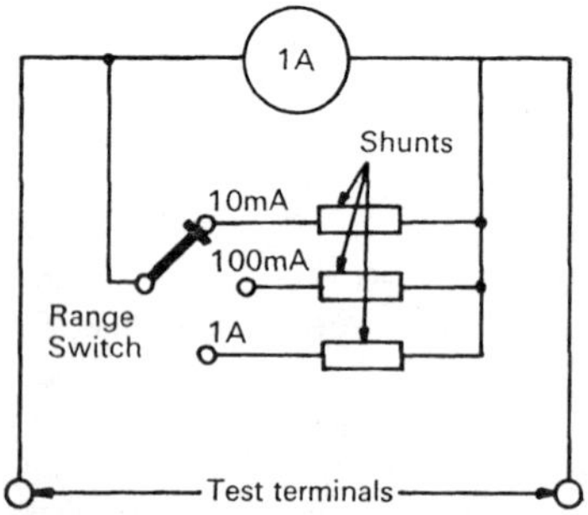

Some meters incorporate what is known as a *universal* shunt.

The formula for calculating the shunt values is different from the one previously given for the simpler type of shunt.

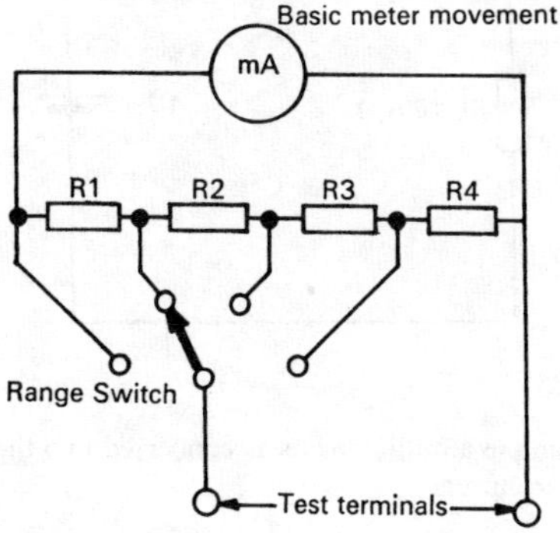

Circuit loading

Use a meter having a low internal resistance. A high resistance meter disturbs the circuit under test and causes inaccurate readings.

The meter resistance should be less than .01 of the resistance of the circuit load.

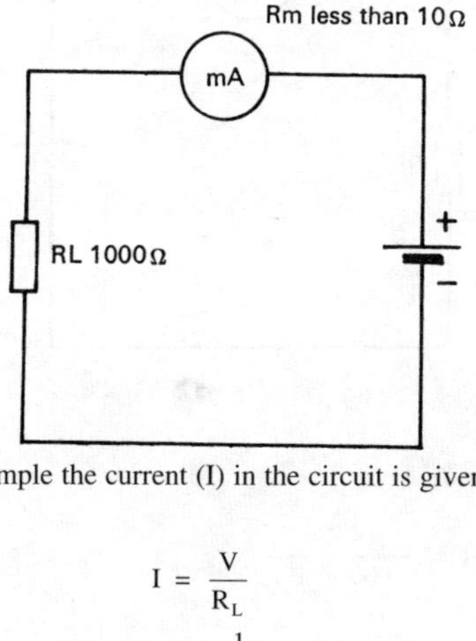

In this example the current (I) in the circuit is given by Ohm's law:

$$I = \frac{V}{R_L}$$
$$= \frac{1}{1{,}000}\ \text{A}$$
$$= 1\,\text{mA}$$

The following examples show how meter resistance affects the current.

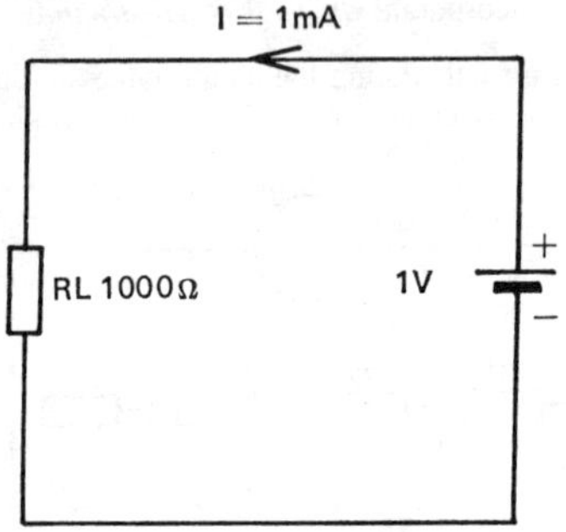

In this example a milliammeter is connected into the circuit to measure the current.

The resistance in the circuit is now:

$R_L + R_m$, where R_m is the meter resistance.

$$\text{Therefore } I = \frac{V}{R_L + R_m}$$

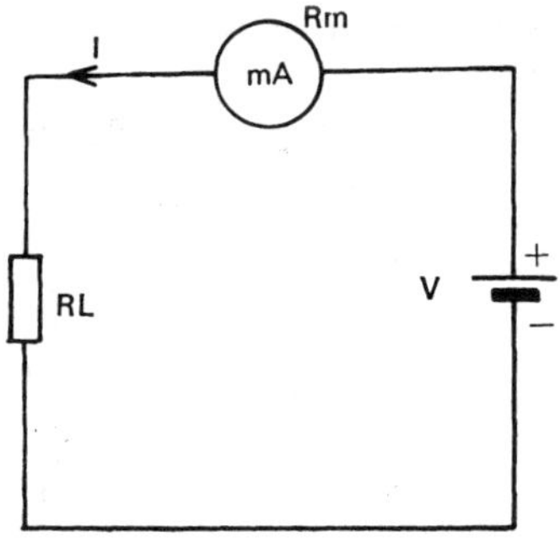

If $R_m = 100\,\Omega$:

$$I = \frac{1}{1{,}000 + 100}$$
$$= \frac{1}{1{,}100}\ \text{A}$$
$$= 0.91\,\text{mA}$$

The error is 9%. The meter has altered the current in the circuit by 9% from 1 mA to 0.91 mA.

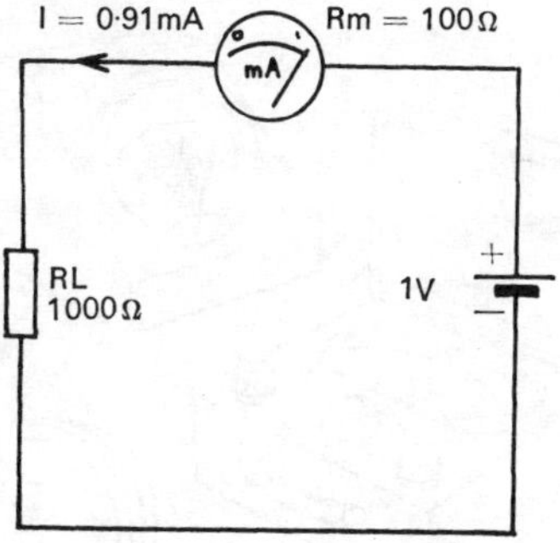

If $R_m = 10\Omega$:

$$I = \frac{1}{1000 + 10}$$
$$= \frac{1}{1{,}010}\ A$$
$$= 0.99\,mA$$

The error is only 1%.

The meter with the larger resistance gives the more accurate measurement of the current flowing through the circuit.

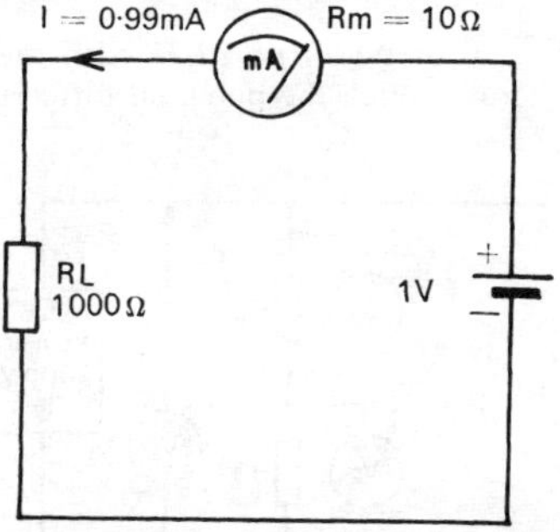

D.C. voltmeters

Use a d.c. *voltmeter* to measure d.c. potential difference.

The simplest voltmeter consists of a moving coil milliammeter or microammeter with a resistor connected in series with the movement.

The resistor is called a *multiplier*.

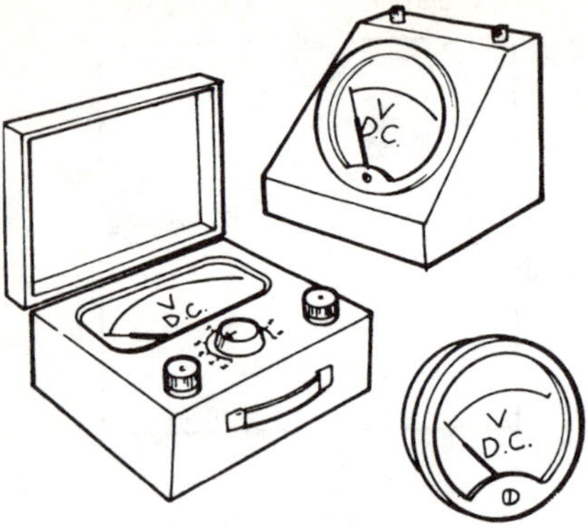

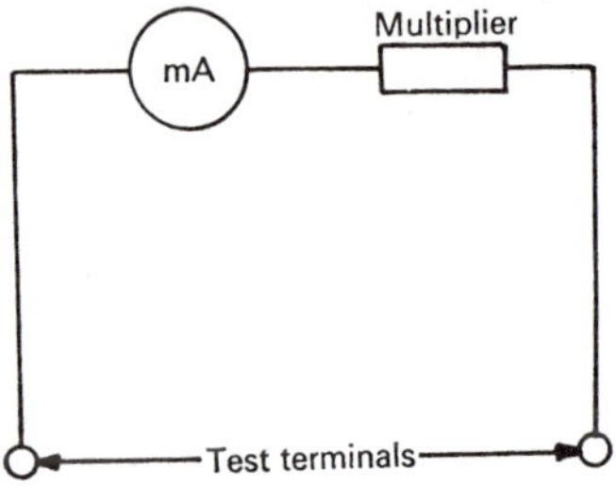

To measure potential difference:

Connect the voltmeter *in parallel* with the circuit or component across which the potential difference is to be measured.

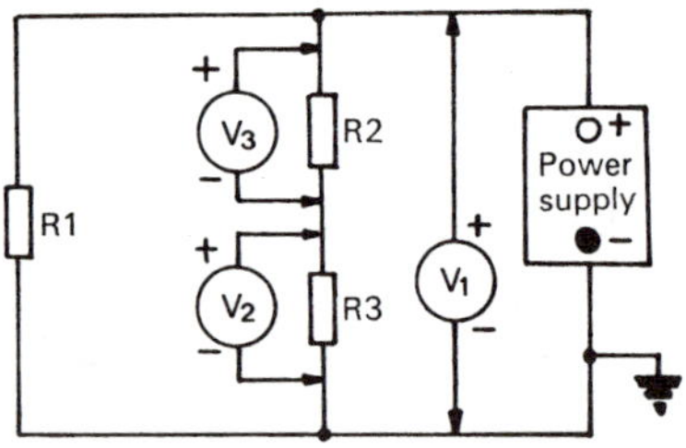

Calculating multiplier values

To calculate the value of multiplier required use the formula:

$$R_x = \frac{V_{f.s.d.}}{I_{f.s.d.}} - R_m,$$ where R_x is the multiplier resistance.

Example 1

A milliammeter has a full-scale deflection ($I_{f.s.d.}$) of 1 mA and a resistance (R_m) of 50 Ω.

Calculate the multiplier resistance (R_x) required to convert the milliammeter to a voltmeter having a full-scale deflection ($V_{f.s.d.}$) of 10 V.

$$R_x = \frac{10}{1.001} - 50$$
$$= 10{,}000 - 50$$
$$= 9{,}950\,\Omega$$

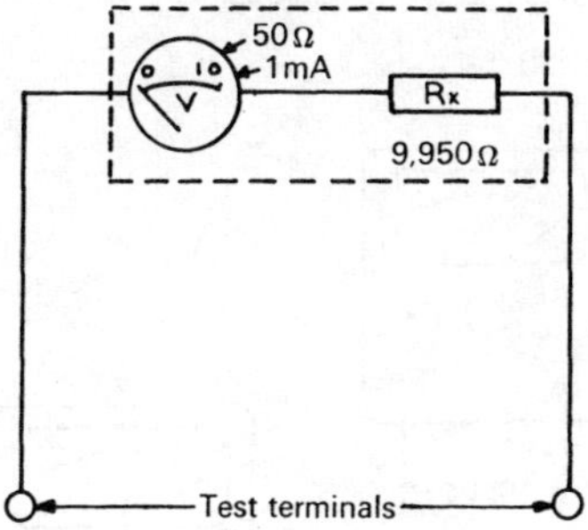

Extending the measuring range

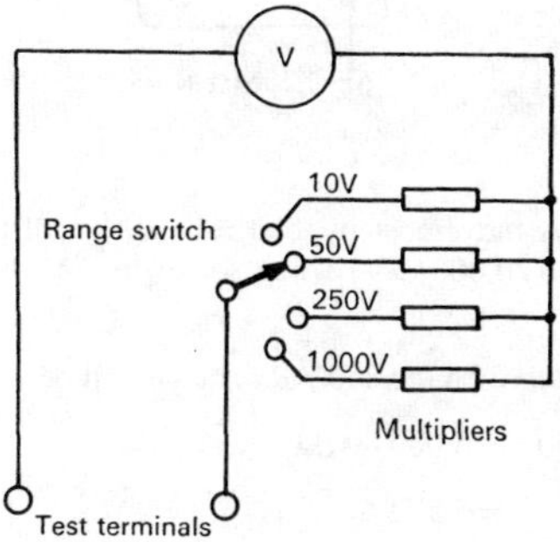

Several voltage ranges can be provided by using different values of multiplier.

The diagrams show two methods of connecting the multipliers.

The voltage ranges and their multipliers are selected by a range switch.

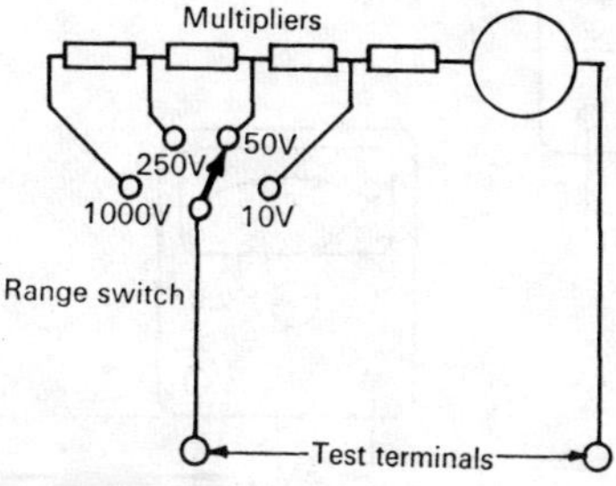

Voltage sensitivity

A multirange voltmeter fitted with a 1 mA movement will have a total resistance of 1,000 ohms when set to its 1 V d.c. range.

Its d.c. voltage sensitivity is 1,000 ohms-per-volt (1,000 Ω/V).

The meter's total resistance on the 250 V d.c. range is:

$$1{,}000 \times 250 = 250{,}000\,\Omega$$

$$= 250\,\text{k}\Omega$$

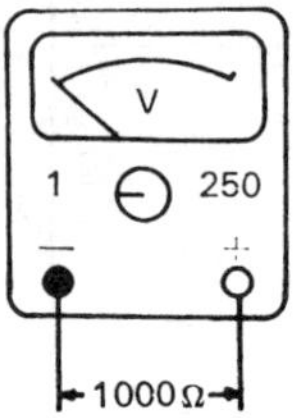

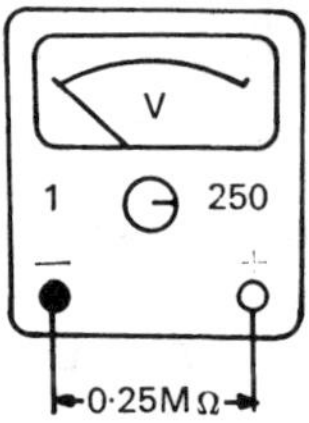

If a meter uses a 50 μA movement its total resistance will be 20,000 ohms-per-volt (20,000 Ω/V) when set to its 1 V d.c. range.

The meter's total resistance on the 250 V d.c. range will be:

$$20{,}000 \times 250 = 5{,}000{,}000\,\Omega$$

$$= 5\,\text{M}\Omega$$

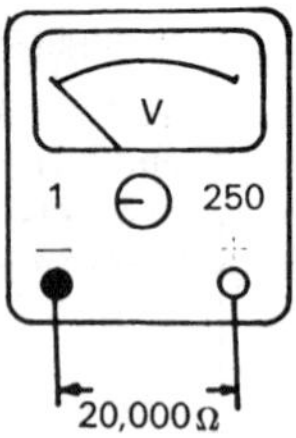

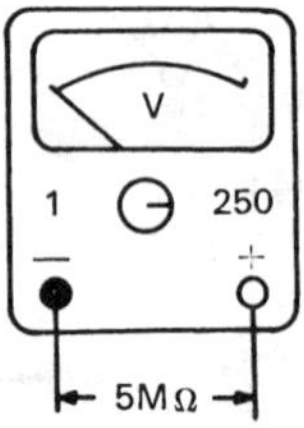

The effect of meter resistance

For accurate measurement a voltmeter must have a high value of resistance compared with the resistance of the circuit under test.

The effect of meter resistance on voltage and current values is shown in the following two examples.

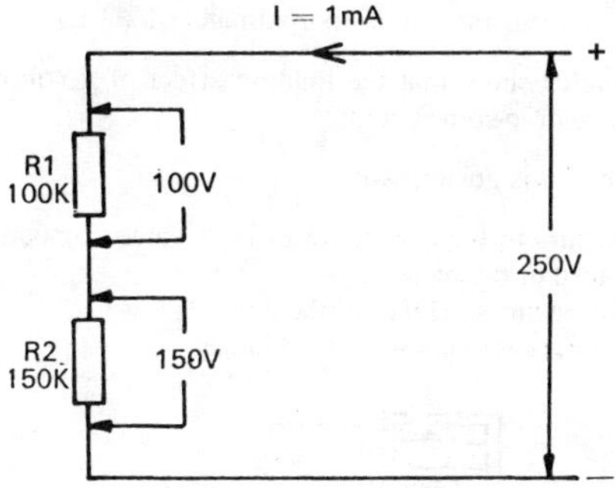

Example 1 shows the effect of connecting a 1,000 Ω/V multirange meter across resistor R1.

The meter is set to the 250 V d.c. range and its total resistance R_m is therefore 250,000 ohms.

Note that:

● due to the parallel resistance R_m of the meter the voltage across R1 is reduced from 100 V to 80 V
● the circuit current has increased from 1 mA to 1.13 mA due to the *loading* effect of the meter.

As the meter reads 80 V the error in measurement is:

$$\frac{20}{100} \times 100 = 20\%$$

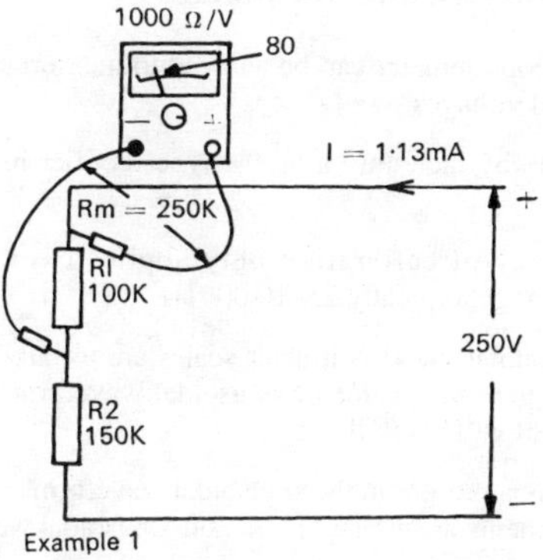

Example 1

In Example 2 a 20,000 Ω/V meter is used to measure the voltage across R1.

The meter is set to its 250 V d.c. range and its total resistance is therefore 5,000,000 ohms (5 MΩ).

In this case the meter reads 98.8 V (1.2 V low).

Circuit current is increased by 0.02 mA.

The error in measurement is approximately 1%.

The examples show that the loading effect of a voltmeter is always present to some extent.

Loading effect is greater when:

- the resistors in the circuit are of high value compared with the resistance of the meter
- the d.c. sensitivity (Ω/V) of the meter is low
- the meter is set to a low voltage range.

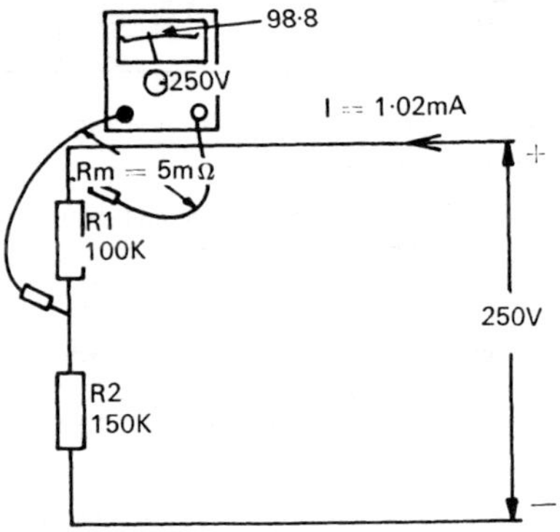

Example 2

Use Ohm's law to obtain the value of R_m from $V_{f.s.d.}$ and

A.C. ammeters and voltmeters

A moving coil ammeter can be adapted to measure alternating currents and voltages.

This is done by including a bridge-type rectifier in the meter circuit.

The accuracy of calibration only applies over a limited frequency range, typically 25–10,000 Hz.

A.C. milliammeter and voltmeter scales are usually calibrated in terms of the r.m.s. value of sinusoidal waveforms. Errors of measurement will result if:

- distortion is present in the sinusoidal waveform
- measurements are attempted on non-sinusoidal waveforms.

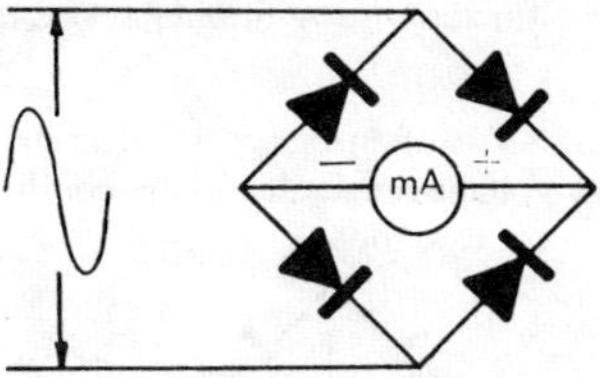

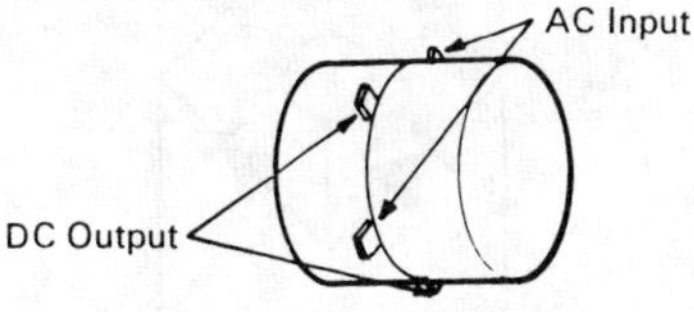

Meter rectifier

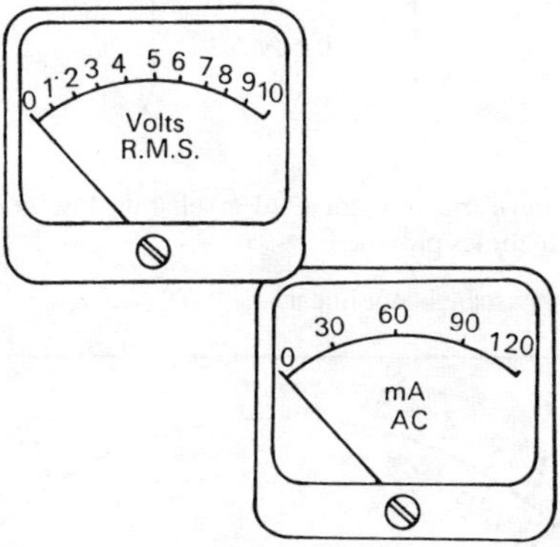

A current transformer is often incorporated in a.c. milliammeters.

The primary winding is tapped to provide several ranges.

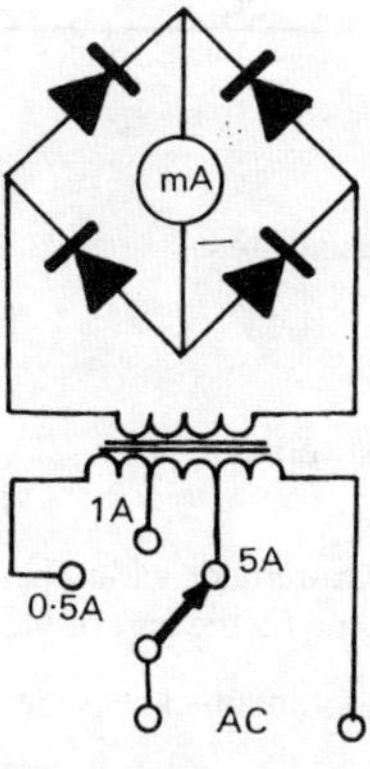

Multirange a.c. voltmeters use multiplier resistors as in d.c. instruments.

A step-up transformer is often incorporated in the instrument for use on the low voltage range, due to the non-linear characteristics of rectifiers at low voltage.

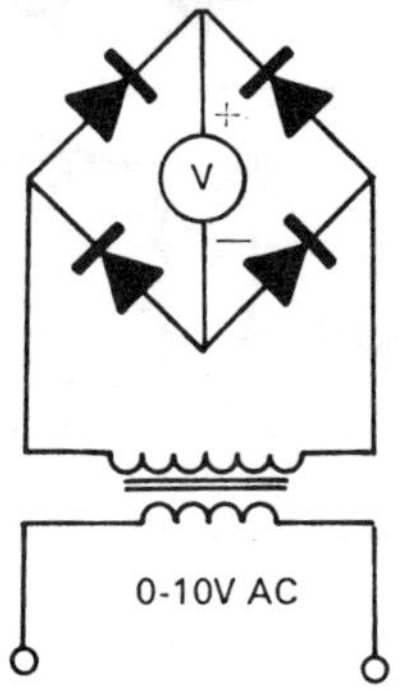

Where a transformer is not used a separate low voltage a.c. scale is sometimes provided.

Note that this scale is not linear.

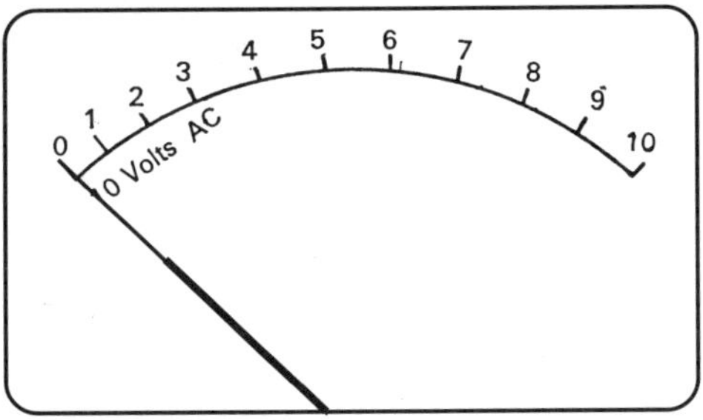

Ohmmeters

Use an *ohmmeter* to measure:

- resistance
- circuit continuity.

Resistance is measured on *ohms* (Ω), *kilohms* (kΩ) and *megohms* (MΩ).

The accuracy of measurement varies between approximately 5% and 10%, depending on the part of the scale used.

The highest accuracy is obtained near the centre of the scale.

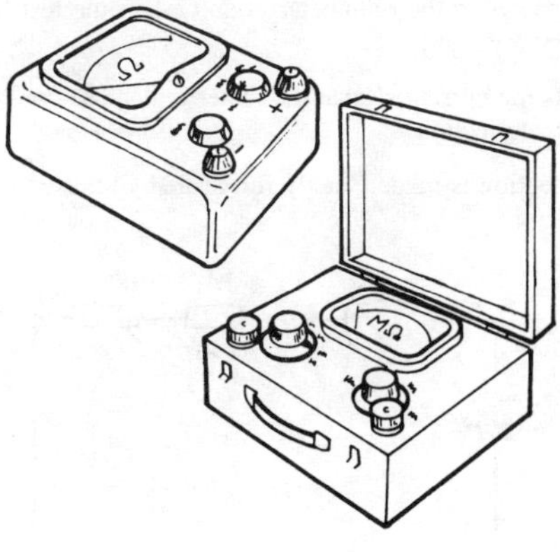

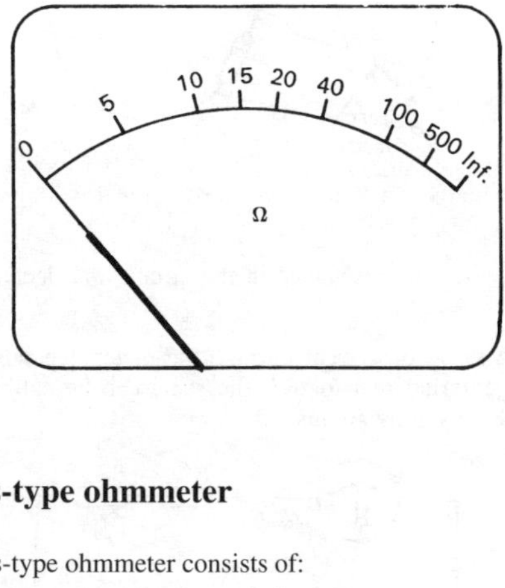

Series-type ohmmeter

A series-type ohmmeter consists of:

- a milliammeter or microammeter
- a fixed resistor R1
- a variable resistor R2
- a battery.

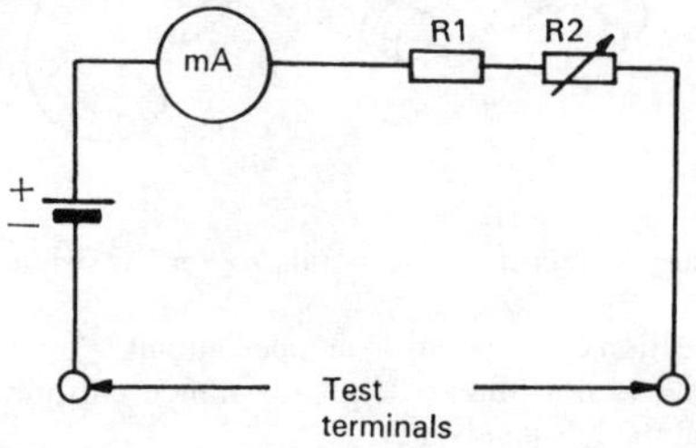

Current flows in the ohmmeter circuit when the test leads are connected together.

R1 limits the current so that the meter indicates approximately full-scale deflection.

The deflection is made exactly full-scale by adjusting the *zero adjust* control, R2.

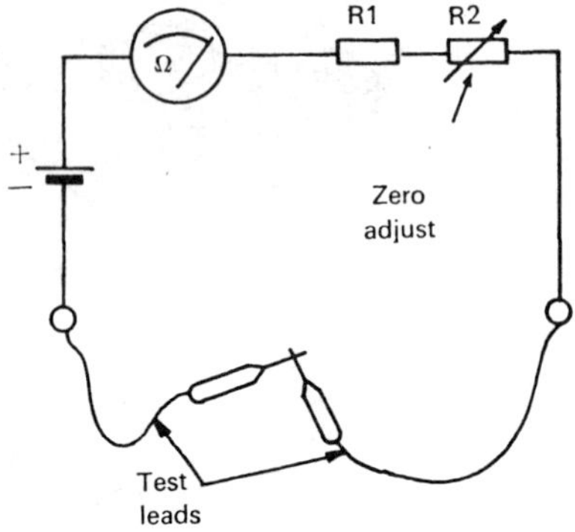

The test leads are connected across the resistance to be measured, R3.

This increases the resistance in the circuit and decreases the current.

Since the value of current through the meter depends on the value of external resistor R3, the scale can be calibrated in ohms, kilohms or megohms.

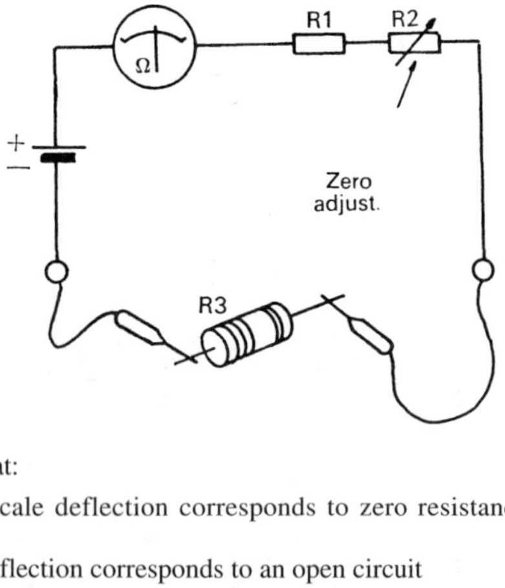

Note that:

- full-scale deflection corresponds to zero resistance (short circuit)
- no deflection corresponds to an open circuit
- the scale is non-linear. High resistance calibrations are cramped at the left-hand end.

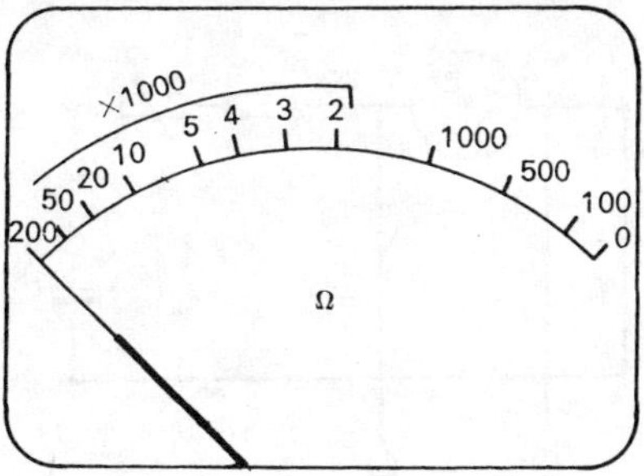

The measuring range can be increased by using a higher battery voltage.

Appropriate batteries and current limiting resistors are switched into circuit as the desired range is selected.

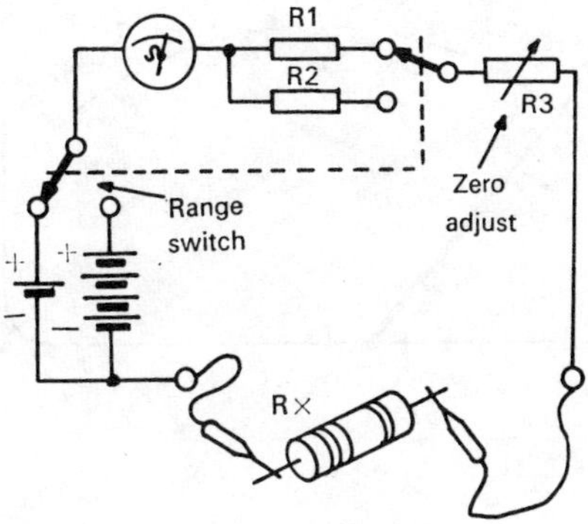

Shunt-type ohmmeter

Use a shunt-type ohmmeter to measure low resistance values.

The resistance to be measured is connected in parallel with the meter.

A battery switch is fitted, since current is always flowing from the battery whether the test terminals are connected or not.

Always switch off the ohmmeter after use.

Unlike the series-type ohmmeter, zero resistance is indicated on the left of the scale.

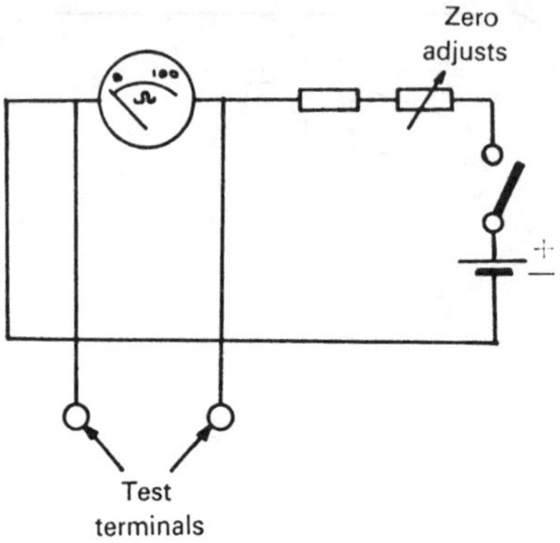

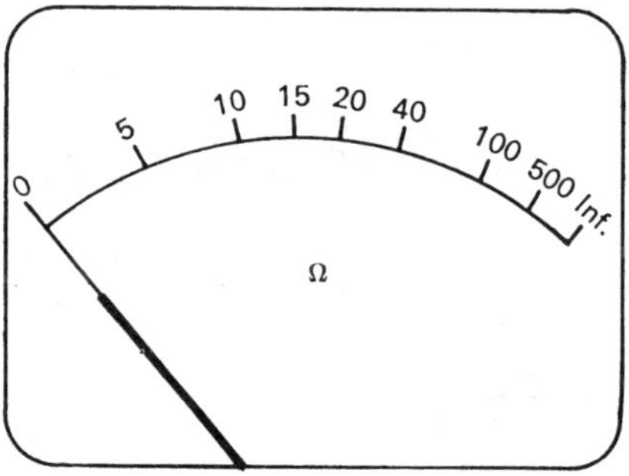

Multirange test meters

A multirange test meter can be used for measuring:

- current (a.c. and d.c.)
- voltage (a.c. and d.c.)
- resistance
- audio frequency power levels.

As a multirange meter combines the functions of a milliammeter, voltmeter (a.c. and d.c.) and ohmmeter it is also known as a:

- universal meter
- multimeter (volt-ohm-milliamp).

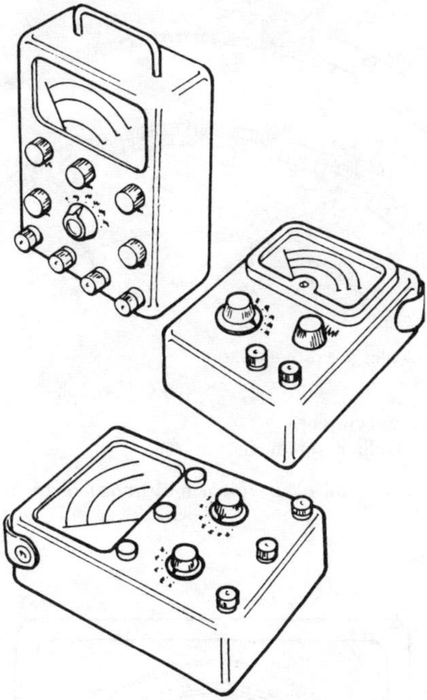

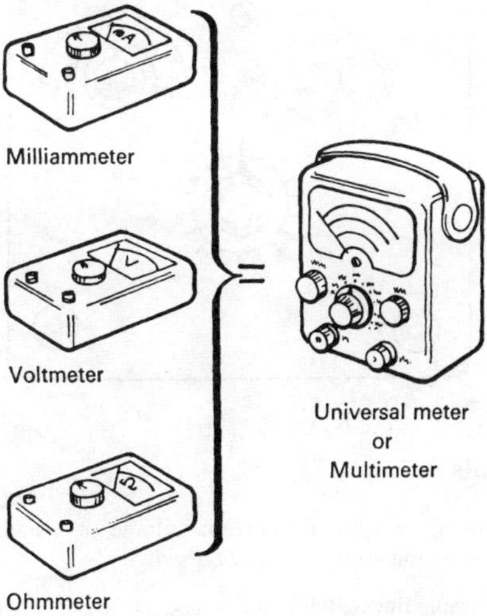
Milliammeter
Voltmeter
Ohmmeter
Universal meter
or
Multimeter

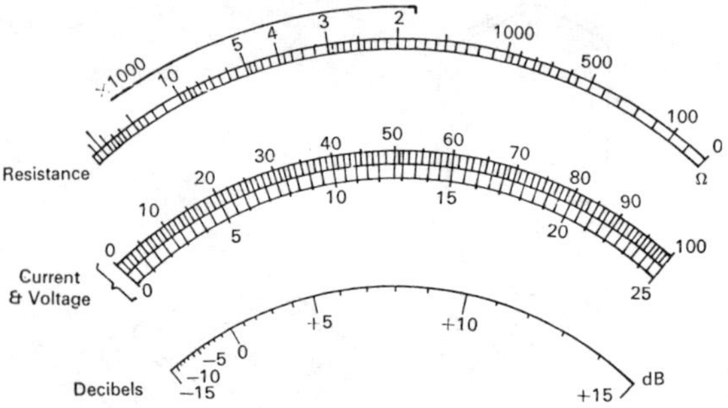

Separate scales are provided for:

- resistance
- voltage and current
- power level measurements.

The control panel of a typical multirange meter is shown below.

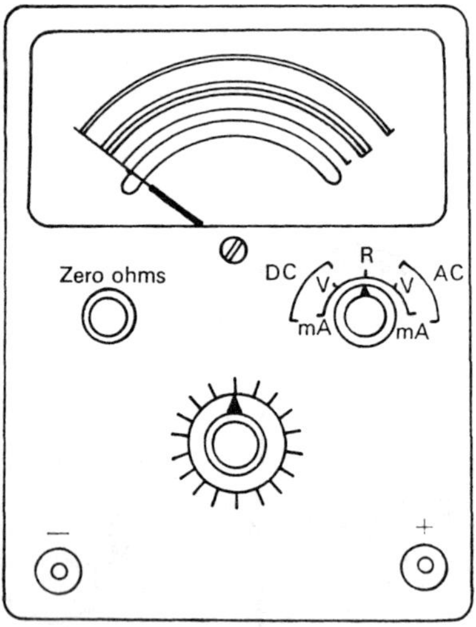

Controls

Set the meter to measure current voltage (a.c. and d.c.) or resistance, by means of the *function* switch.

In the example the switch is set to mA, a.c.

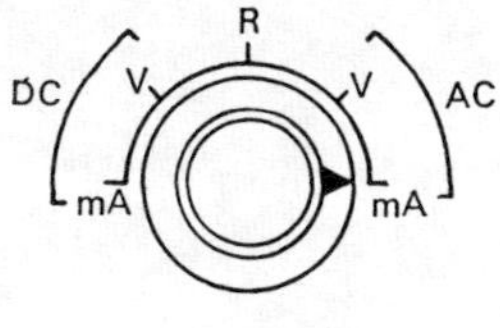

Set the meter to the required current, voltage or resistance range by means of the *range* switch.

In the example the switch is set to 2.5—volts or mA, depending on the setting of the *function* switch.

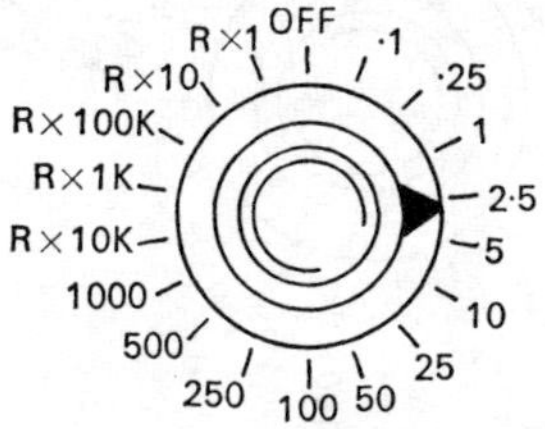

In some meters the *function* and *range* are selected by a single switch.

The example shows the switch set to 25 V d.c.

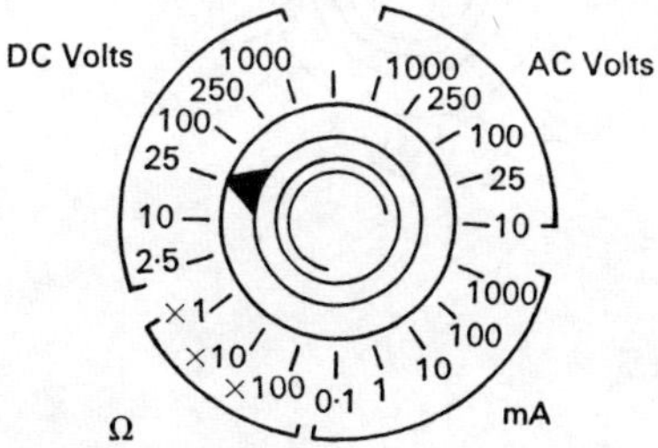

Some meters use two *function/range* switches, one for d.c. ranges and the other for a.c. and resistance (ohms) ranges.

The examples on following pages show switches set to 250 V a.c., 100mA d.c. and the last is set to *resistance*, Ω x 100 range.

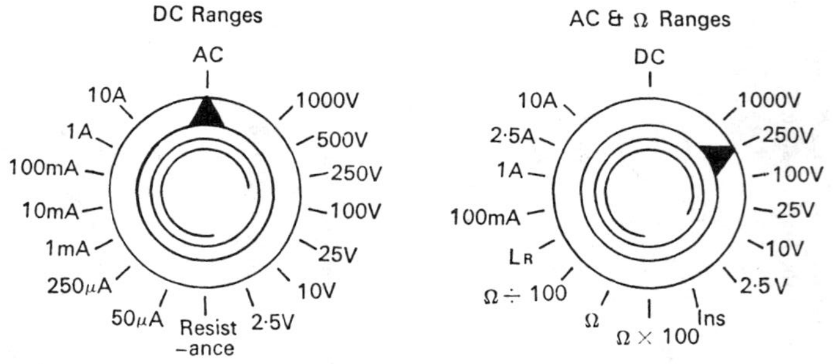

Switches set to 250 V a.c.

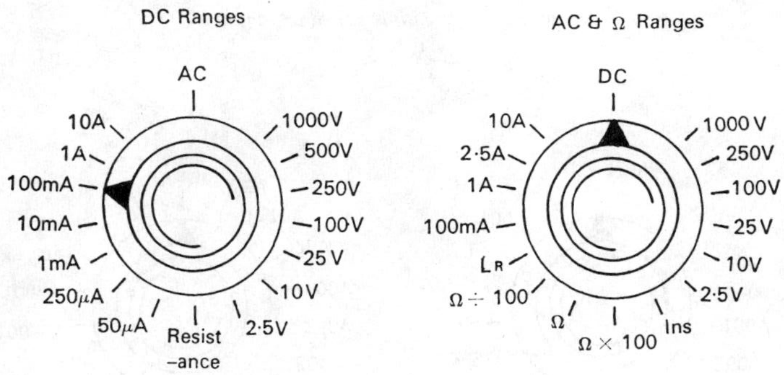

Switches set to 100 mA d.c.

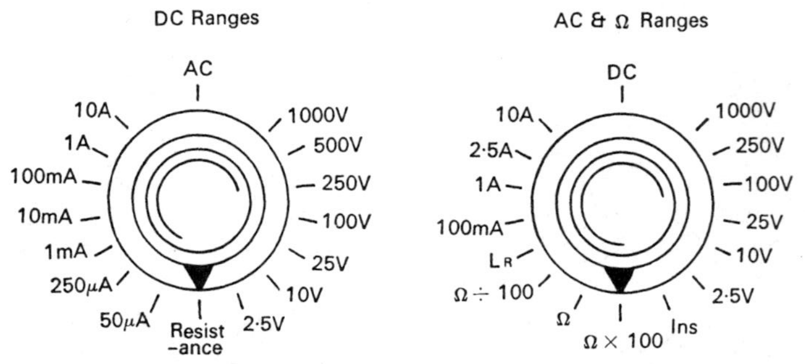

Switches set to *resistance*, Ω x 100 range

The *ohms adjust* or *zero* Ω control compensates, within limits, for variation of the battery voltage.

Some meters use separate *ohms adjust* controls, one for each resistance range (R/100, R, and 100R).

The controls may be inter-dependent — read the manufacturer's instruction book.

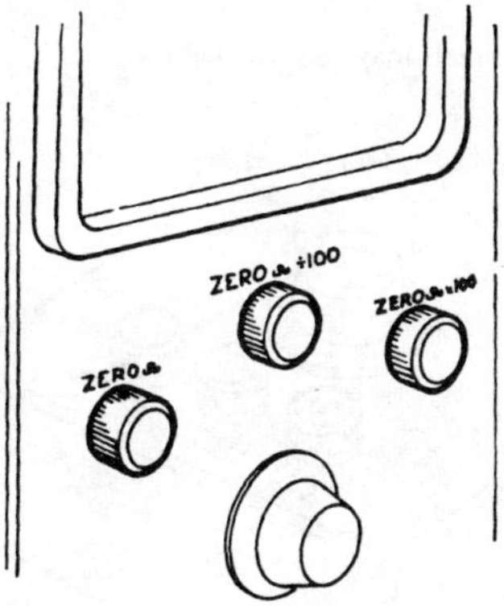

The *reverse meter* or *rev. m.c.* switch avoids the need to reverse test leads should the polarity of current or voltage change in the circuit under test.

Meter terminals

Two terminals are used for most measurements.

One of these is marked — *negative* or *common negative*. It may be colour coded *black*.

The other is marked +, or *test pos*. Sometimes colour coded *red*.

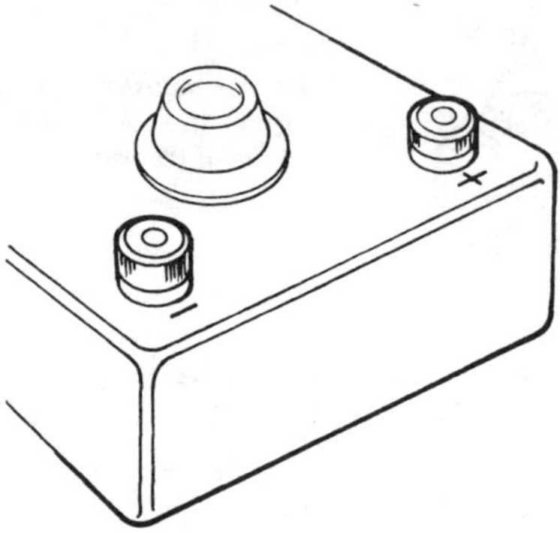

Separate terminals may be provided for:

- high voltage ranges
- audio frequency *output* measurements.

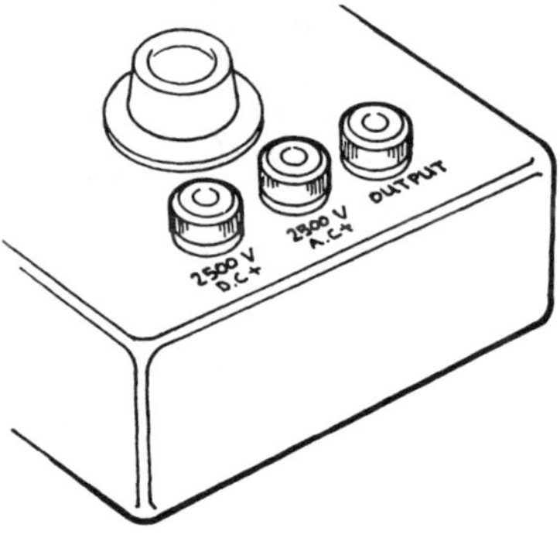

E.H.T. multiplier

To enable *extra high tension* (E.H.T.) voltage to be measured, some meters are provided with a special adaptor called an E.H.T. multiplier.

Meter accuracy

To ensure accurate readings, position the meter as specified by the instrument manufacturer.

Check that the pointer is set to zero.

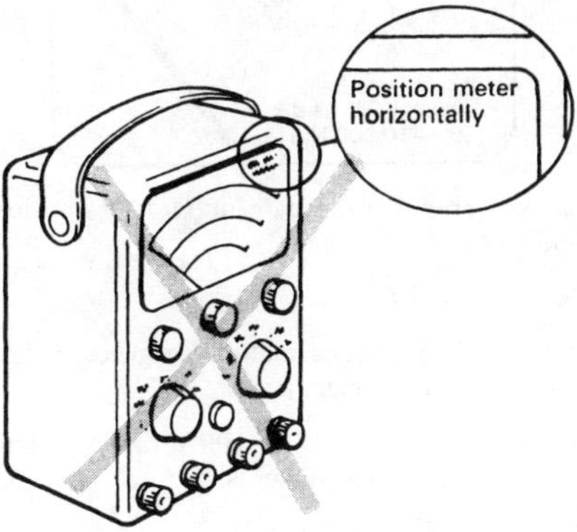

The accuracy of a meter is usually specified as a percentage of the reading at full-scale deflection (f.s.d.).

Accuracy is typically 2% (f.s.d.) on d.c. ranges.

For the meter shown, the *true* voltage at *full-scale deflection* falls between the limits 98 V and 102 V.

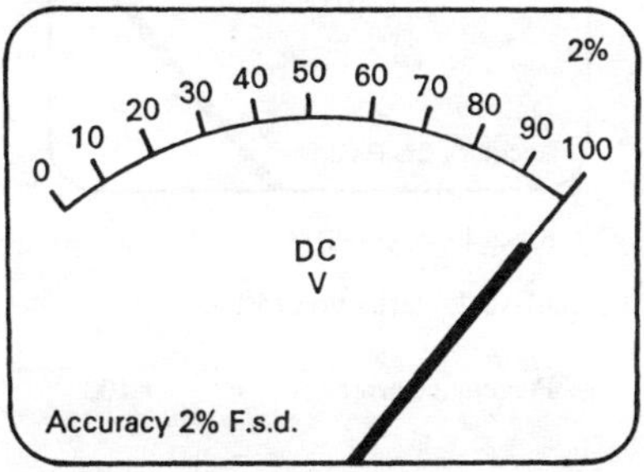

On a.c. ranges, accuracy is usually between 2% and 3% (f.s.d.), depending on the meter.

The specified accuracy applies only when the voltage or current being measured is:

- of sinusoidal waveform
- within a specified frequency range.

For the meter shown, the *true* voltage at *full-scale deflection* falls between the limits 242.5 V and 257.5 V for voltages within the frequency range 25–10,000 Hz.

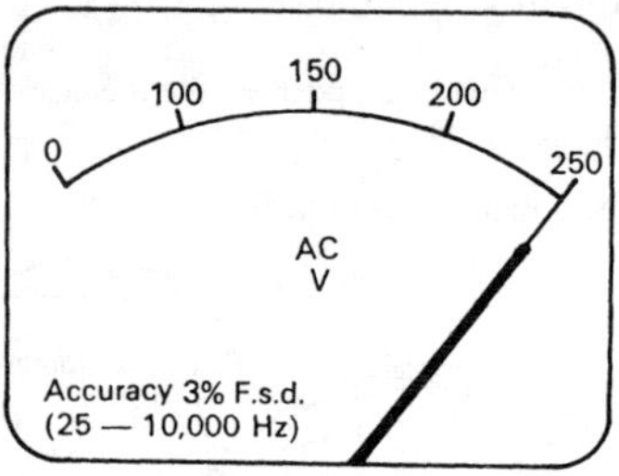

The *voltage* error at full-scale deflection applies to *any deflection* over a *given range*.

Example:

The accuracy of a particular multirange meter is specified as 2% (f.s.d.) on its d.c. voltage ranges.

If the meter is switched to its 100 V range, calculate the percentage error when the pointer is at:

- half-scale deflection (50 V)
- one-tenth of full-scale deflection (10 V).

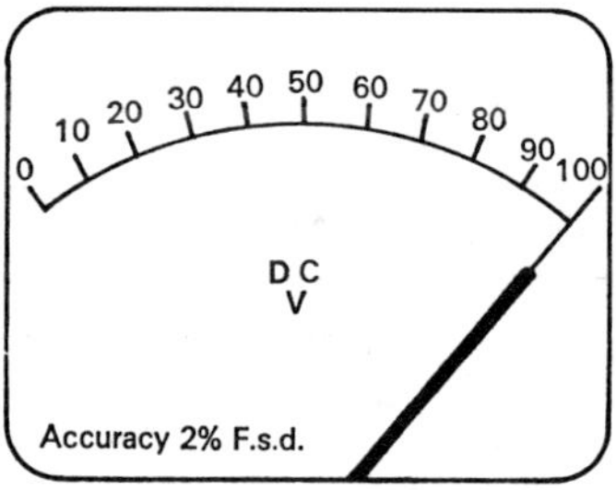

The *voltage* error at f.s.d. is ± 2 V.

Therefore at half-scale deflection (50 V):

$$\text{Percentage error} = \frac{2}{50} \times 100$$
$$= 4\%$$

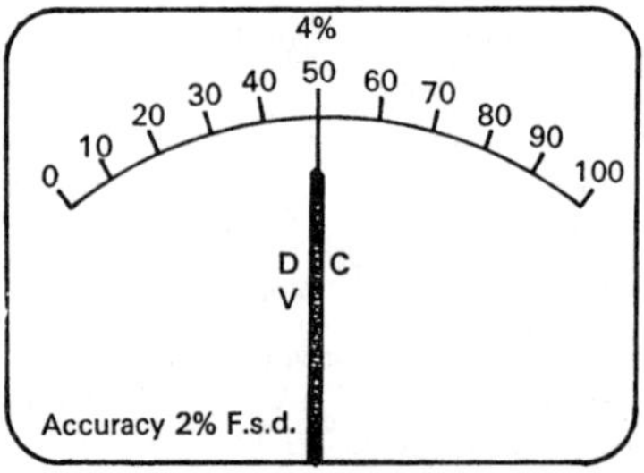

At one-tenth of full-scale deflection (10 V):

$$\text{Percentage error} = \frac{2}{10} \times 100 = 20\%$$

The above example indicates that the accuracy of measurement decreases as the pointer deflection decreases.

To obtain more accurate measurements:

- set the range switch to its highest range to avoid overloading the meter
- re-set the range switch to obtain a deflection on the right-hand half of the scale.

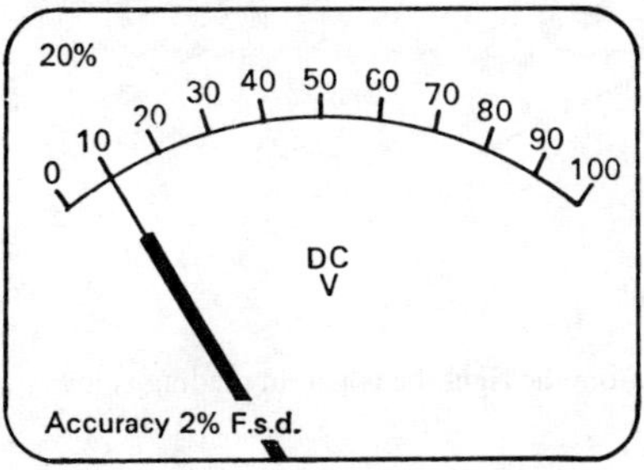

Anti-parallax mirror

Use the anti-parallax mirror, if fitted, to avoid errors in reading the scale due to oblique viewing.

Position yourself so that the reflection of the pointer is hidden by the pointer itself.

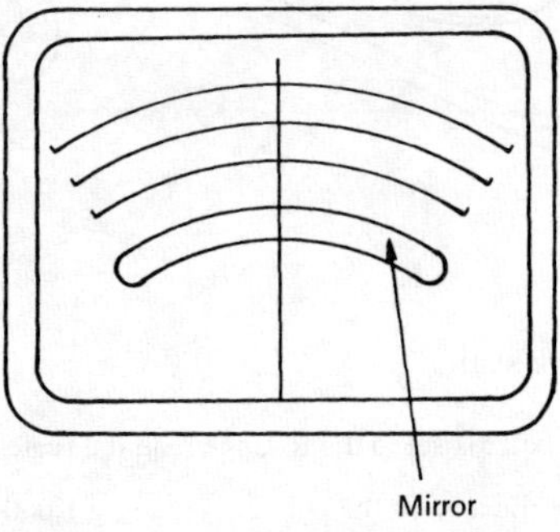

Viewed from the left the apparent reading is high.

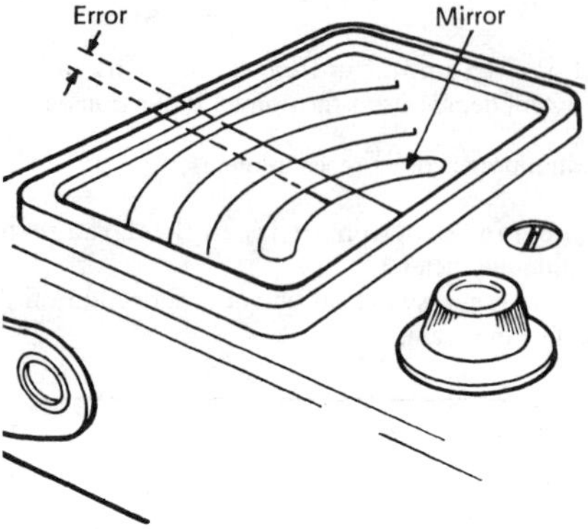

Viewed from the right the apparent reading is low.

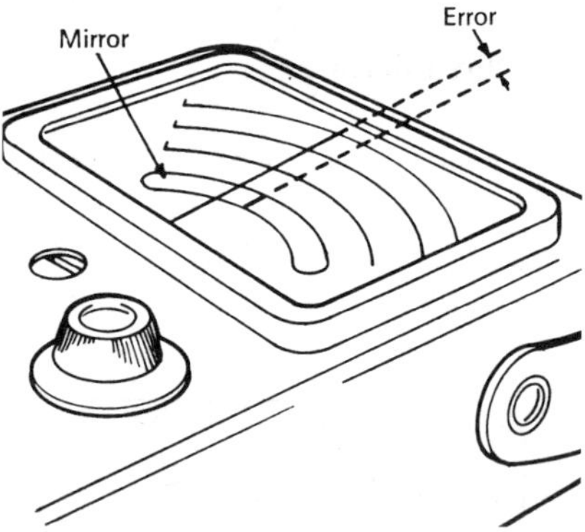

Resistance scale

When the meter is set to the R range read the scale directly.

In the scale illustration the pointer is reading 1,000 ohms.

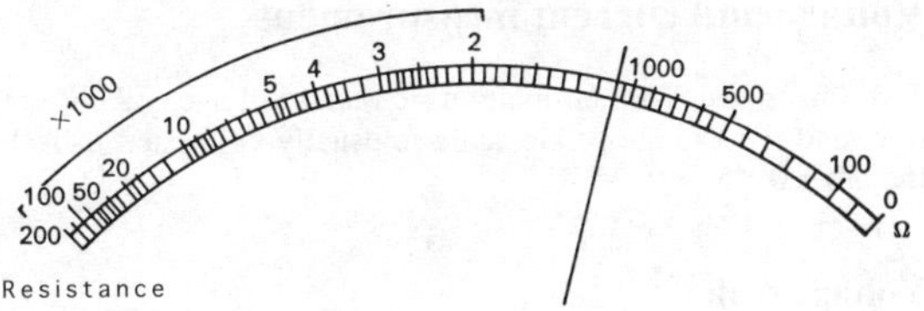

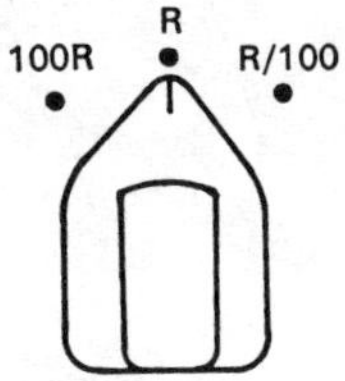

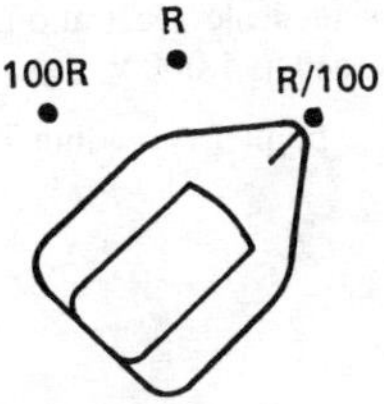

When set to the R/100 range: divide the scale reading by 100.

In the scale illustration the pointer is reading 10 ohms.

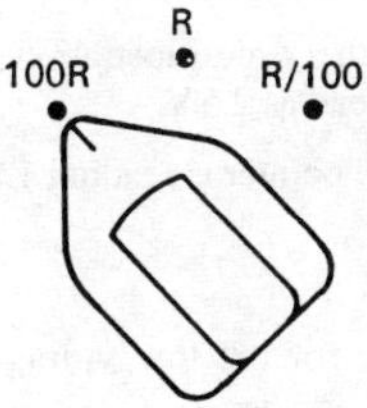

When set to the 100R range: multiply the scale reading by 100.

In the scale illustration the pointer is reading 100,000 ohms.

Voltage and current measurement

Two linear scales are normally used on all voltage and current (a.c. and d.c.) ranges. One scale is usually calibrated 0–100, the other 0–25 or 0–30.

Voltage scale

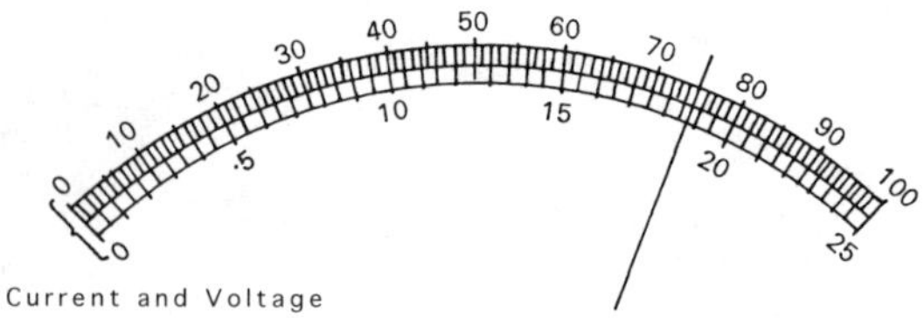

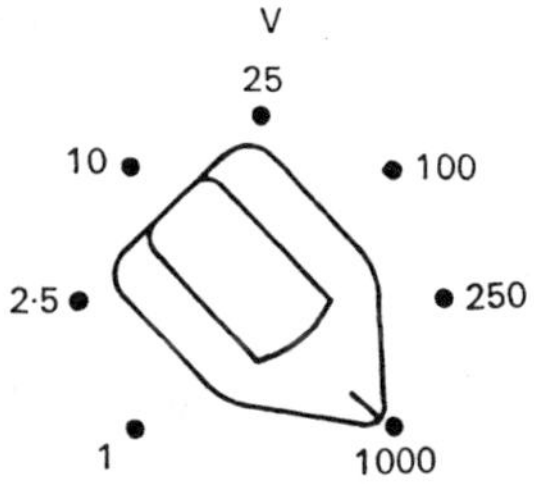

When the meter is set to the 1,000 V (a.c. or d.c.) range:

- use the 0–100 scale
- multiply the reading by 10.

In the scale illustration, f.s.d. is read as 1,000 V.

The pointer is reading 750V.

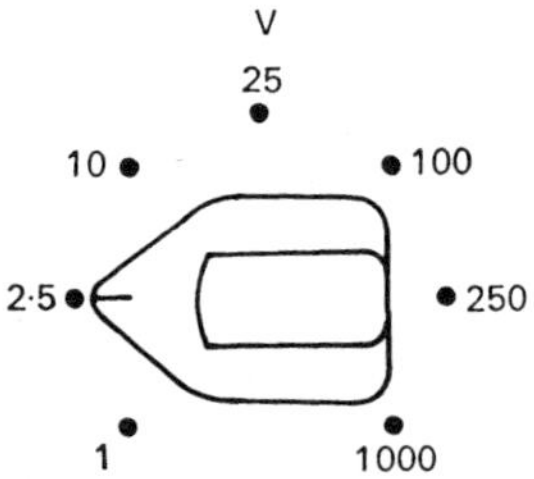

When the meter is set to the 2.5 V (a.c. or d.c.) range:

- use the 0–25 scale
- divide the reading by 10.

In the scale illustration, f.s.d. is read as 2.5 V.

The pointer is reading 1.87 V.

Some meters use a separate scale for the low voltage a.c. range.

Current scale

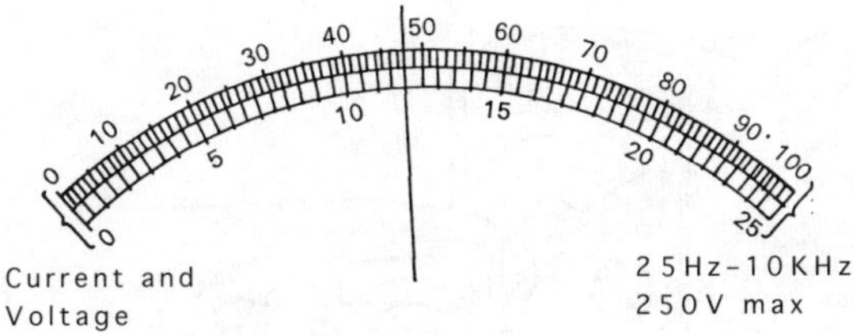

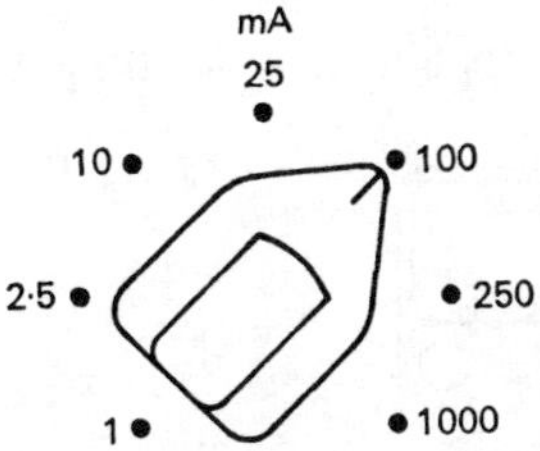

When the meter is set to the 100 mA (a.c. or d.c.) range:

- use the 0–100 scale
- read the scale directly.

In the scale illustration the pointer is reading 48 mA.

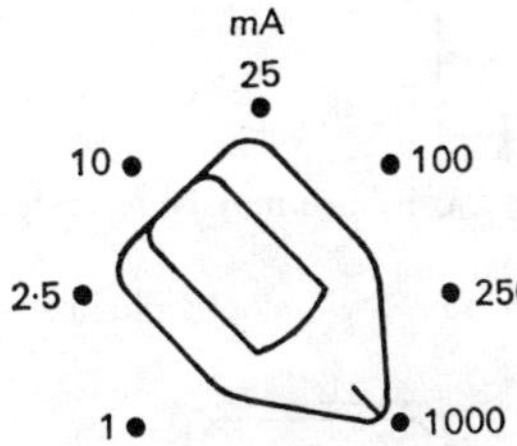

When the meter is set to the 1 A range:

- use the 0–100 scale
- divide the reading by 100.

In the scale illustration f.s.d. is read as 1 A, and the pointer is reading 0.48 A.

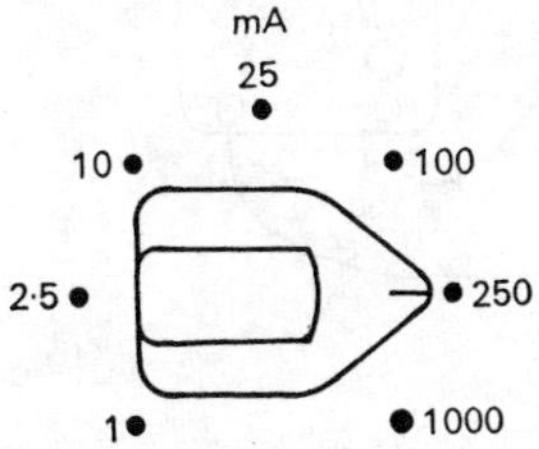

When the meter is set to the 250 μA range:

- use the 0–25 scale
- multiply the reading by 10.

In the scale illustration f.s.d. is read as 250 μA, and the pointer is reading 120 μA.

Electronic voltmeters

Electronic voltmeters, either analogue or digital, are used to measure a.c. and d.c. voltages. Basics of both types of electronic voltmeters are the same.

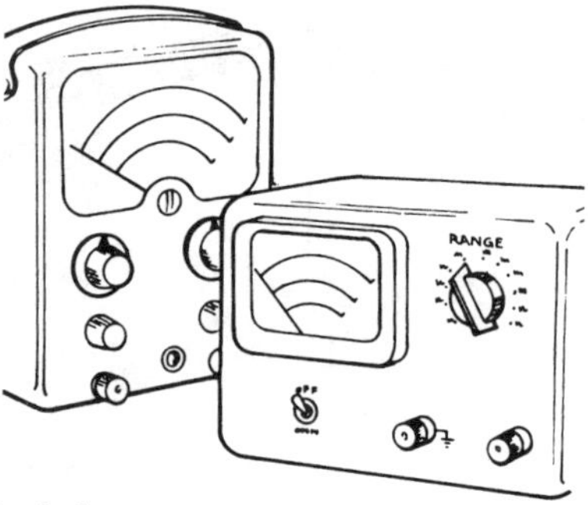

Characteristics

An electronic voltmeter has high input resistance (R) and a low input capacitance (C).

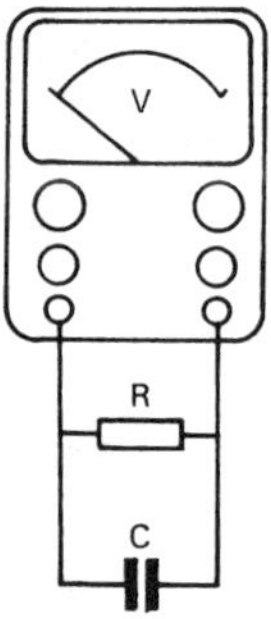

For d.c. measurements, the input resistance may be as high as 100 MΩ.

This enables measurements to be made with negligible loading effect on the circuit under test.

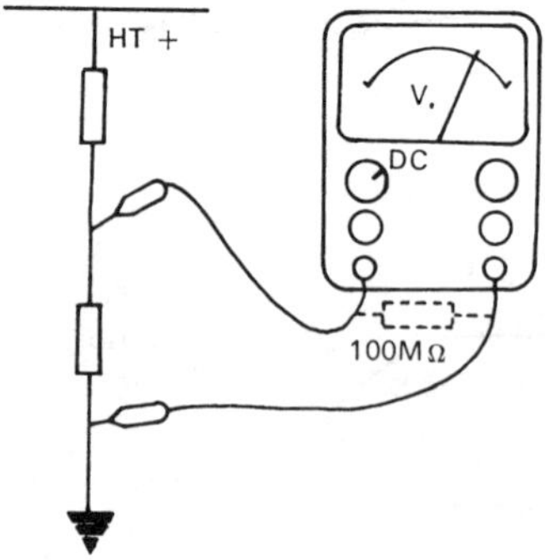

For a.c. measurements, a high input resistance and low capacitance combine to provide a high input impedance.

The input capacitance is usually within the range 20–30 pF but may be as low as 1 pF in some instruments.

A high input impedance enables accurate a.c. measurements to be made at audio and radio frequencies.

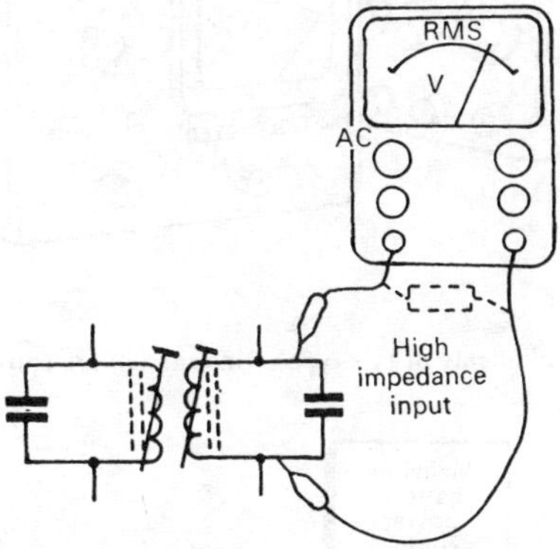

A typical frequency range for an electronic voltmeter is 10 Hz to 5 MHz. Many instruments operate up to 1,000 MHz or more.

A frequency response curve is usually provided in the instrument instruction book.

The curve indicates the frequency range over which measurements may be made with a specific accuracy.

Accuracy is usually between 1% and 5% of full-scale deflection, depending on the instrument.

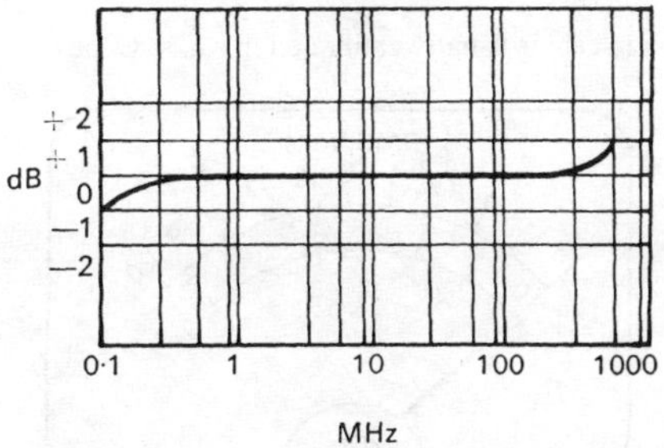

Response curve

Microvoltmeters and millivoltmeters

Sensitive electronic voltmeters are often known as microvoltmeters and millivoltmeters.

These instruments are generally designed to measure a.c. voltage only.

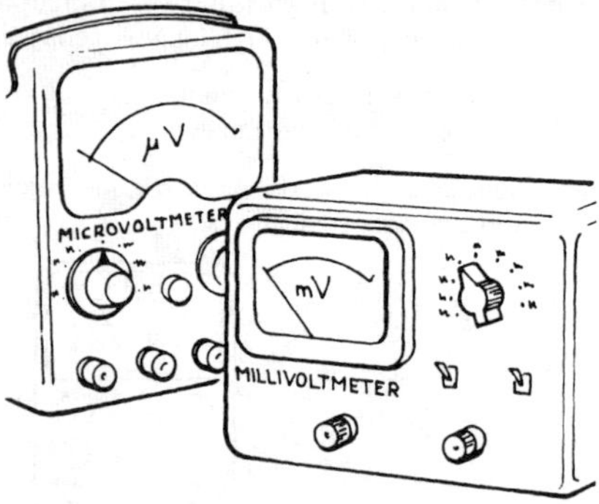

The main elements of this type of instrument are shown in the block diagram.

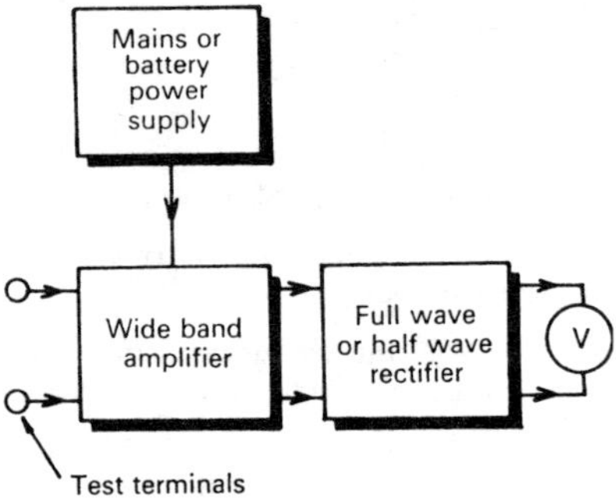

The meter scale is usually calibrated in r.m.s. values.

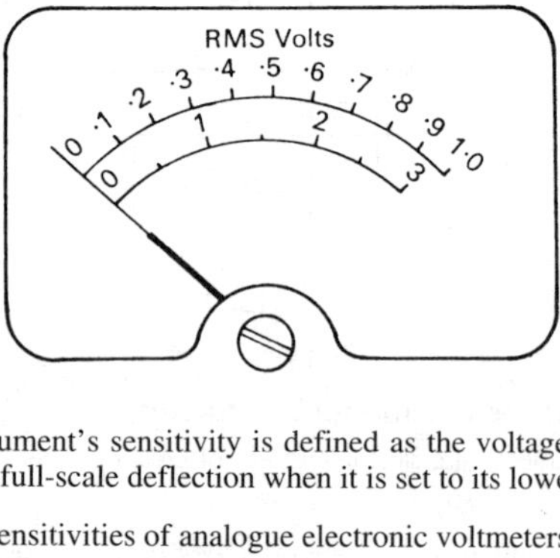

The instrument's sensitivity is defined as the voltage required to obtain full-scale deflection when it is set to its lowest range.

Typical sensitivities of analogue electronic voltmeters are:

100 μV, 1 mV and 10 mV.

The voltage and frequency ranges vary from one instrument to another.

Always make sure that both the amplitude and the frequency of the signal to be measured are within the range of the instrument.

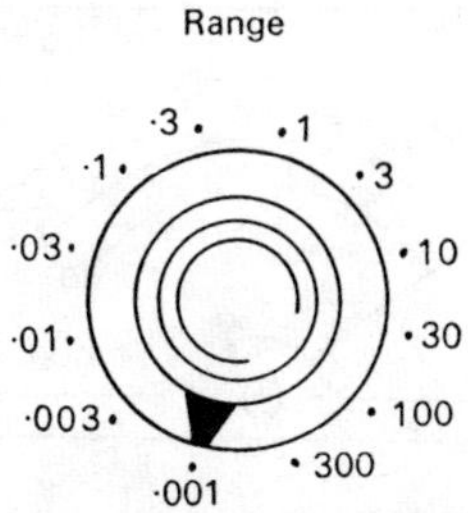

Modern instruments are usually transistorised and powered by internal batteries.

A built-in battery test facility is often provided.

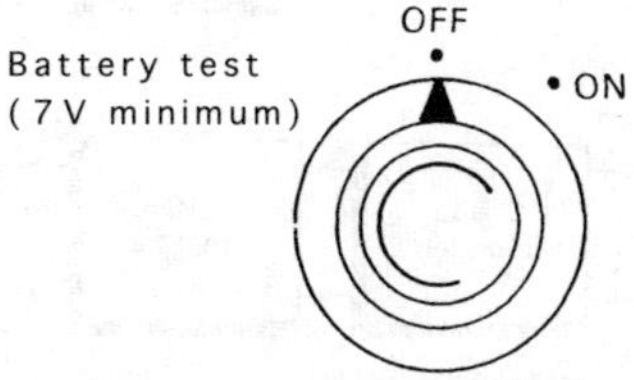

General-purpose electronic voltmeters

Many general-purpose electronic voltmeters measure:

- a.c. voltage over a specified frequency range
- d.c. voltage
- resistance
- decibels
- current.

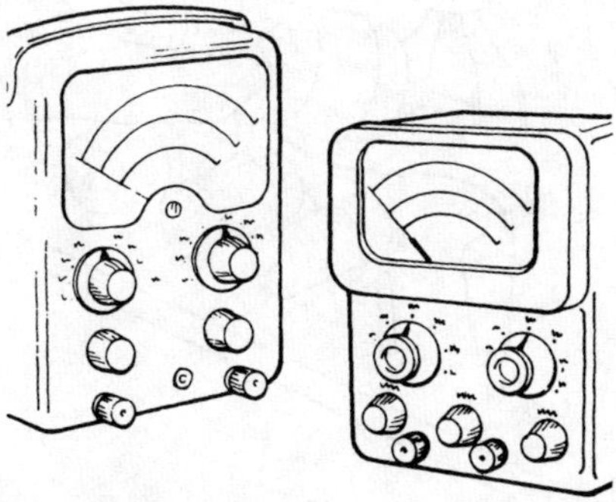

Separate scales are provided for:

- resistance
- a.c. and d.c. voltage/current
- decibels.

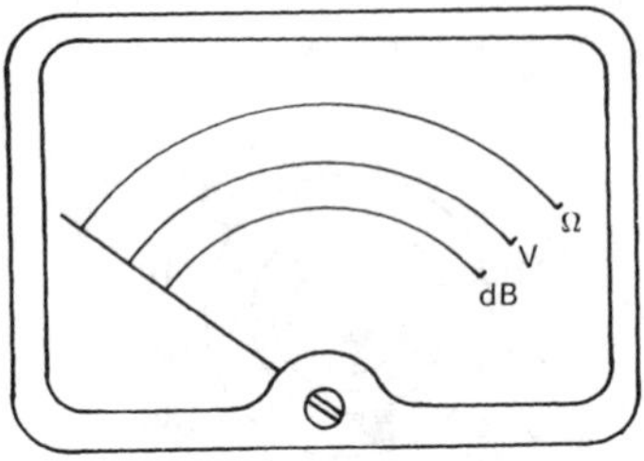

Mode of operation

The elements of this type of instrument are shown in this block diagram.

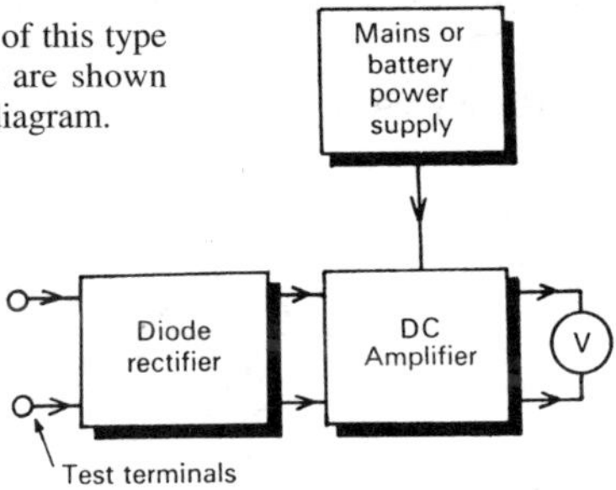

Measurements

The probe is brought up to the equipment under test.

An earth connection between the instrument and the equipment under test is provided by a short lead attached to the probe unit.

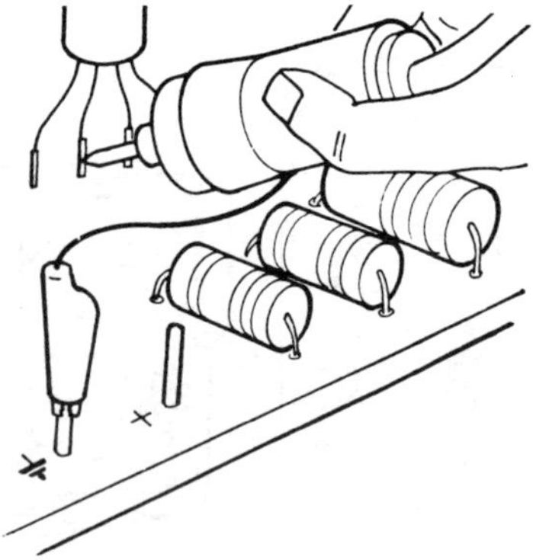

Most electronic voltmeters are scaled to measure the r.m.s. value of sinusoidal waveforms. Obtain the peak value of a sinusoidal voltage by multiplying the r.m.s. value by 1.414.

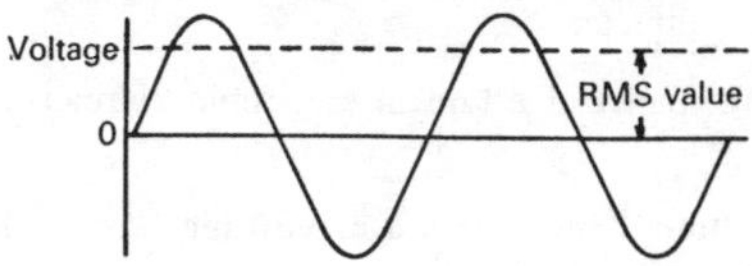

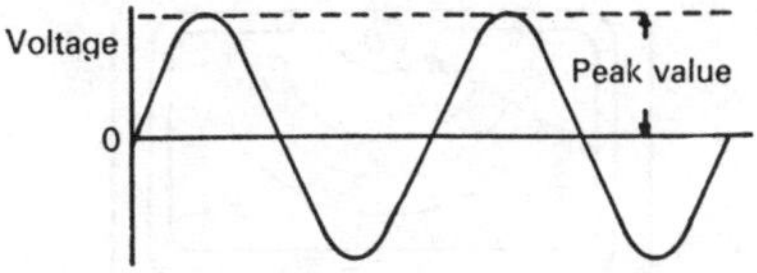

Errors in measurement will result if:

- distortion is present in the sinusoidal waveform
- measurements are attempted on non-sinusoidal waveforms.

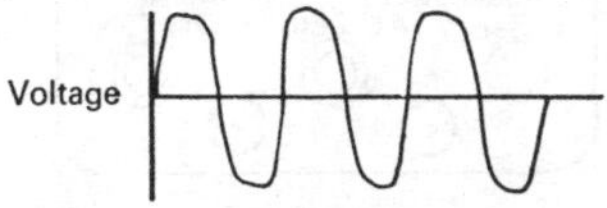

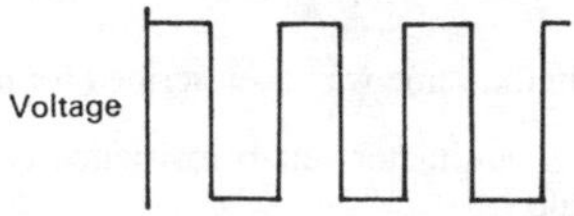

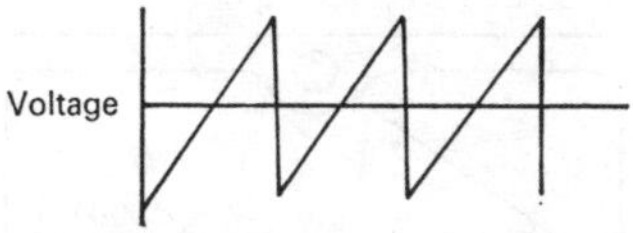

Use an oscilloscope to measure the amplitude of non-sinusoidal waveforms.

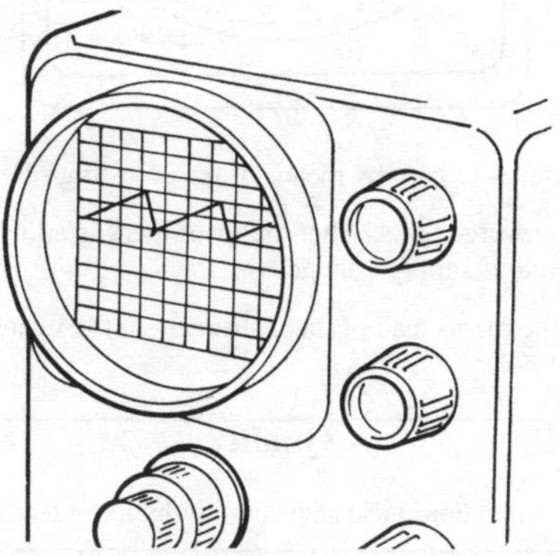

Using an electronic voltmeter

Read the manufacturer's instruction book before using an electronic voltmeter.

The control panel of a typical electronic voltmeter is shown below.

The instrument measures a.c. voltage, d.c. voltage and resistance in several ranges.

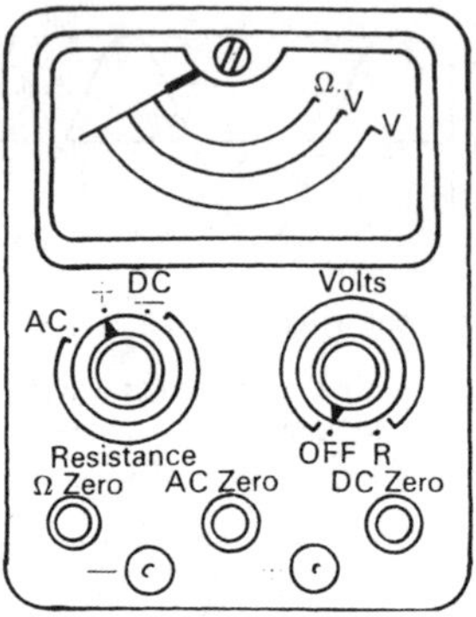

Read the scales in the same way as described for multimeters.

Note that, unlike multimeters, zero resistance is indicated on the left of the scale.

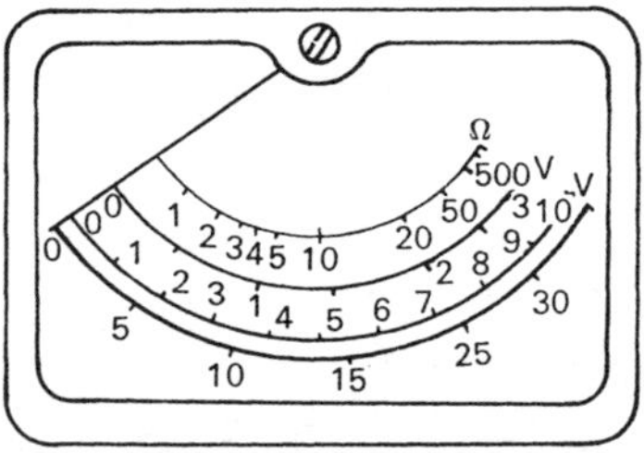

Set the pointer to zero by means of the adjusting screw.

If mains powered, check that the instrument is adjusted to suit the local mains supply voltage.

Connect the mains lead of the instrument to the mains.

Quality

When not in use short circuit the input leads.

A.C. measurements

Set the *a.c./d.c./resistance* switch to a.c.

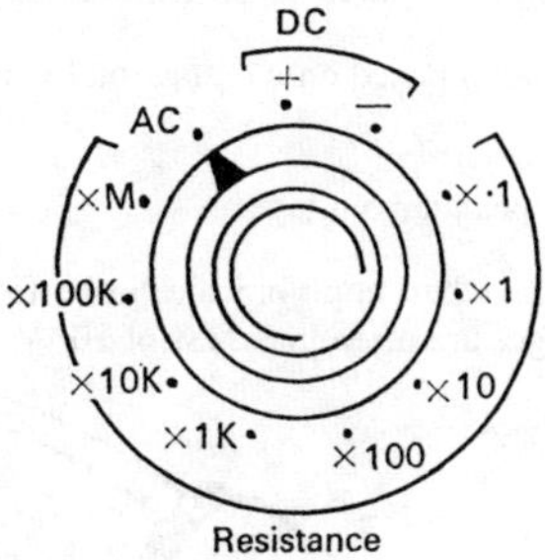

Switch on the meter by rotating the *volts* range switch clockwise from the *off* position.

Set the range switch to maximum (300 V).

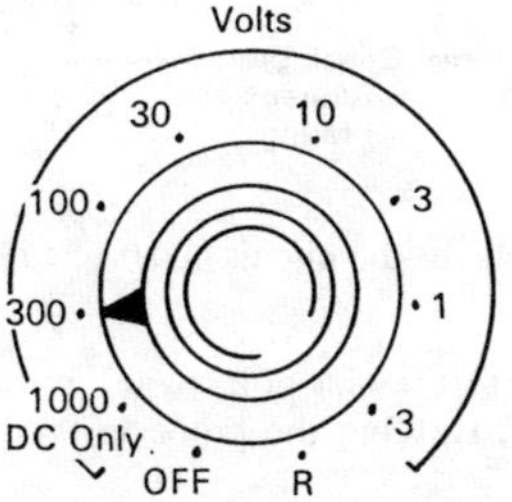

Connect the tip of the a.c. probe and its earth lead together.

Set the pointer to zero by adjusting the *a.c. zero* control.

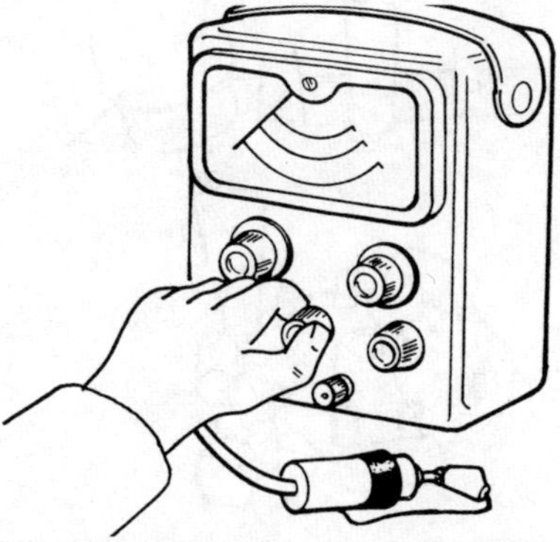

Before connecting the instrument to the circuit under test:

Check the maximum voltage rating of the probe. This rating is indicated in the manufacturer's instruction book.

Where a.c. is superimposed on d.c., the total voltage applied to the probe is:

d.c. voltage + peak a.c. voltage.

Take special care when applying the probe to valve circuits where d.c. voltages are often in excess of 100 V.

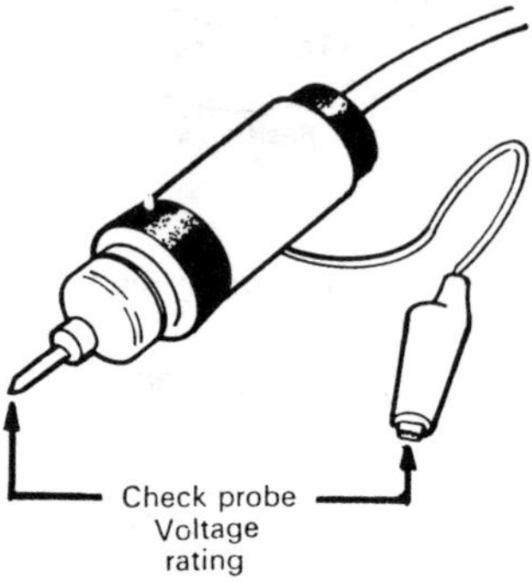

Connect the probe to the circuit under test as indicated in the diagram.

Clip the probe to the test point when measuring very low signal voltages. Holding the probe in the hand is liable to cause errors.

Observe any special instructions given in the instrument's handbook when connecting the probe to equipment operating at UHF.

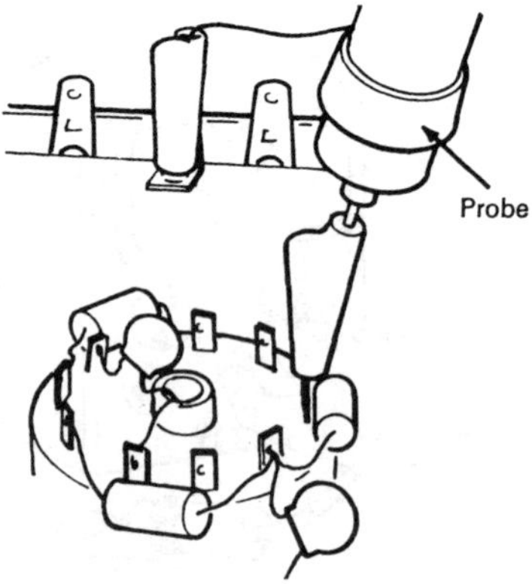

Re-set the *volts* range switch to obtain a suitable pointer deflection.

Note the reading.

Re-set the meter to its highest voltage range before disconnecting the meter from the equipment.

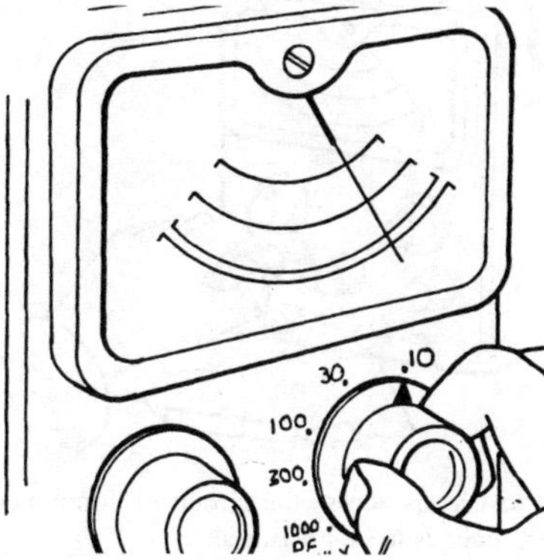

D.C. measurements

Set the *a.c./d.c./resistance* switch to d.c. positive or negative, depending on the polarity of the voltage being measured.

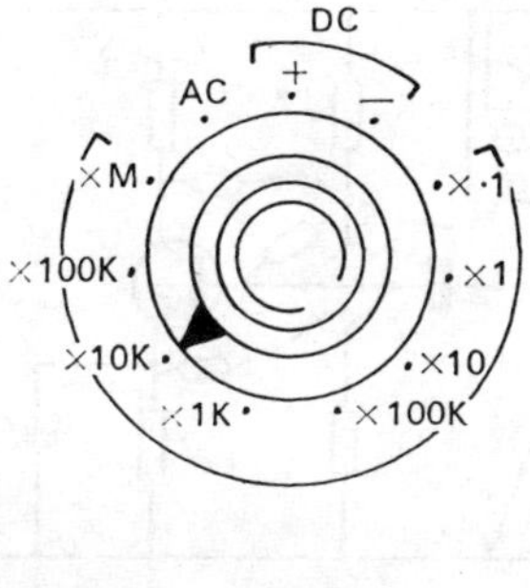

Set the *volts* range switch to 1,000 V.

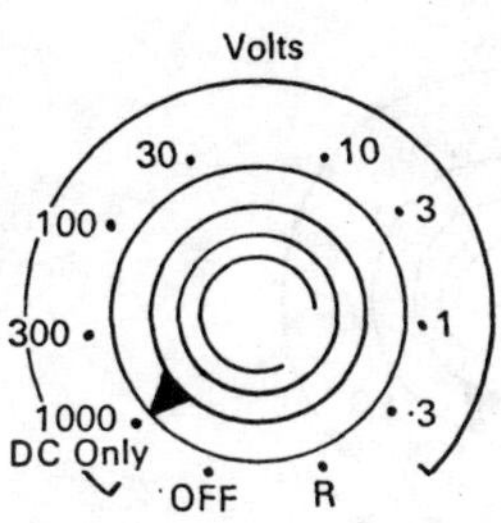

Connect the test leads to the negative (–) and positive (+) sockets.

Connect the free ends of the leads together.

Set the pointer to zero by adjusting the *d.c. zero* control.

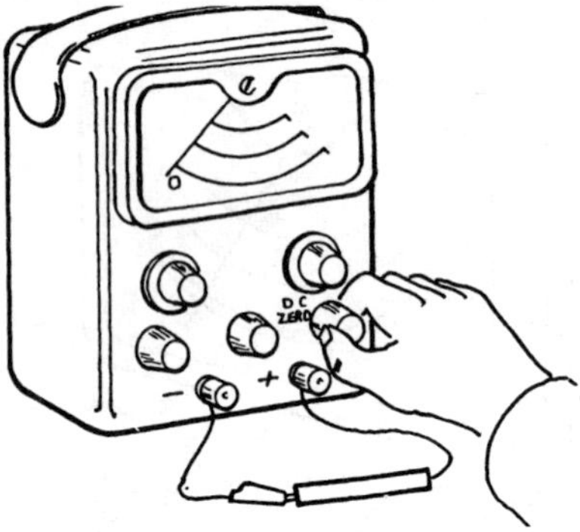

Connect the test leads across the circuit or component whose potential difference is to be measured.

Re-set the *volts* range switch to obtain a suitable deflection.

Note the reading.

Re-set the meter to a high voltage range before disconnecting it.

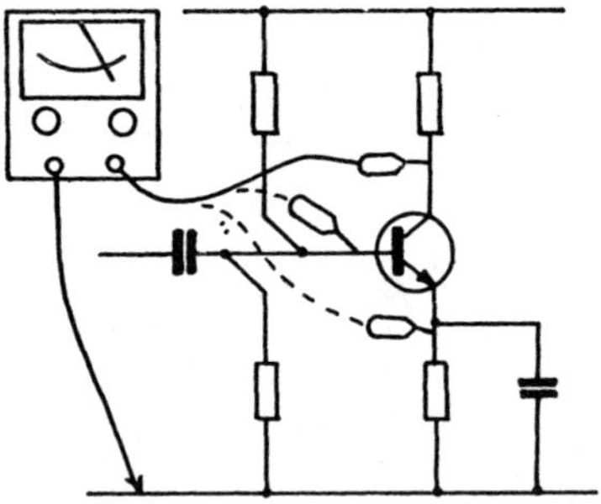

Resistance measurements

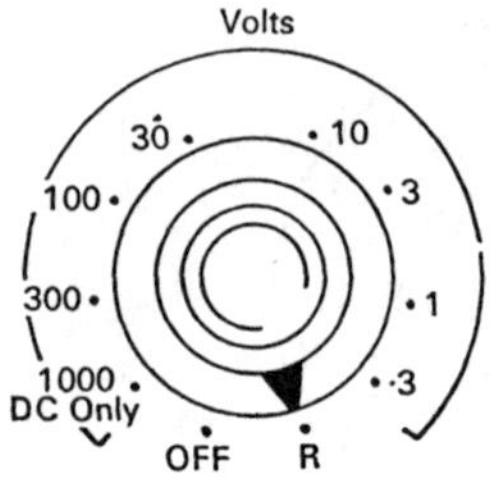

Set the *volts* range switch to R.

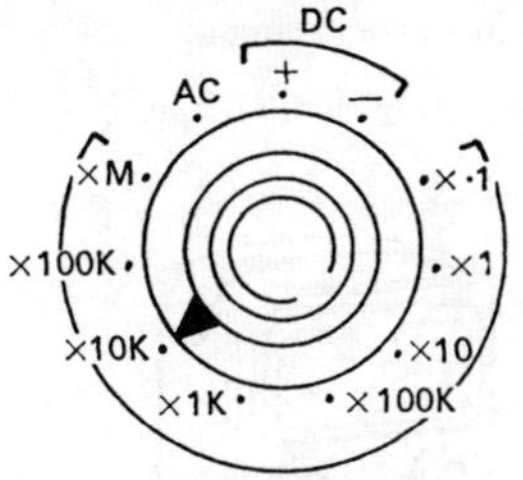

Set the a.c./d.c./resistance switch to an appropriate resistance range.

Connect the positive (+) and negative (–) test leads together.

Adjust the Ω *zero* control to obtain a zero resistance reading.

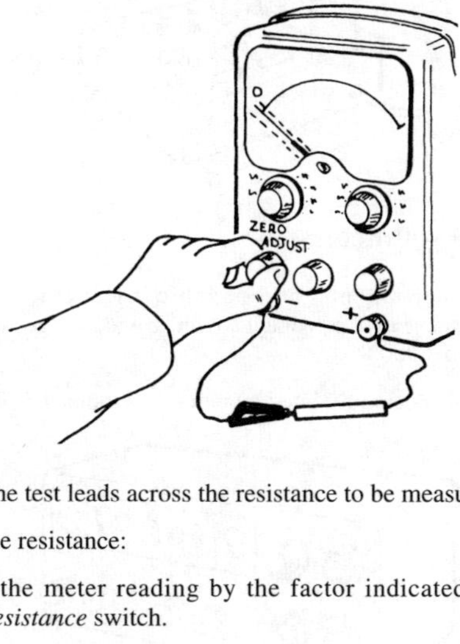

Connect the test leads across the resistance to be measured.

To read the resistance:

Multiply the meter reading by the factor indicated on the *a.c./d.c./resistance* switch.

After use, set the meter to a high voltage range.

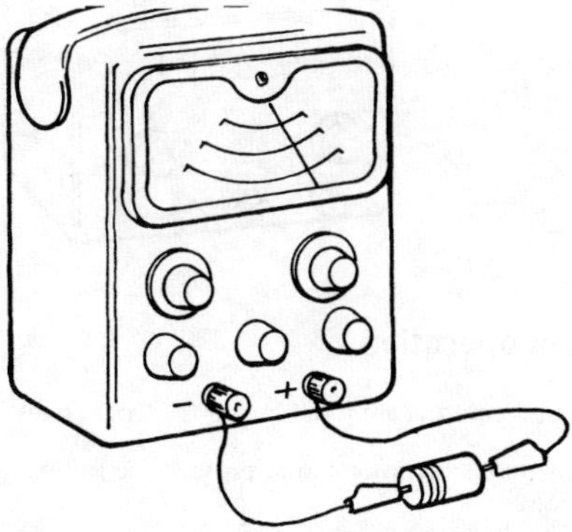

To measure the resistance of a component in circuit:

- disconnect the equipment under test from the supply
- proceed as shown above.

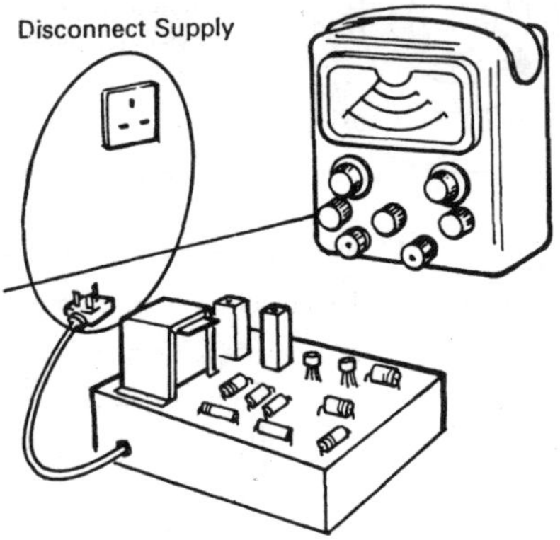

Digital voltmeters

These instruments are capable of making accurate measurement of d.c. voltages over a wide range of typically 100 μV to 2,000 V.

The readings are presented on numerical indicators.

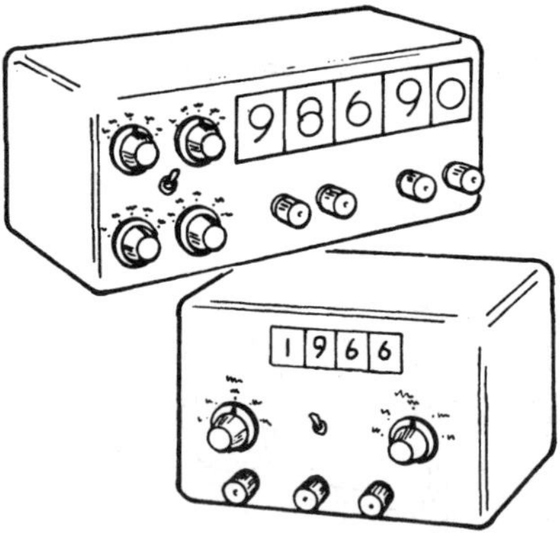

Modes of operation

All operating controls are mounted on the front panel.

A mode switch is incorporated to provide the following modes of operation:

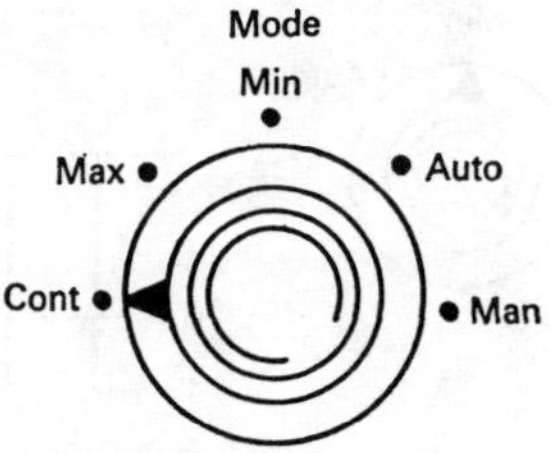

Continuous reading

Readings are repeated 50 times per second. Operation of a *trip* button will freeze a particular reading.

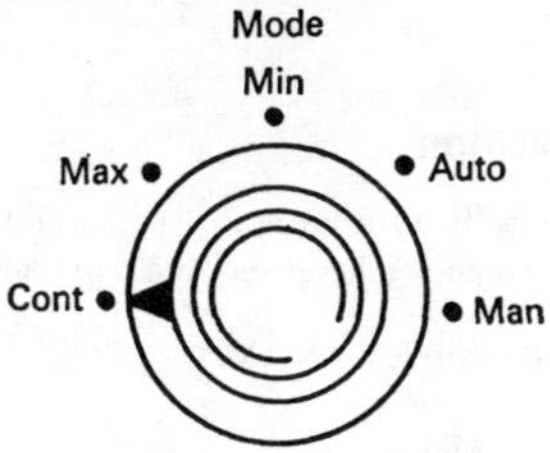

Maximum readings

The instrument will register the most positive excursion of a changing input voltage.

The readings recorded in the two examples would be:

- + 6 V
- – 2 V.

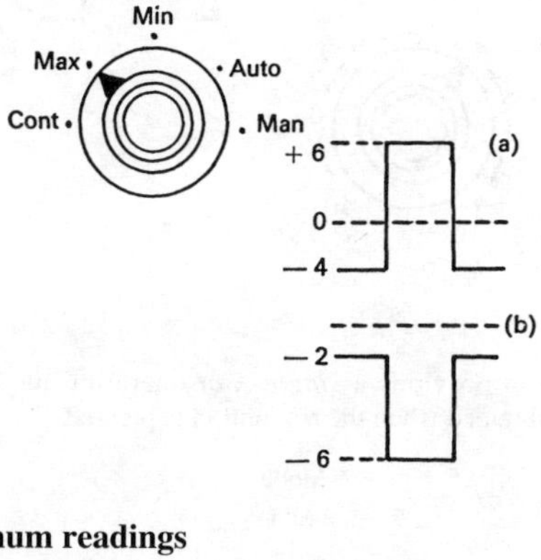

Minimum readings

The instrument will register the most negative excursion of a changing input voltage.

In this case the readings would be:

- – 4 V
- – 6 V.

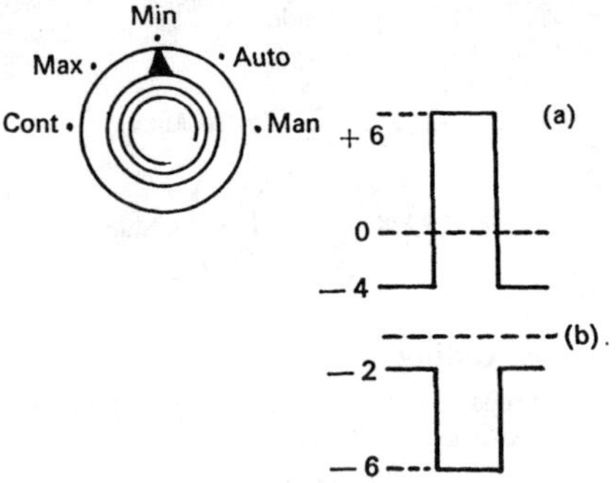

Automatic tracking

The instrument will automatically register the amount by which the input exceeds a level set on a *trip* control.

In this example the trip level = 4 V.

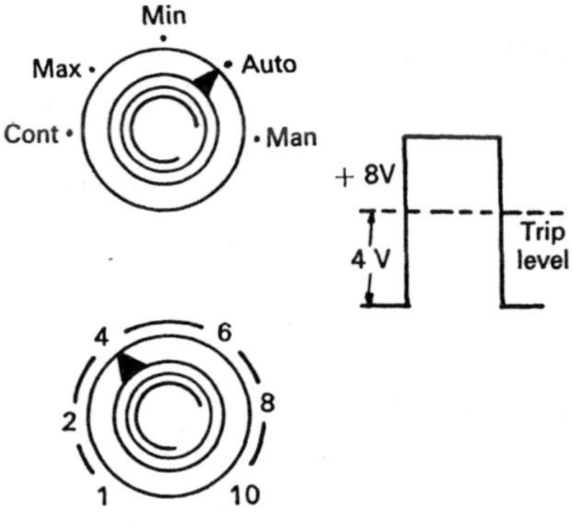

Manual

This position provides a *single-shot* operation in which a reading is obtained when the *trip* button is pressed.

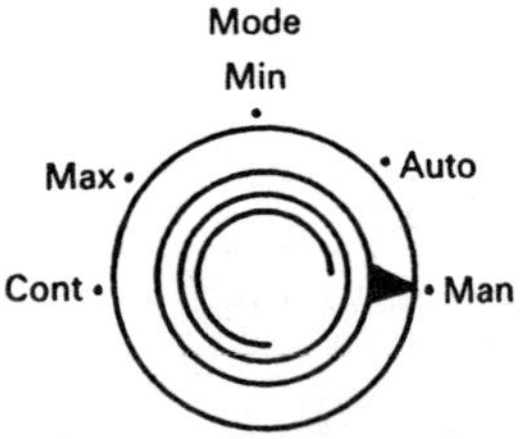

Calibration

This facility is provided to switch an internal standard voltage to the input so that the accuracy of the indications may be checked. Zero is also checked. Set zero first.

If the instrument is found to be inaccurate, refer to the test equipment engineer or manufacturer's instruction manual.

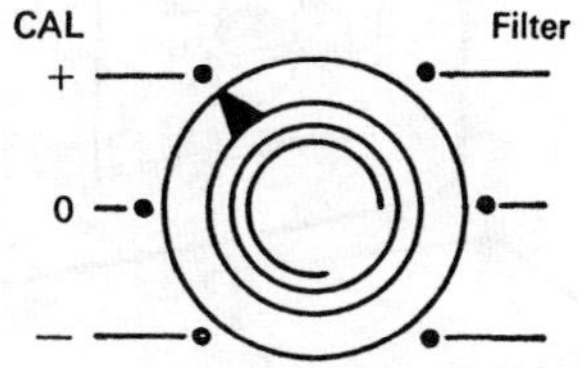

Filter

This facility connects a filter to the instrument's input circuits and may be used to offset the effects of noise and hum on the input from the unit under test.

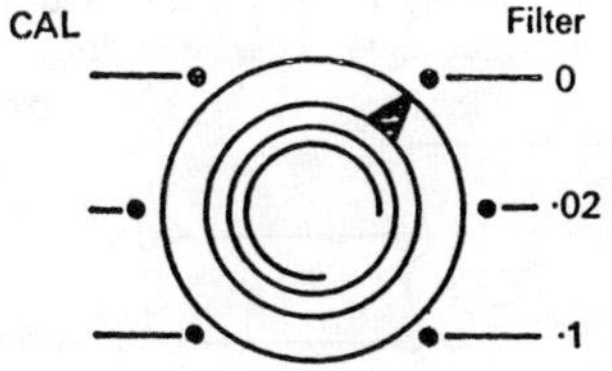

Using the digital voltmeter

Do not stand the unit near sources of heat.

Allow adequate space for ventilation.

Do not obstruct cooling louvres.

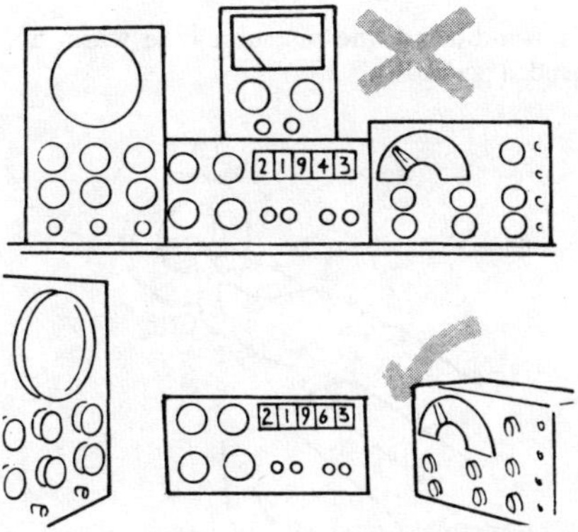

Ensure that the unit is set to operate from the appropriate mains supply.

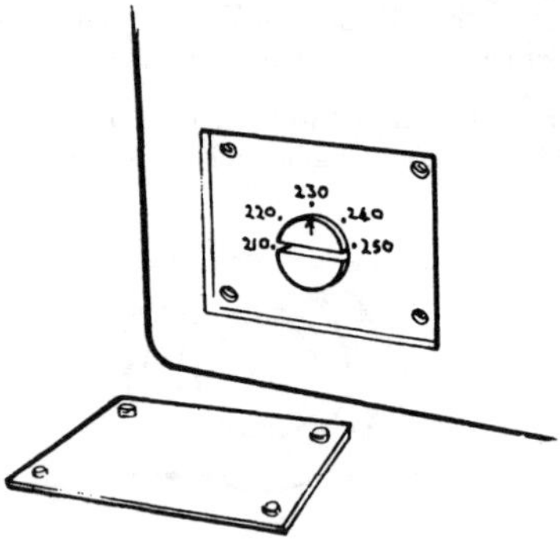

Make sure that instructions in the manufacturer's handbook regarding earthing arrangements are followed. Failure to do this may result in erratic and inaccurate readings.

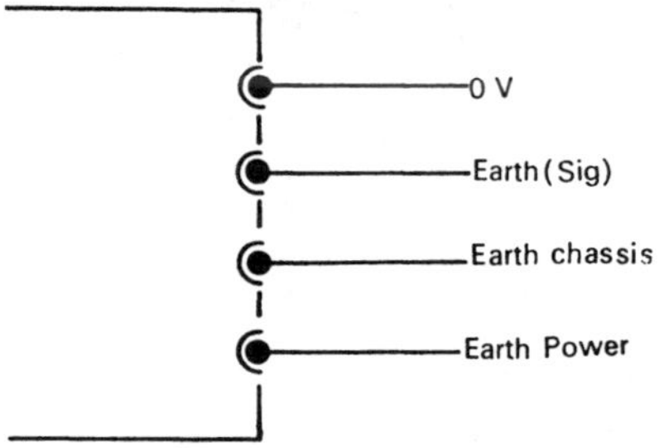

Switch on and allow the instrument to warm up for the recommended period.

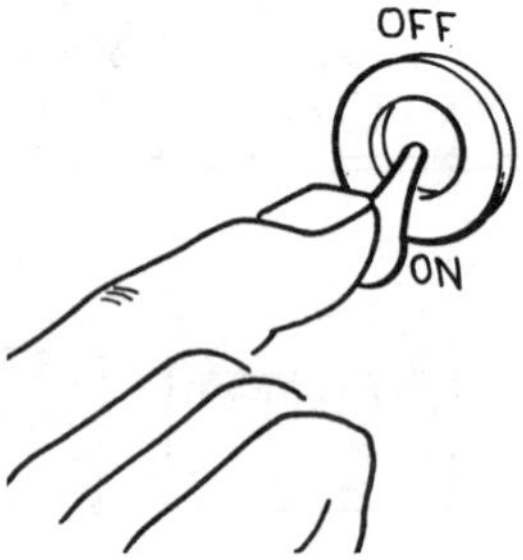

Always use the leads provided with the instrument.

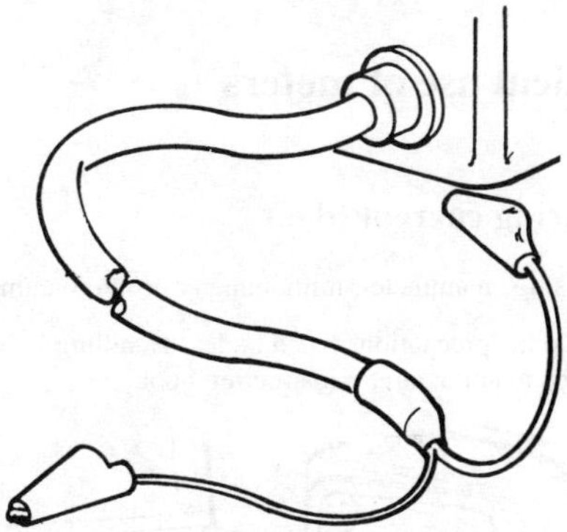

Do not connect a DVM to a completely unknown voltage. Determine the approximate reading you will obtain and select the appropriate range on the DVM.

Check that the maximum voltage which may be applied between either of the terminals and earth will not be exceeded.

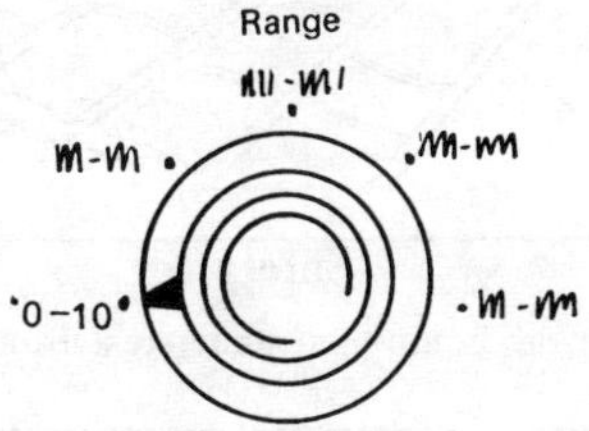

Practical use of meters

Measuring current (d.c.)

Before using an ammeter, milliammeter or microammeter:

- observe the precautions listed under "Handling Meters"
- read the manufacturer's instruction book.

Safety

Disconnect the equipment under test from the power supply.

Set the meter to its highest range.

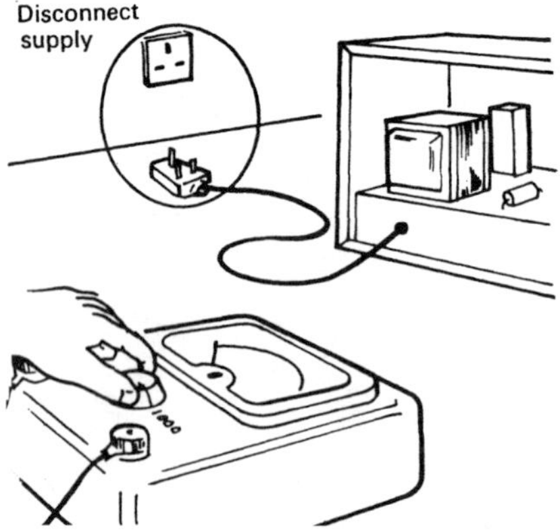

Connect the meter *in series* with that part of the circuit where the current value is to be measured.

Connect the meter in the correct polarity — *positive* meter terminal to the *positive* side of the circuit.

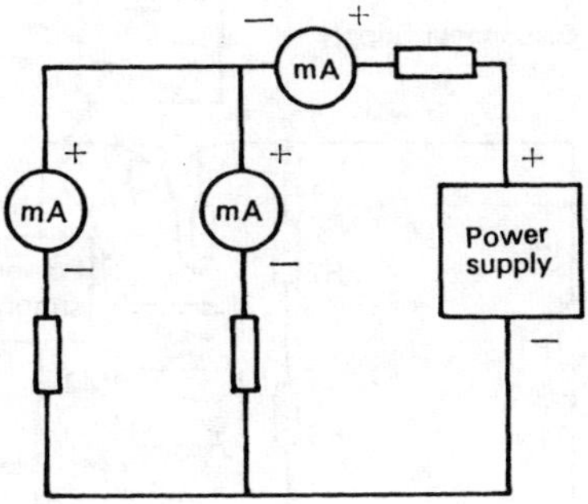

Remember that, to connect an ammeter, milliammeter or microammeter correctly *some part of the circuit must be disconnected.*

Quality

Never connect a milliammeter:

- in parallel with a source of voltage such as the mains supply, a battery or power supply unit
- in parallel with a circuit component.

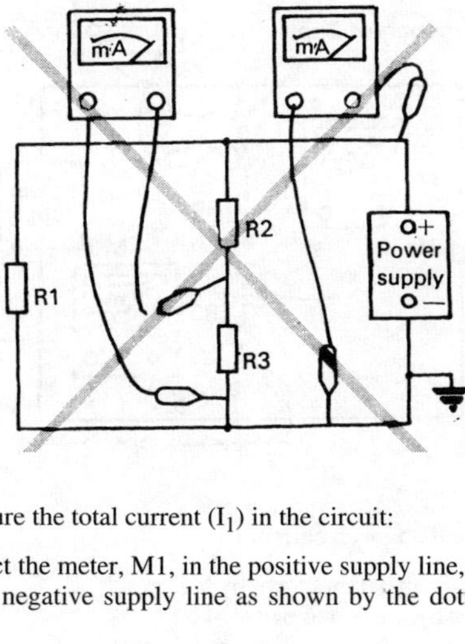

To measure the total current (I_1) in the circuit:

- connect the meter, M1, in the positive supply line, or
- in the negative supply line as shown by the dotted meter, M2.

Note the polarity of the meter terminals for each method of connection.

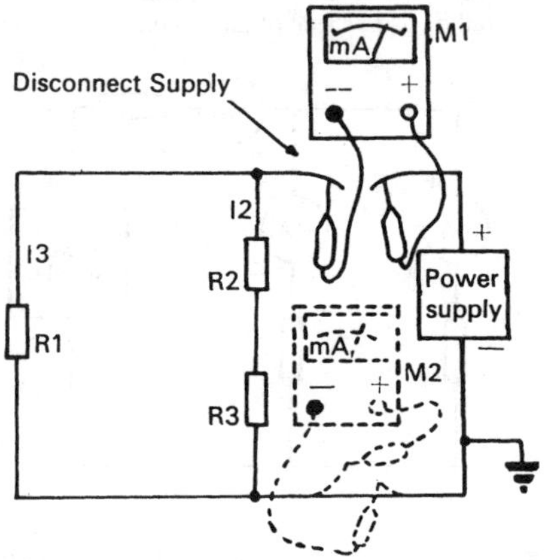

To measure the branch current I_2:

- connect the meter as shown by M1, or
- as shown by the dotted meters, M2 and M3.

Note the polarity of the meter terminals.

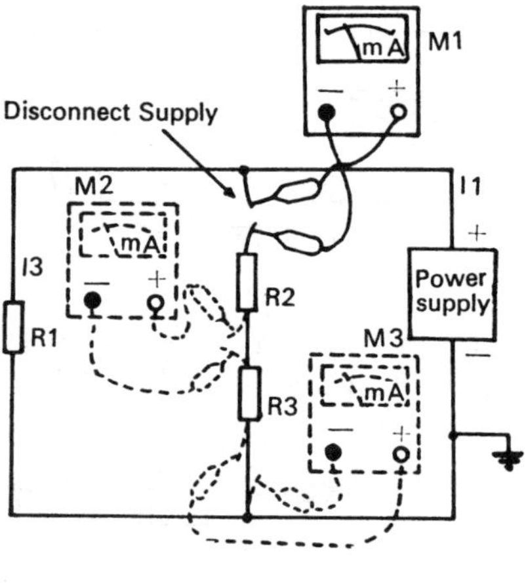

To measure the branch current I_3:

- connect the meter as shown by M1, or
- as shown by the dotted meter M2.

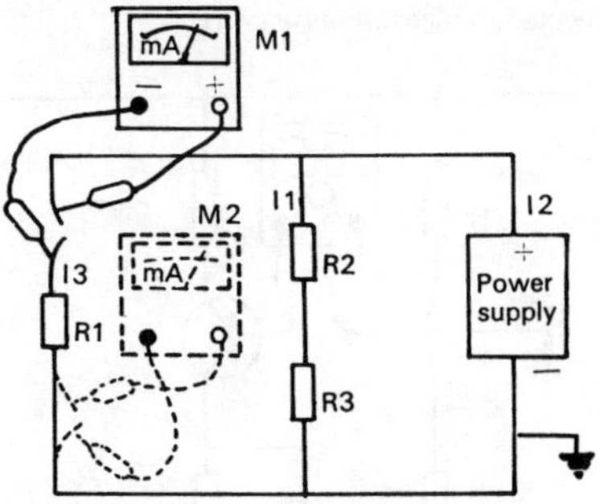

Switch on the equipment.

Adjust the range switch to obtain a suitable pointer deflection.

Note the reading.

Safety

Disconnect the power supply to the equipment under test.

Set the meter to its highest range before disconnecting it. Reconnect the circuit component.

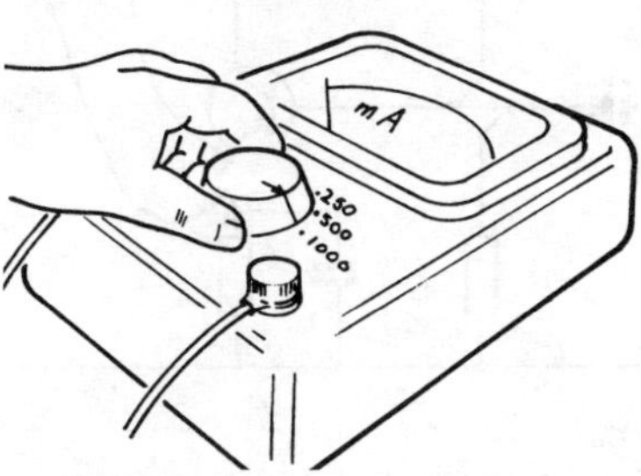

Measuring transistor currents

Meter connected to measure *collector* current.

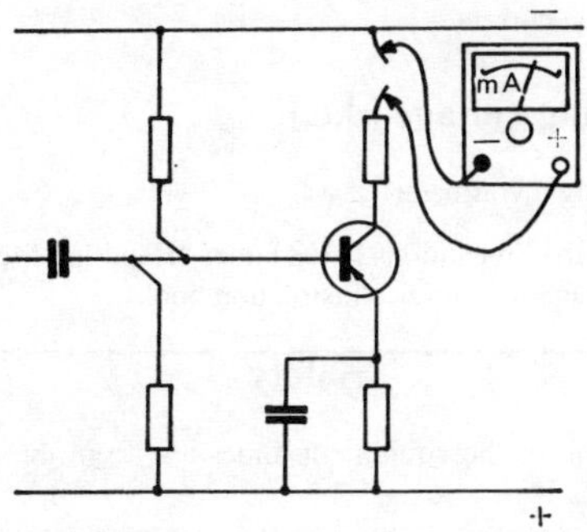

Meter connected to measure *base* current.

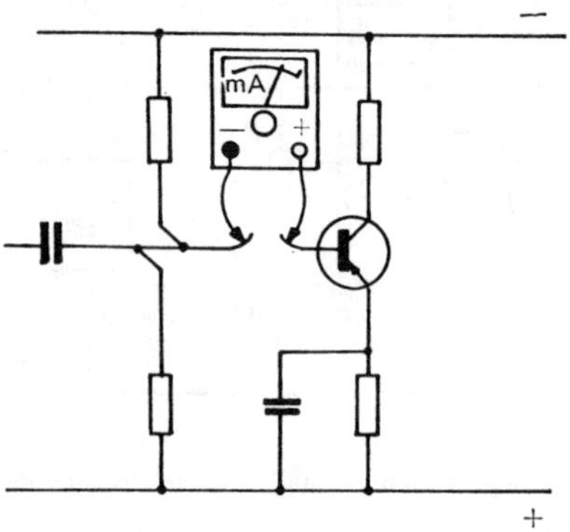

Meter connected to measure *emitter* current.

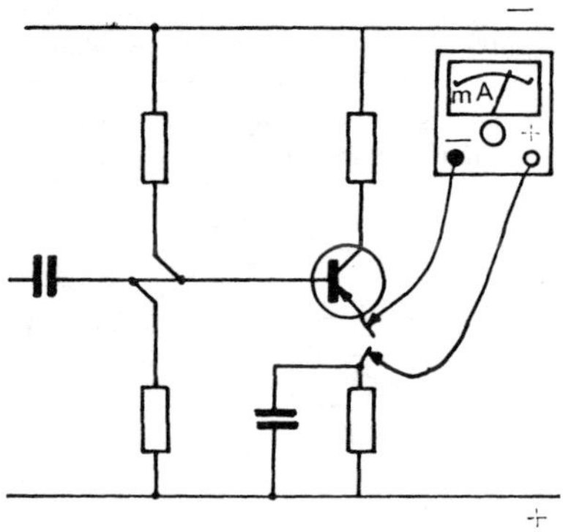

It is often impractical to disconnect components to measure current in the manner shown previously. The current can be determined approximately by measuring the voltage developed across a resistor in the current path. Use Ohm's law to calculate the current.

Measuring voltage (d.c.)

Before using a voltmeter:

- observe the precautions listed under "Handling Meters"
- read the manufacturer's instruction book.

Safety

Disconnect the equipment under test from the power supply.

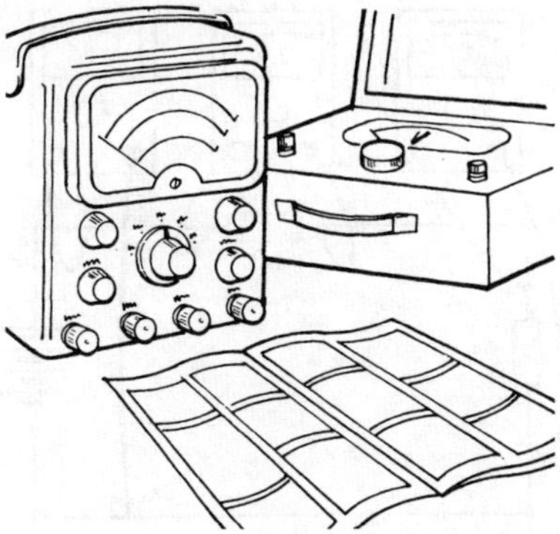

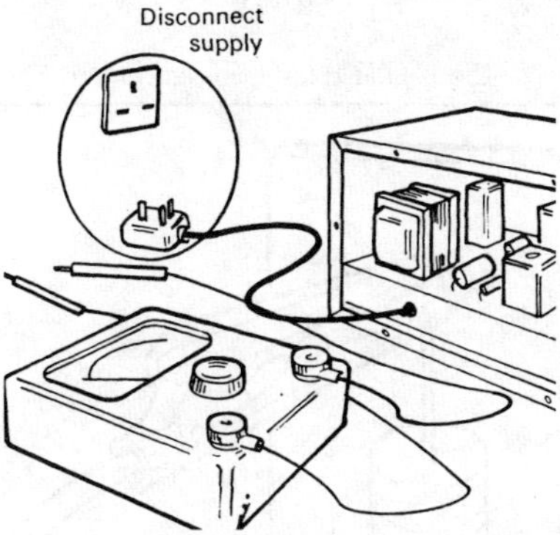

Set the meter to its highest range.

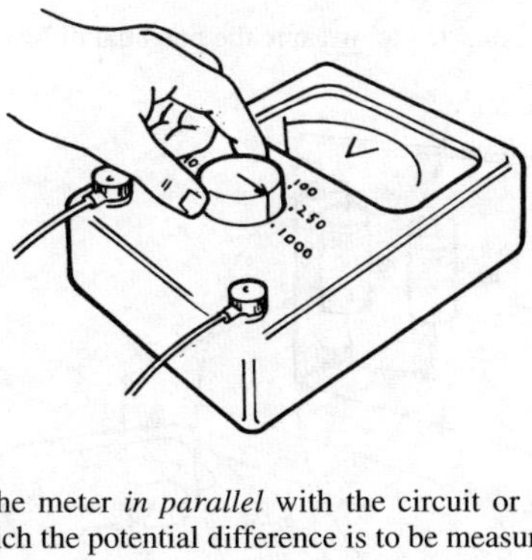

Connect the meter *in parallel* with the circuit or component across which the potential difference is to be measured.

Connect the meter in the correct polarity — *positive* meter terminal to the *positive* side of the circuit under test.

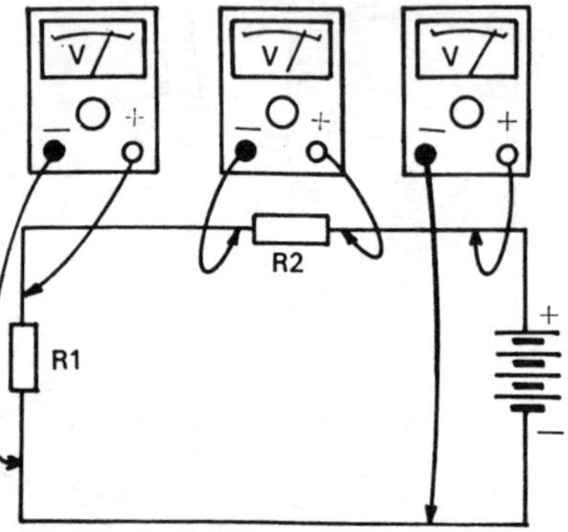

Safety

Use insulated test leads and clips.

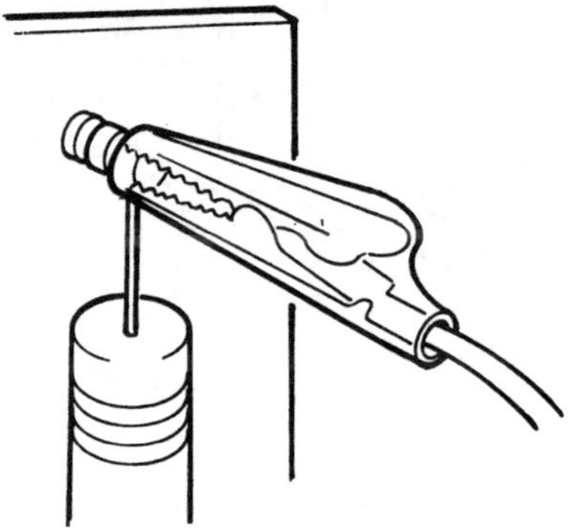

Voltmeter connected to measure the potential difference across R2.

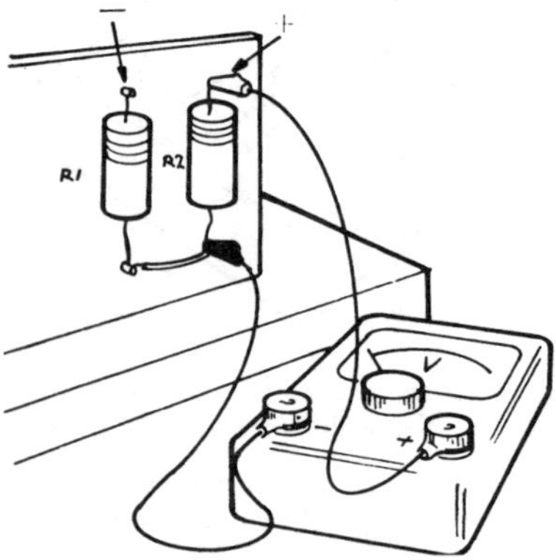

Connect the equipment under test to the supply.

Adjust the meter range switch to obtain a suitable pointer deflection.

Note the reading.

Disconnect the equipment from the supply.

Set the meter to its highest range before disconnecting it.

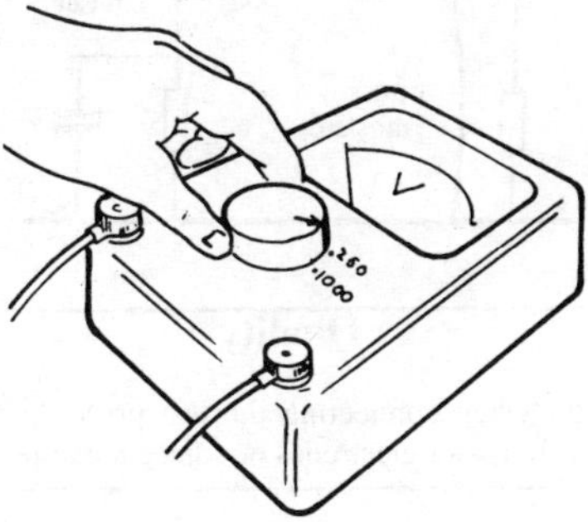

Measuring transistor voltages (d.c.)

Use a high resistance meter such as an electronic voltmeter.

Connect the voltmeter as indicated in the test specification.

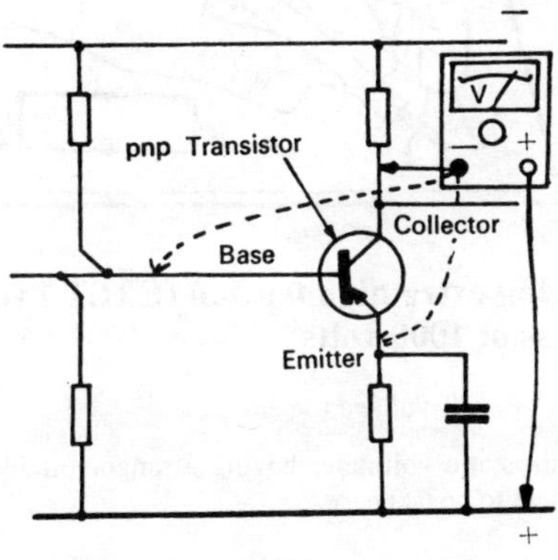

Collector, base and emitter voltages are usually measured with respect to the low or the high potential side of the circuit, depending on the type of transistor.

The diagrams show methods of connecting the voltmeter to circuit using:

- a pnp transistor
- an npn transistor.

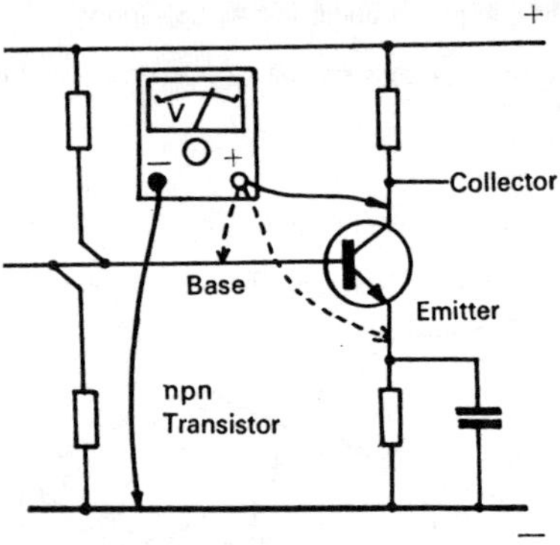

Quality
Be careful when connecting the test prod. Avoid short circuiting transistor electrodes or adjacent components.

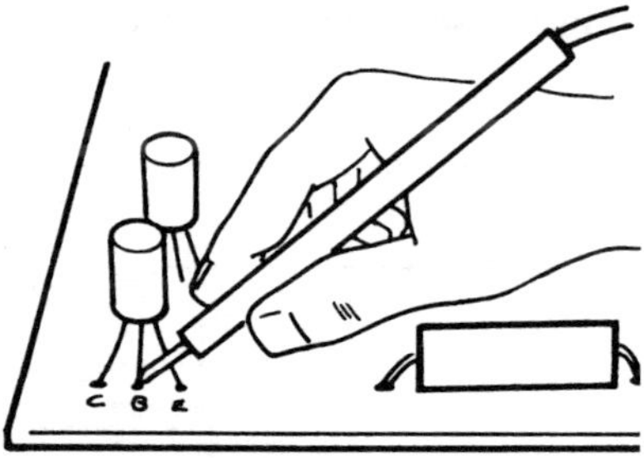

Measuring extra high tension (E.H.T.) voltage (in excess of 1000 volts)

To measure E.H.T. voltage use either:

● an electrostatic voltmeter having a range suitable for the measurement to be taken, or

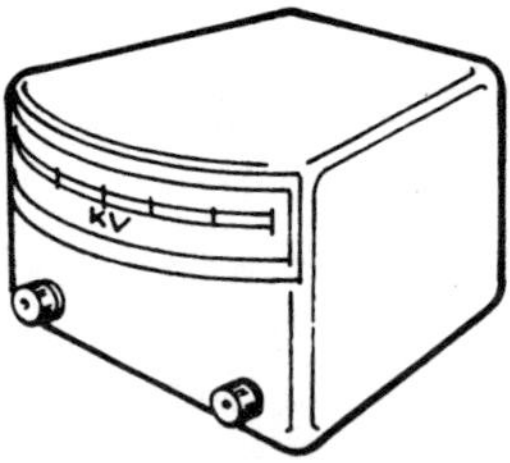

● a multimeter switched to a d.c. voltage range and fitted with a special adaptor called an *E.H.T. multiplier*.

Use the meter as specified in the manufacturer's instruction book.

Safety

E.H.T. voltages can be lethal — take great care.
Use the recommended high voltage test leads, specially insulated terminal connectors and test clips.

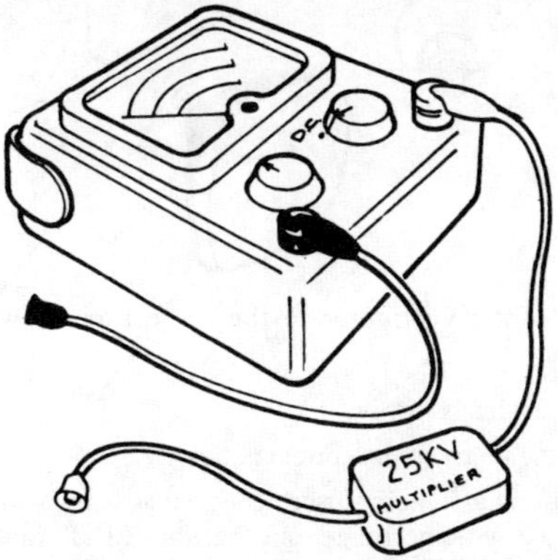

To measure E.H.T. voltage proceed as follows:

Safety

● disconnect the equipment under test from the supply

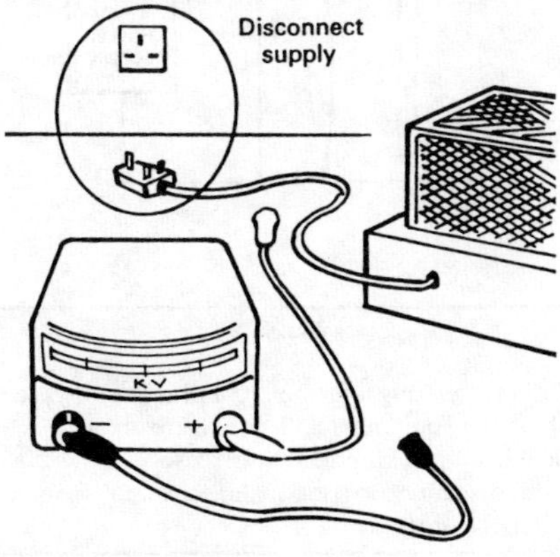

Safety

● according to company procedure, discharge components such as capacitors and cathode-ray tubes. These can hold a charge for some time after the equipment is switched off.

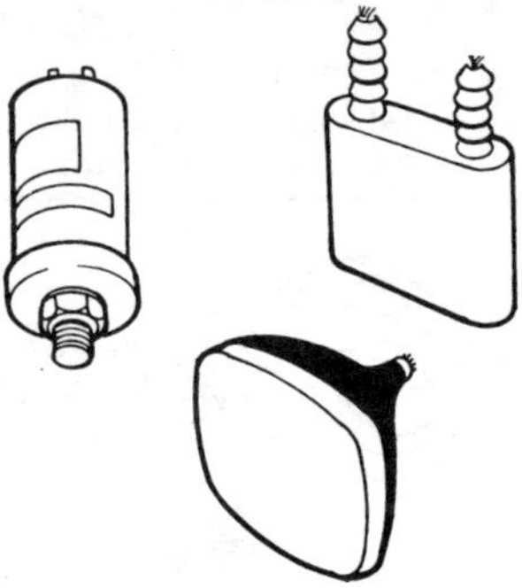

● connect the voltmeter to the circuit as shown in the diagrams.

Connect test leads firmly.

Connect the meter in the correct polarity.

If a multimeter is being used, check that it is switched to a suitable d.c. voltage range and that the E.H.T. multiplier is connected as specified.

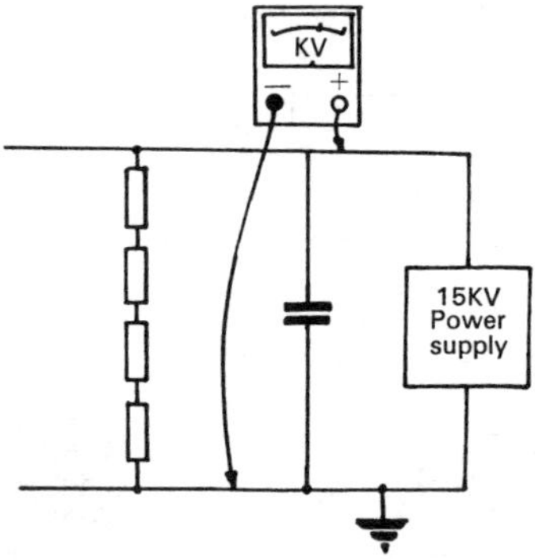

Safety

● without touching the meter, multiplier or test leads, switch on the equipment and take the reading

● disconnect the equipment under test

● before disconnecting the meter, ensure that its reading has fallen to zero.

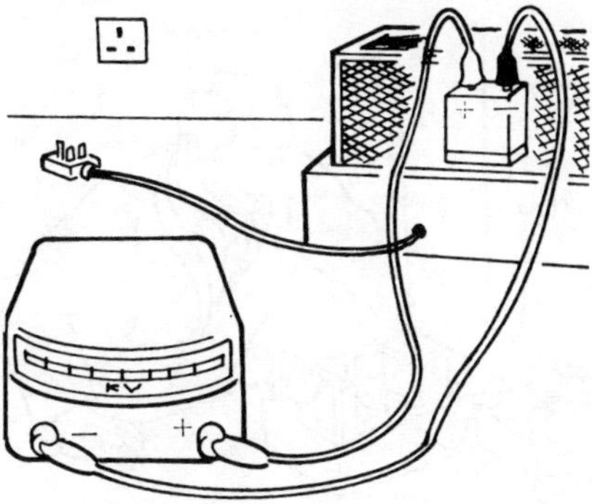

Measuring a.c. voltage

Safety

Take extra care when measuring a.c. voltage.

● disconnect the equipment under test from the mains before connecting (or disconnecting) the meter

● use insulated terminal connectors, test leads and clips.

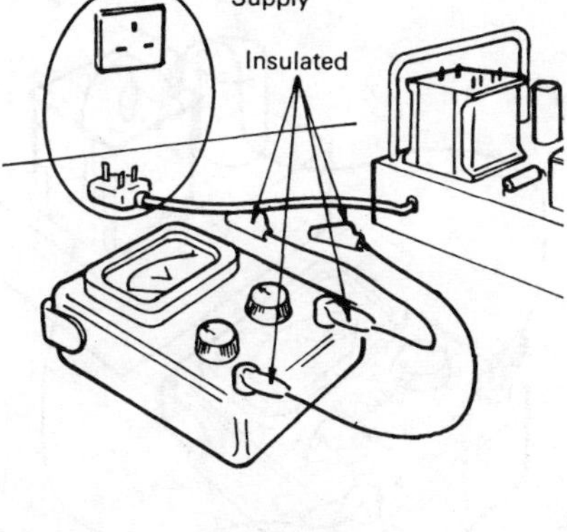

Safety

Position the equipment to avoid accidental contact with live points.

In the case of an a.c. voltmeter connected to measure the voltage across the secondary winding of the mains transformer T1, the polarity of the meter terminals is immaterial for a.c. measurements.

To measure the voltage proceed as follows:

Safety

Disconnect the power supply unit from the mains

- switch the voltmeter to its 1,000 V a.c. range
- *using insulated test clips*, connect the meter across the secondary winding (points 1 and 2)

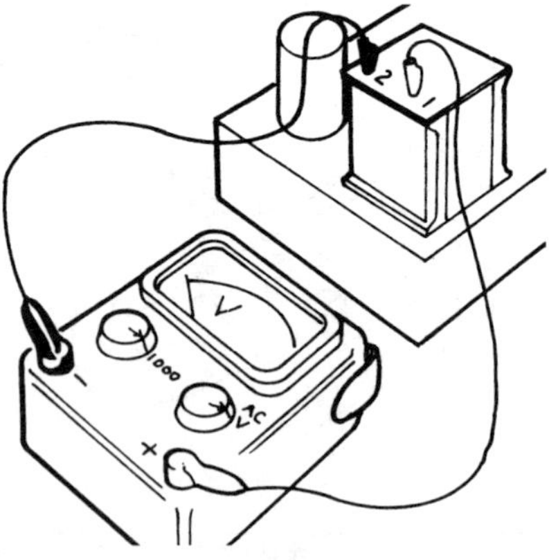

- connect the power supply unit to the mains
- re-set the range switch to provide a suitable deflection
- note the reading

Safety

Disconnect the unit from the mains

● set the meter to its highest a.c. voltage range and disconnect the meter.

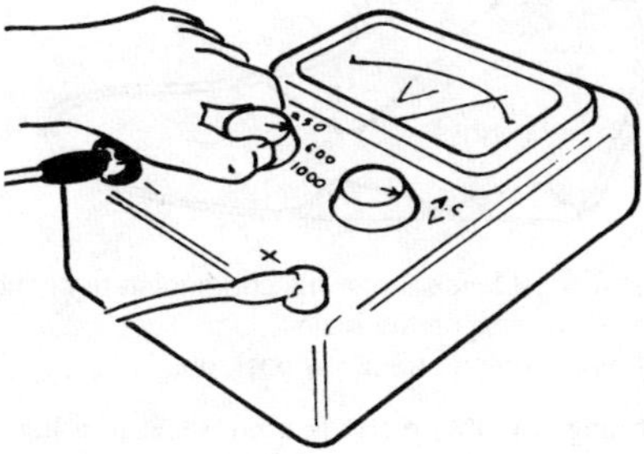

Measuring resistance

Use an *ohmmeter* to measure resistance and circuit continuity.

To measure the d.c. resistance of a component, proceed as follows:

● check that the pointer reads maximum resistance (series-type ohmmeter) or zero resistance (shunt-type ohmmeter).

If necessary, adjust pointer.

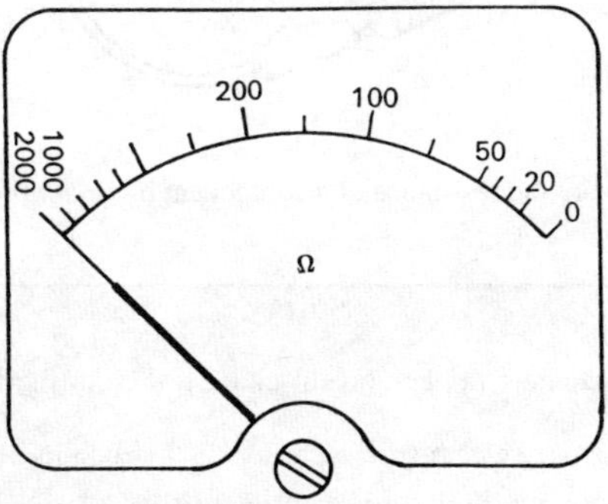

● set the meter to a suitable resistance range
● connect the test leads together
● set the *zero adjust* control to obtain a zero resistance reading

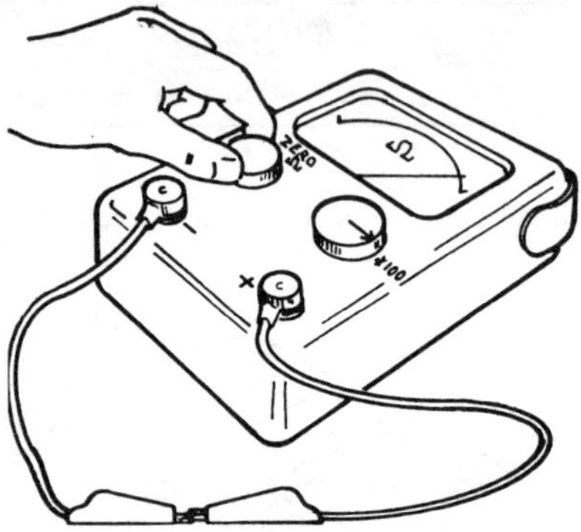

- connect the test leads across the component to be measured
- read the resistance on the scale
- switch the ohmmeter to the *off* position.

If a multirange (VOM) meter is used, set it to a high voltage range.

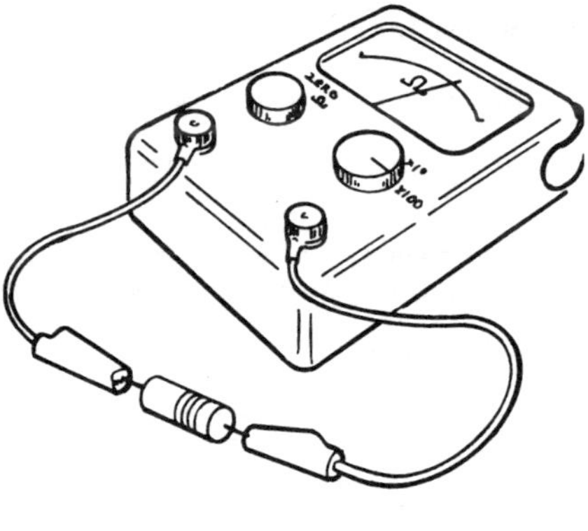

To measure the resistance of a component in *circuit* proceed as follows.

Safety

- disconnect the equipment under test from the power supply
- discharge electrolytic capacitors in the equipment.

Various methods are used to do this → **ask your supervisor**.

The ohmmeter will be damaged if measurements are attempted with the equipment switched on, or residual voltages left on large value capacitors.

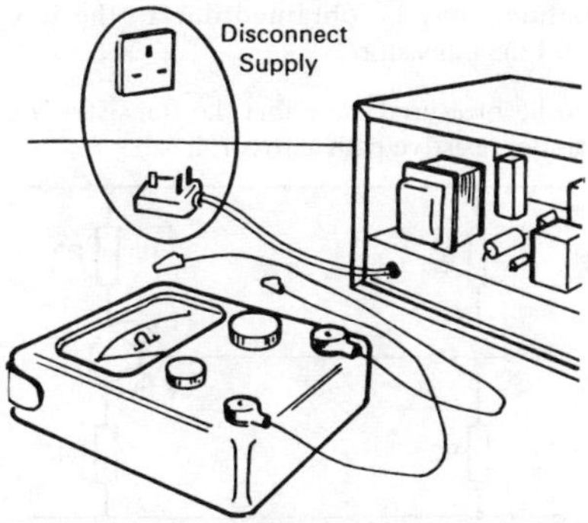

Quality

Transistors are easily damaged by the meter's internal batteries.

Do not apply an ohmmeter between any leads of a transistor — in or out of circuit — unless you are given specific instructions on how to do so.

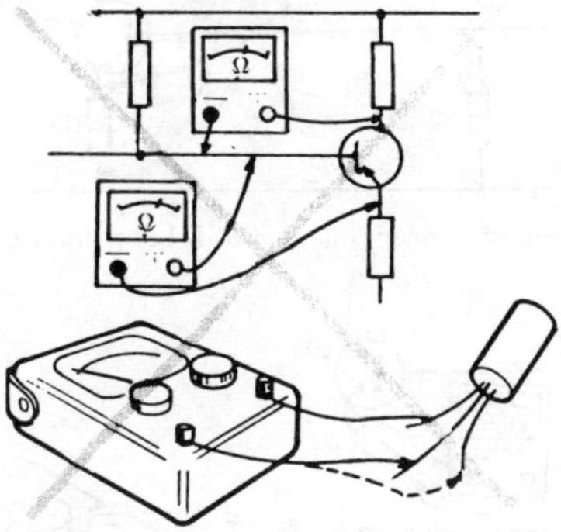

● connect the ohmmeter across the components to be measured.

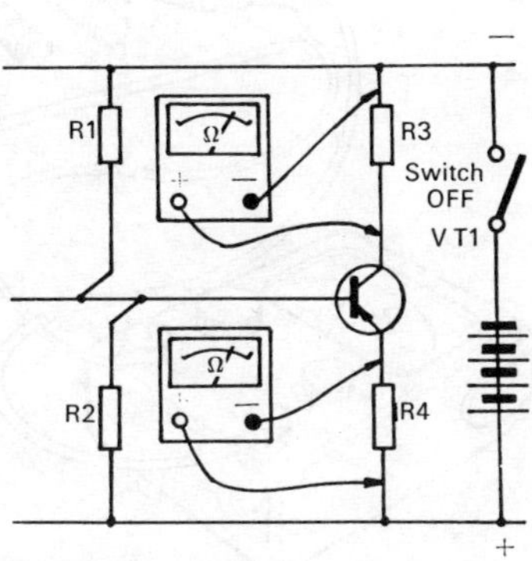

False readings may be obtained due to the low forward resistance of the transistor.

If R4 is to be measured, note that the transistor VT1 and R2 form a parallel resistive path across R4.

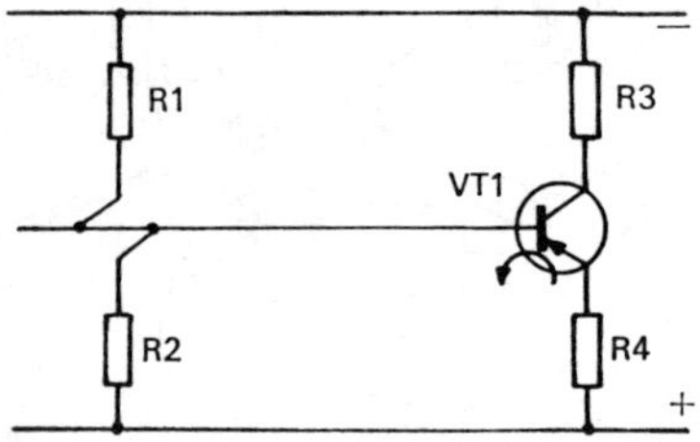

In this case either:

● reverse the meter connections to the circuit, when the correct reading will be obtained, or

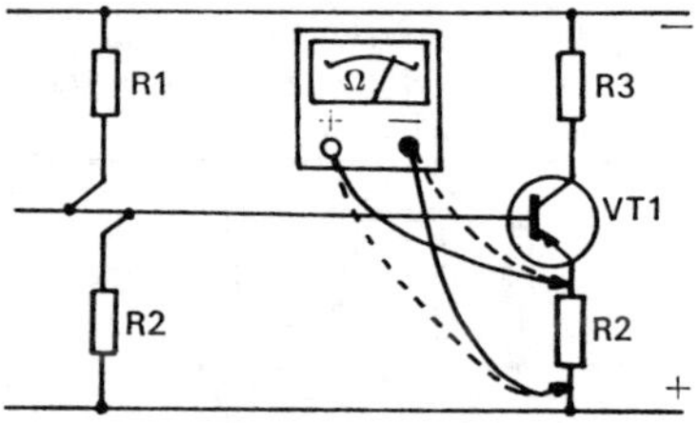

● better still, disconnect one end of R4 and connect the meter across it.

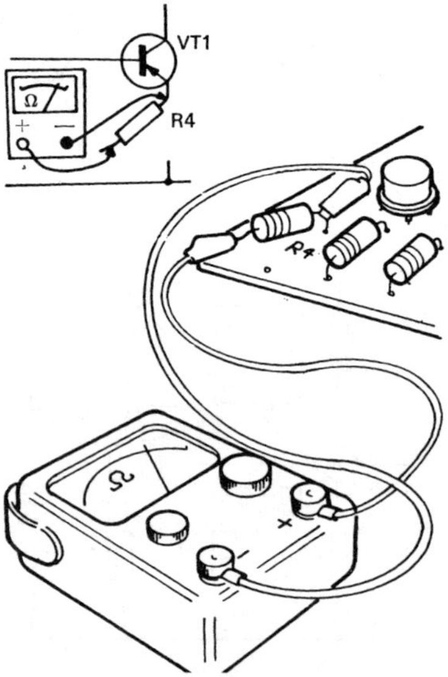

Connecting the ohmmeter as shown measures the effective resistance of R1 and R2 in parallel.

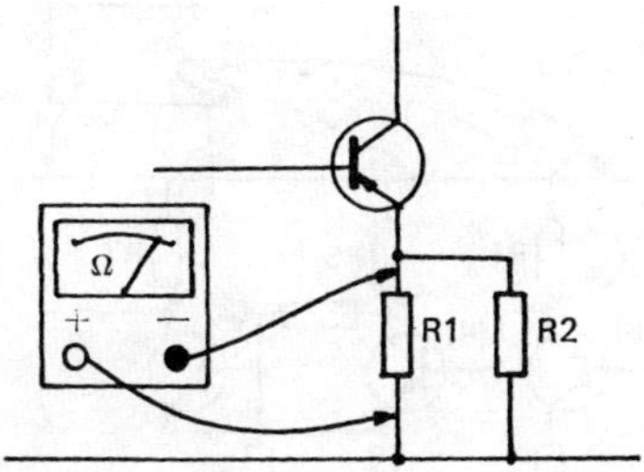

To measure R1:

Disconnect one end of the resistor as shown and connect the meter across it.

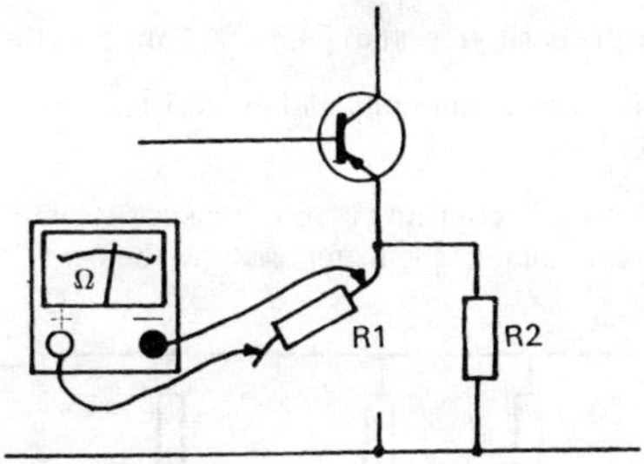

To measure R6 or R7:

Disconnect the appropriate resistor at the points shown and connect the meter across it.

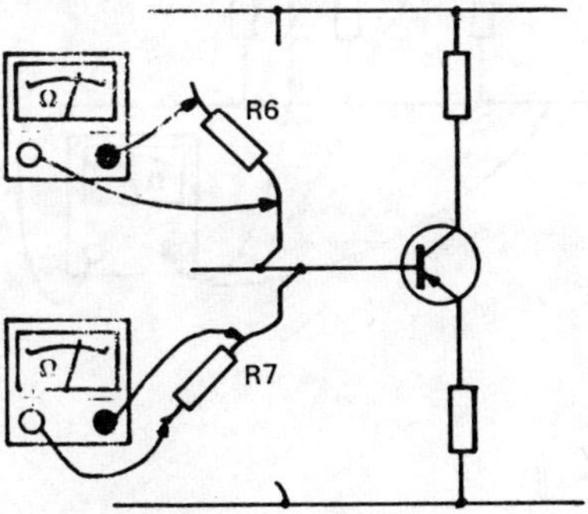

To check circuit continuity and resistance in the collector circuit proceed as follows:

- connect the negative test lead to point 1
- connect the positive test lead to point 2.

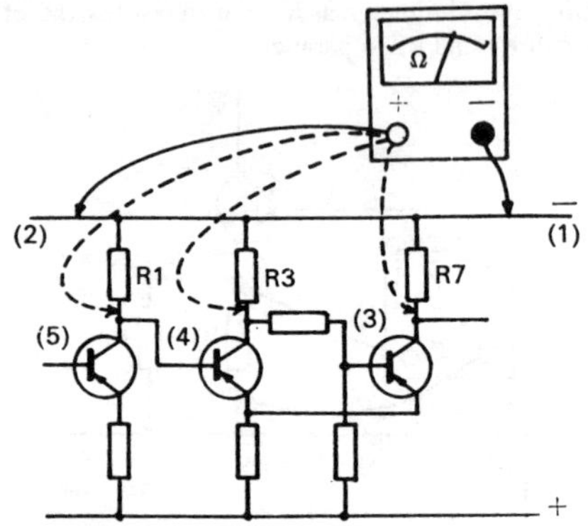

This checks the continuity of the negative supply line.

● transfer the positive lead to points 3, 4 and 5 in turn.

This checks circuit continuity and the resistance of R7, R3 and R1, respectively.

To check circuit continuity and measure resistance in the emitter circuit connect the meter as shown below.

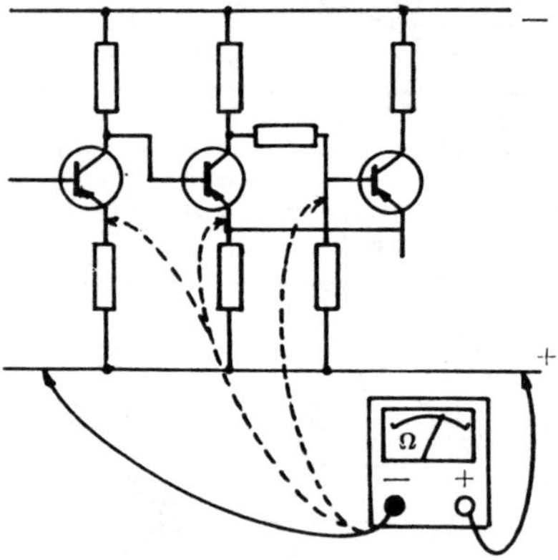

Signal generators

General

A signal generator provides a signal of known frequency, amplitude and waveform to test, service and measure the performance of equipment.

Signal generators are classified as general-purpose or standard-signal types according to:

- the accuracy of frequency and output level (amplitude)
- the stability of frequency and output level
- the amount of distortion present in the signal.

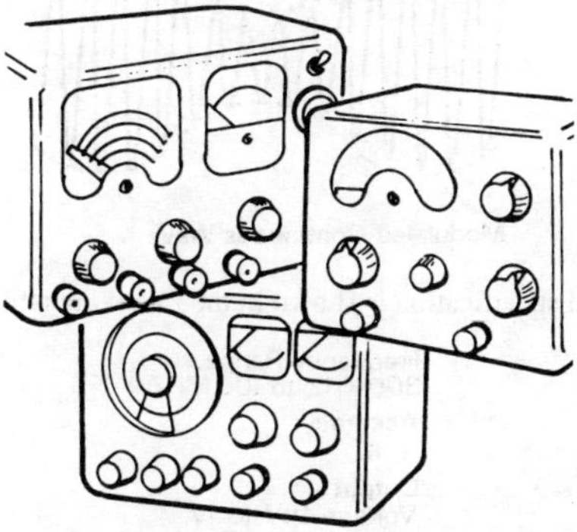

Check in the manufacturer's instruction book that the instrument's performance is suitable for the measurement to be made.

Use a general-purpose instrument where a high degree of accuracy and stability is not important.

Use a standard-signal generator where highly accurate measurements are to be made.

Radio frequency signal generators

A radio frequency (R.F.) signal generator provides:

- a continuous-wave R.F. signal
- a modulated continuous-wave R.F. signal.

These signals are described later in this chapter.

Instruments are available to cover the frequency range 100 kHz to 1,000 MHz or more.

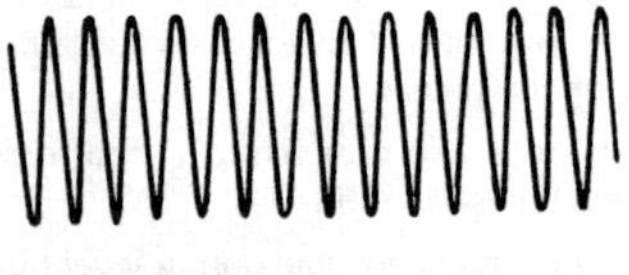

Continuous Wave

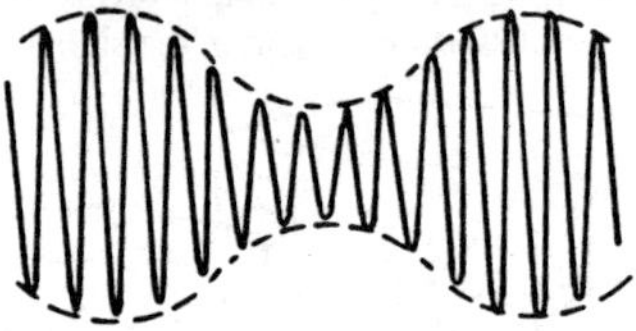

Modulated Continuous Wave

A typical specification is shown in the following table.

Frequency Range:-
300 KHZ to 100 MHZ

Accuracy:-
± 1%

Output:-
Voltage 1µV to 1V
Impedance 75Ω

Modulation:-
Internal AM: 400 Hz
Depth: 30%

Use an R.F. signal generator to test, service and measure the performance of equipment operating in R.F.

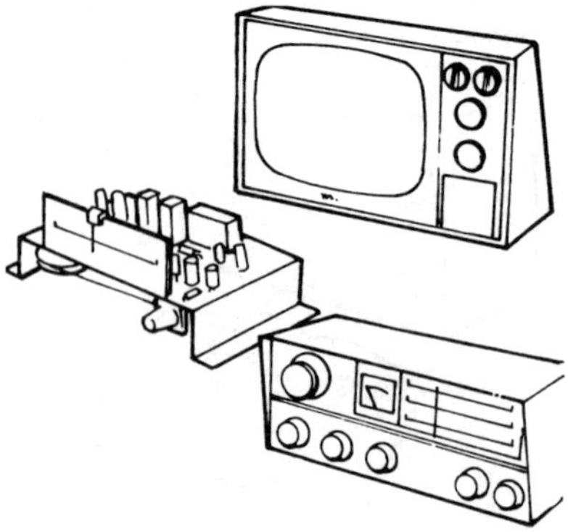

Audio frequency signal generators

Audio frequency (A.F.) signal generators provide an A.F. signal of:

- sine-wave output, or
- square-wave output
- or both.

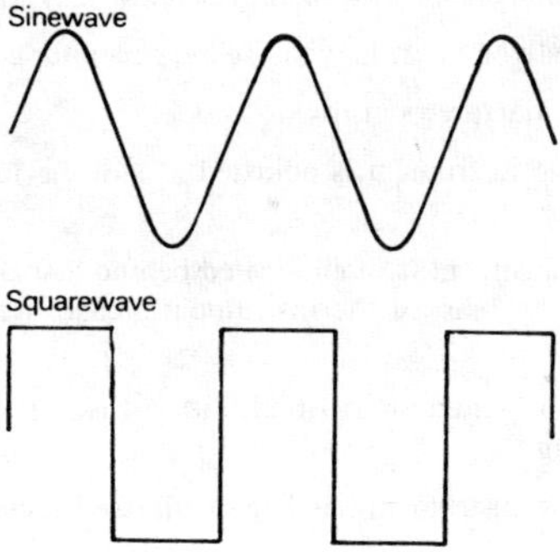

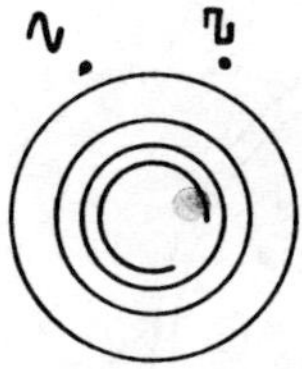

The required waveform is selected by a switch.

A typical A.F. signal generator covers the frequency range 5 Hz–100 kHz at output voltage adjustable from 0–30 V, this being the r.m.s. value for a sine-wave signal and peak-to-peak value for a square-wave signal.

Frequency selection is provided by a range switch, a frequency control and calibrated dial. Often, signal generators are of a digital nature but their operation is simple.

The output level is selected by attenuator (volts) controls.

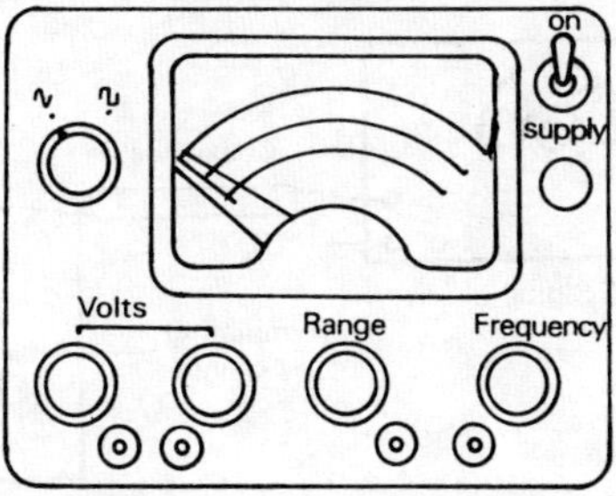

Use the instrument to test and measure the performance of equipment operating in the A.F. band.

Care and handling of signal generators

Handle signal generators carefully.

Their calibration can be disturbed by careless handling.

Make sure that you know how to use the generator correctly.

Read the manufacturer's instruction book.

Check that the instrument is adjusted to suit the local mains supply voltage.

The mains adjustment is often located behind a small panel at the base, or the rear, of the instrument's case. Refer to the instruction book.

Check that the generator is fitted with a fuse of the correct type and rating.

Check that the instrument has been calibrated recently. **If in doubt → ask.**

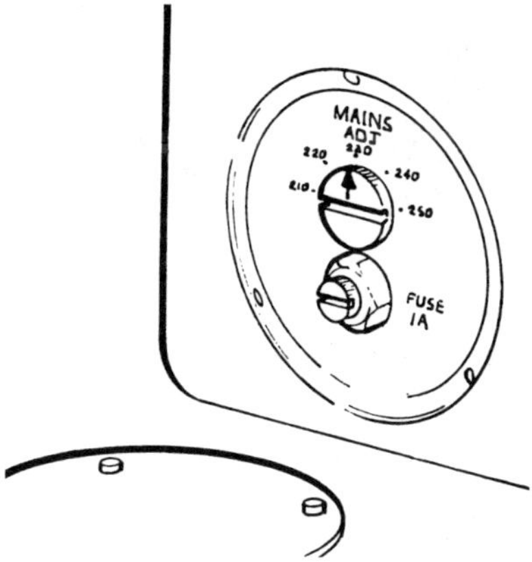

Use the specified test cable.

Terminate the cable with the recommended impedance.

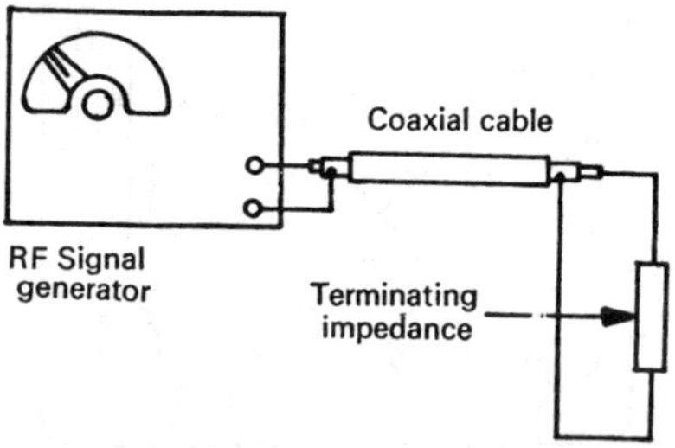

Use a blocking capacitor of specified value if the generator is to be connected to a point where a d.c. voltage is present.

Many signal generators are provided with a d.c. isolating unit for this purpose.

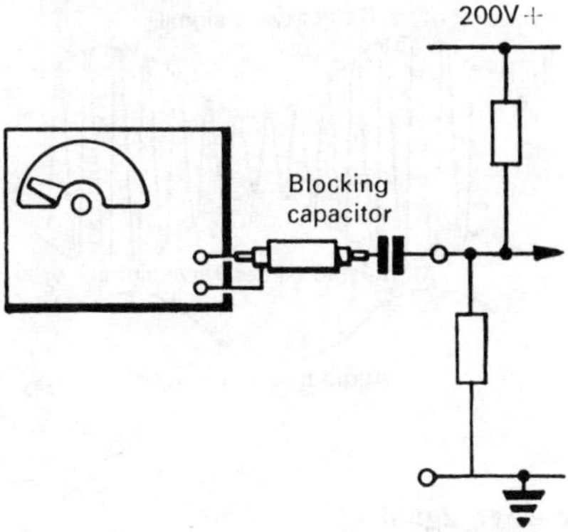

R.F. signal generators — mode of operation

The basic elements of a general purpose R.F. signal generator are shown in the schematic diagram.

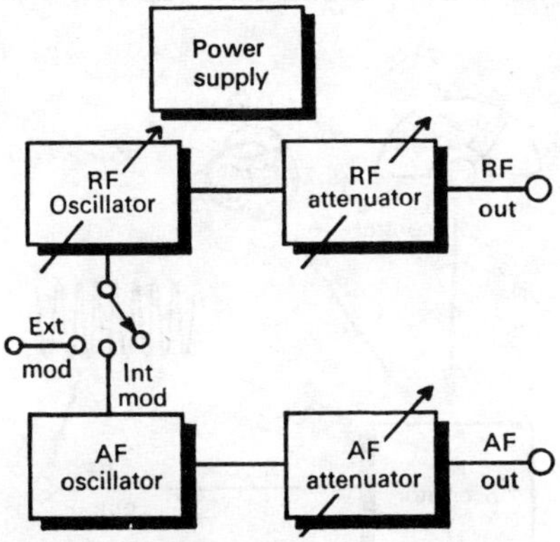

Three different types of signal are provided by the instrument:

- an unmodulated radio frequency (R.F.) signal of selected frequency and amplitude. This is a continuous wave (C.W.) signal and when used to align and test radio receivers, is known as a carrier wave
- an amplitude modulated R.F. signal of selected frequency and amplitude
- an audio frequency (A.F.) signal of fixed frequency but of variable amplitude.

The signals are sinusoidal.

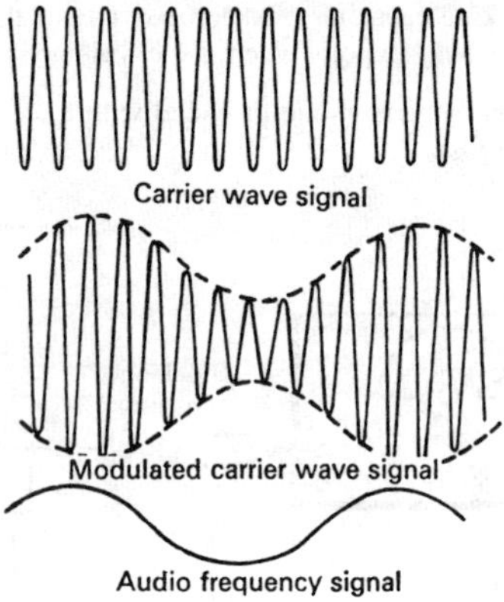

Carrier wave signal

The unmodulated carrier wave signal is derived from an R.F. oscillator.

The frequency of the oscillator is varied over the range of the instrument by means of a tuning control and range switch.

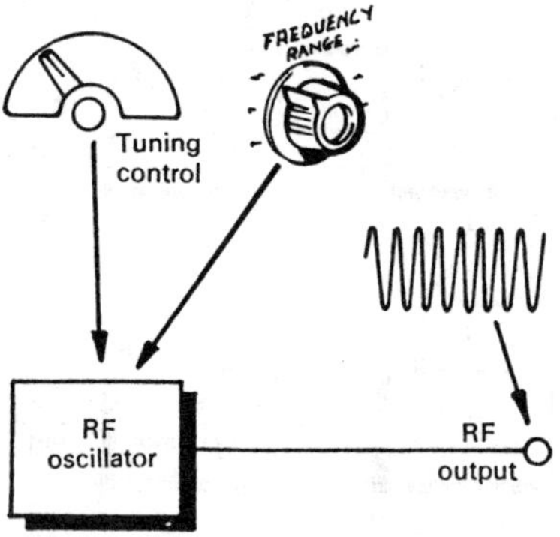

The amplitude of the signal at the *R.F. output* socket is controlled by an attenuator.

Signal amplitude is expressed in Volts r.m.s.

The attenuation is varied by means of a continuously variable *voltage* control and a *multiplier* switch.

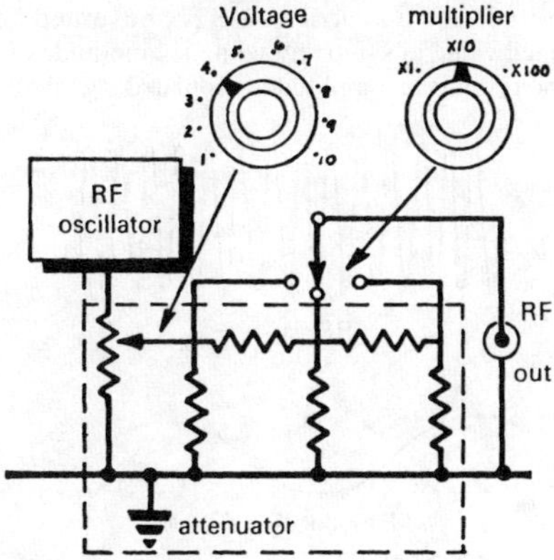

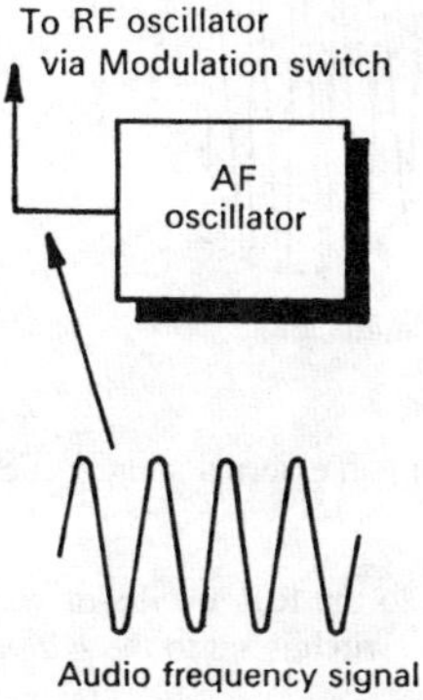

Audio frequency signal

The audio frequency (A.F.) oscillator provides a sinusoidal signal of fixed frequency (usually 400 Hz).

Modulated continuous wave signal

A modulated R.F. signal is obtained by setting the modulation switch to *Int. mod.*

This connects the A.F. oscillator to the R.F. oscillator.

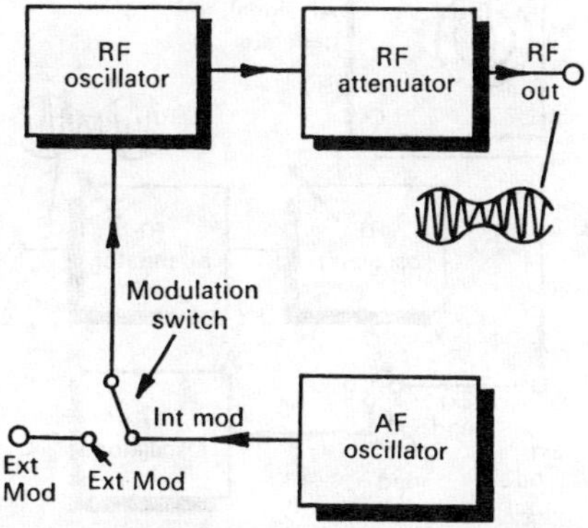

The amplitude of the carrier swings above and below its unmodulated value in sympathy with the amplitude of the A.F. signal. The result is an amplitude-modulated signal.

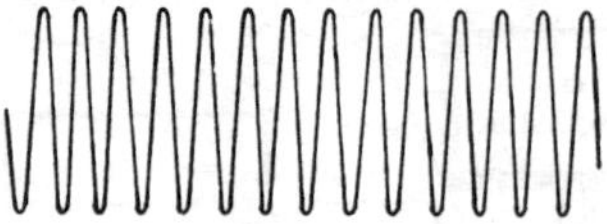

RF signal

AF modulating signal

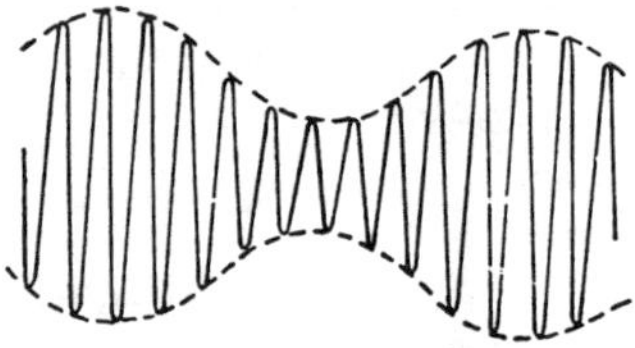

Amplitude modulated signal

The carrier may be modulated from an external source, such as an A.F. signal generator.

The modulating signal is applied to the R.F. oscillator via the *Ext. mod.* socket. The modulation switch is set to the *Ext. mod.* position.

In many general-purpose R.F. signal generators a separate attenuator and output socket enables the A.F. signal to be used for testing A.F. equipment.

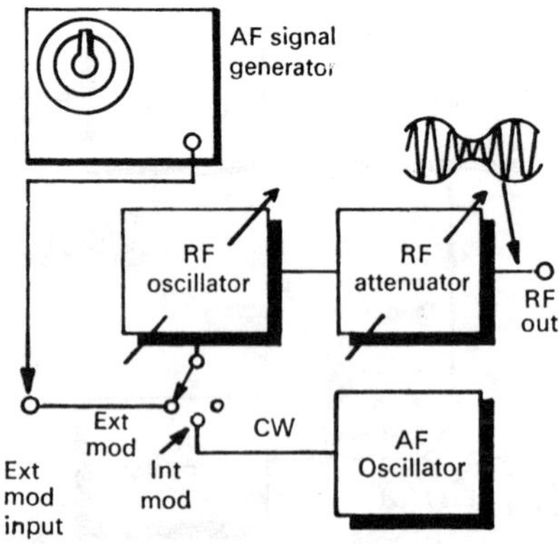

The R.F. signal is applied to the equipment under test via a co-axial cable having a specified characteristic.

The A.F. signal is applied to the equipment by a 2-core unscreened cable or a co-axial cable, depending on the equipment.

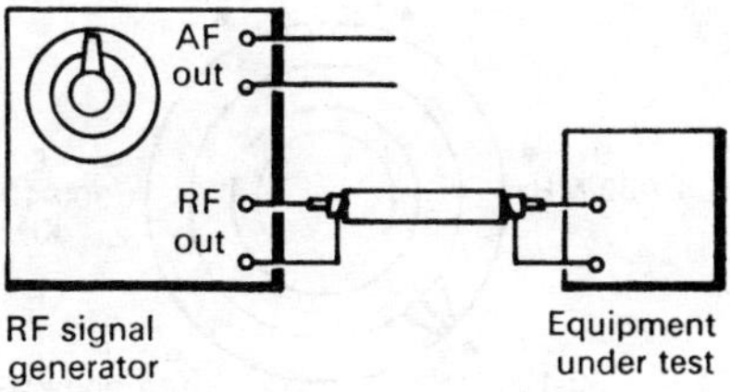

Controls

The control panel of a typical general-purpose R.F. signal generator is shown below.

The generator's frequency range is 100 kHz–100 MHz.

This range is divided into six bands as follows:

Band		
A	30– 100	MHz (Mc/s)
B	10– 30	MHz (Mc/s)
C	3– 10	MHz (Mc/s)
D	1– 3	MHz (Mc/s)
E	300–1,000	kHz (kc/s)
F	100– 300	kHz (kc/s)

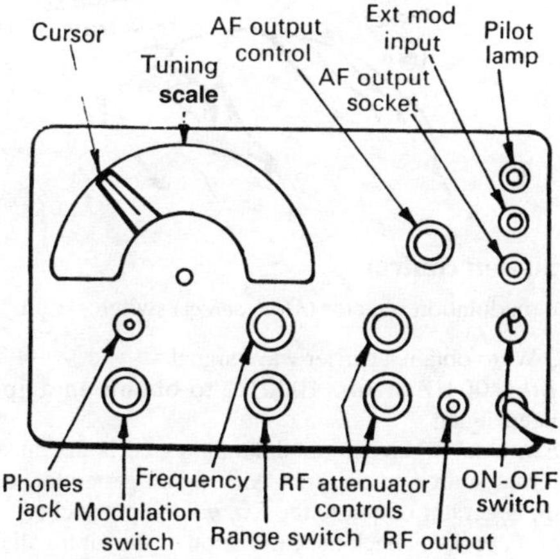

Frequency controls

Set the *frequency range* switch to the required band.

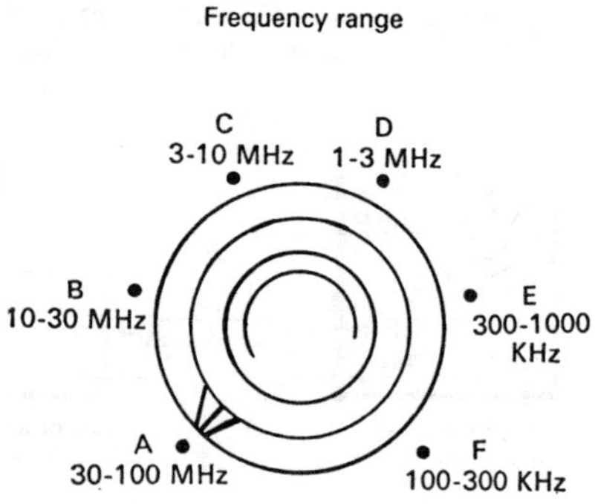

The tuning dial is marked with six scales, one for each band.

The appropriate scale for each band is identified by a letter A-F (or by a frequency range) printed on the panel.

Adjust the *frequency* control until the cursor coincides with the required frequency calibration on the appropriate scale.

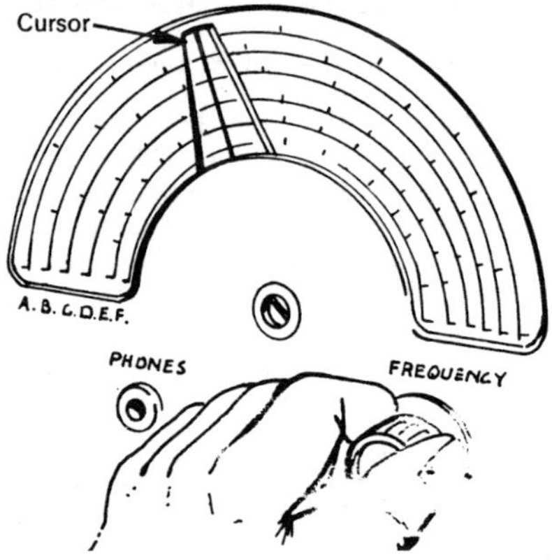

Modulation control

Set the modulation selector (*Mod. select*) switch:

- to C.W. to obtain a carrier wave signal
- to *Int*. 400 Hz or *Int*. 1000 Hz to obtain an amplitude modulated signal
- to *Ext. mod.* where the modulating signal is obtained from an external source, such as an A.F. signal generator, connect the A.F. generator output to the *Ext. mod.* input socket
- to *Xtal check* to check the calibration of the tuning dial.

Mod select

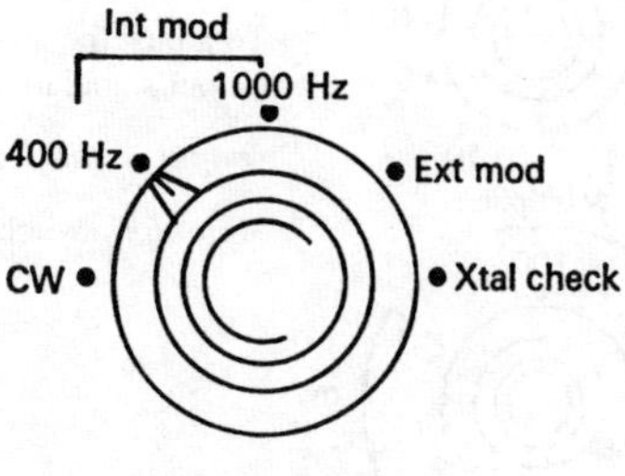

Attenuator controls

The output attenuator is operated by two controls:

- a potentiometer (*output voltage*), calibrated 1–10

Output voltage

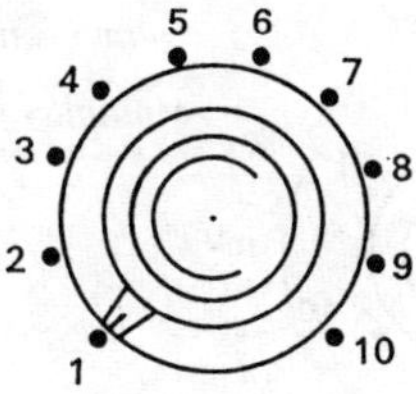

- 5-position *multiplier* switch, calibrated X1 μV, X10 μV and X100 μV and X1 mV, X10 mV.

Multiplier

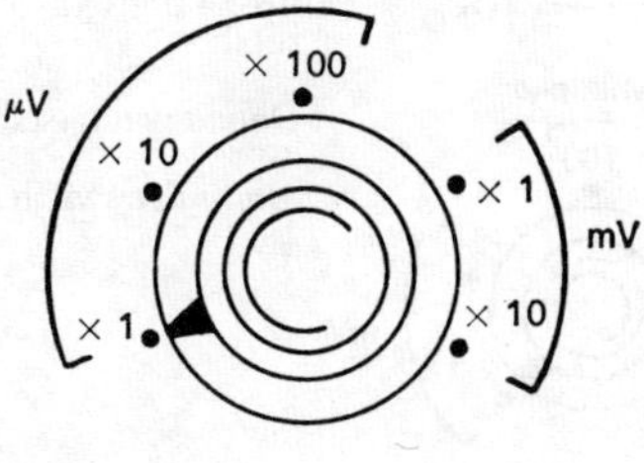

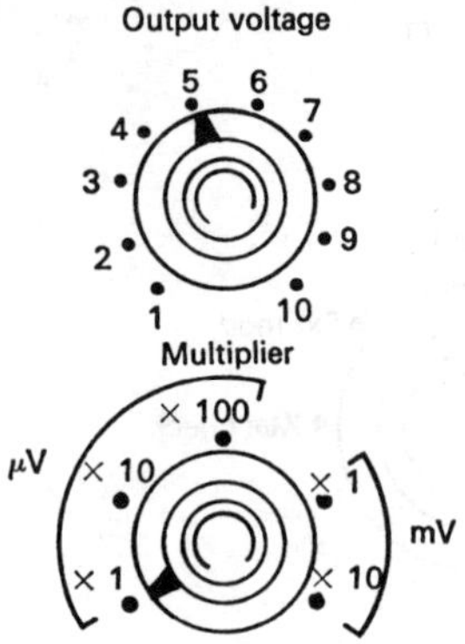

To adjust the amplitude of the output signal to 5 μV, set the:

Output voltage control to 5.

Multiplier switch to X1 μV.

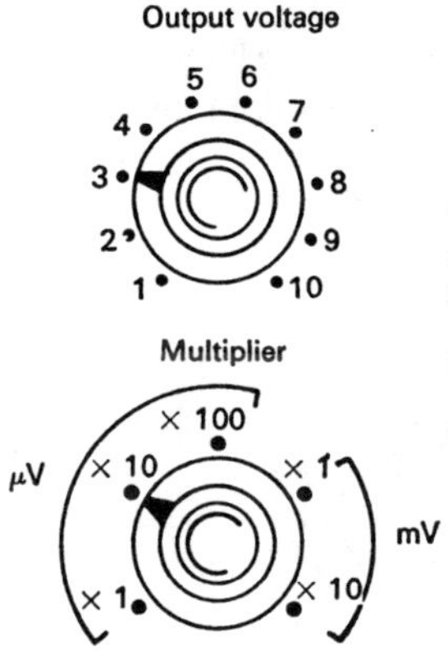

To adjust the amplitude of the output signal to 30 μV, set the:

Output voltage control to 3.

Multiplier switch to X10 μV.

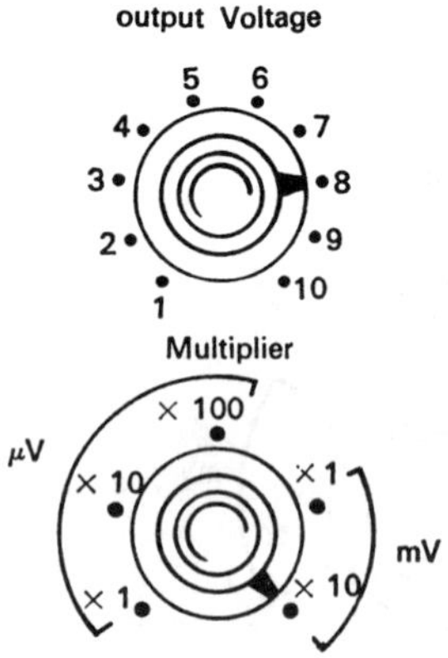

To adjust the amplitude of the output signal to 80 mV, set the:

Output voltage control to 8.

Multiplier switch to X10 mV.

Test leads and terminations

Generally, the attenuator calibration only applies when:

- the correct type of cable is used
- the cable is terminated with an impedance of the same value as the generator output impedance.

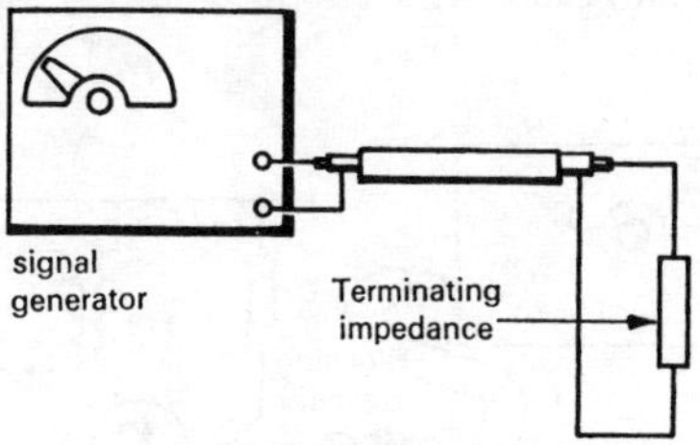

When the generator is connected to equipment under test, the terminating impedance is the input impedance of the equipment.

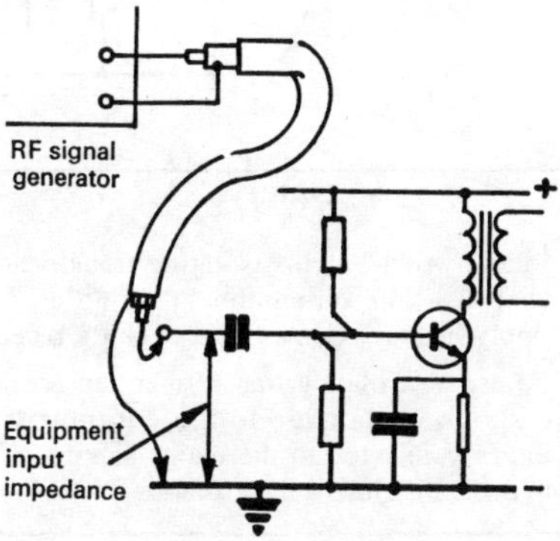

When the generator output impedance and equipment input impedance differ, use an impedance matching pad.

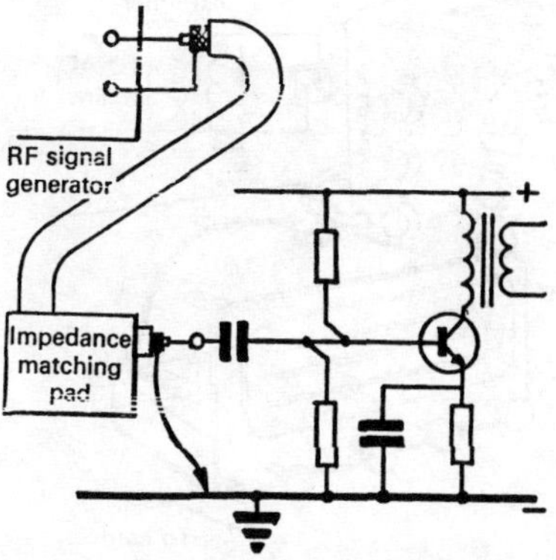

If the generator is to be connected to a point where a d.c. voltage is present, always connect a blocking capacitor of specified value in series with screened lead of the cable.

Failure to use a blocking capacitor can result in damage to:

- components in the equipment under test
- the signal generator, due to d.c. flowing through the attenuator
- or both.

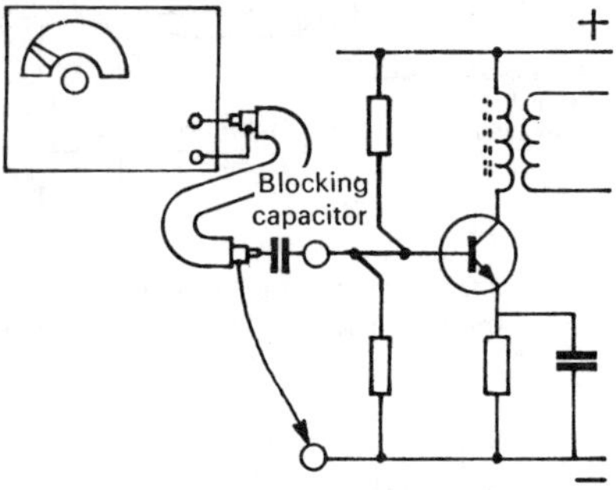

Safety

Use a double-wound mains isolating transformer (1:1 ratio) when working on equipment using an a.c./d.c. power supply such as a domestic radio or TV receiver.

Failure to use an isolating transformer can result in an electric shock or damage to the equipment if the equipment is connected to the mains supply in such a way as to make the chassis live to earth.

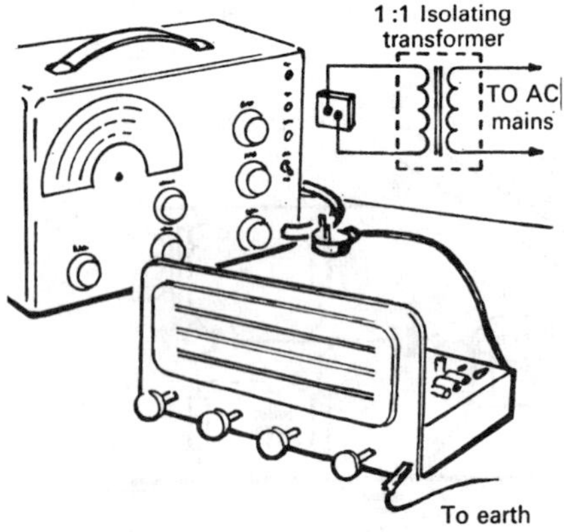

When connecting a signal generator to the aerial circuit of a receiver:

Use an *artificial aerial matching unit*, commonly known as a dummy aerial.

The dummy aerial circuit shown is suitable for use with ordinary broadcast receivers.

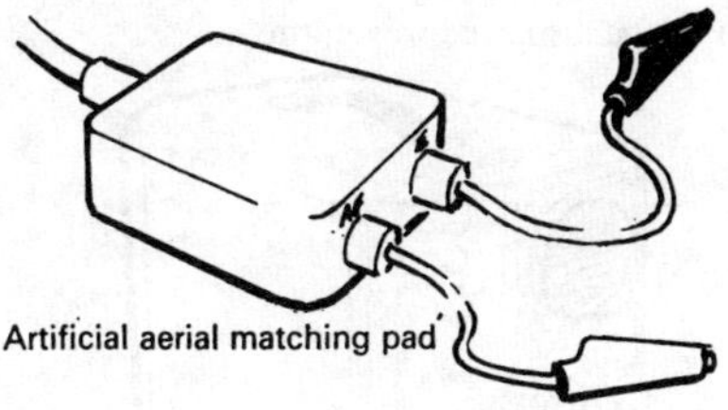

Artificial aerial matching pad

Cathode-ray oscilloscopes

An oscilloscope is used to display and measure the characteristics of the waveforms in electronic circuits.

Oscilloscopes are calibrated to measure the amplitude and frequency of the displayed waveform.

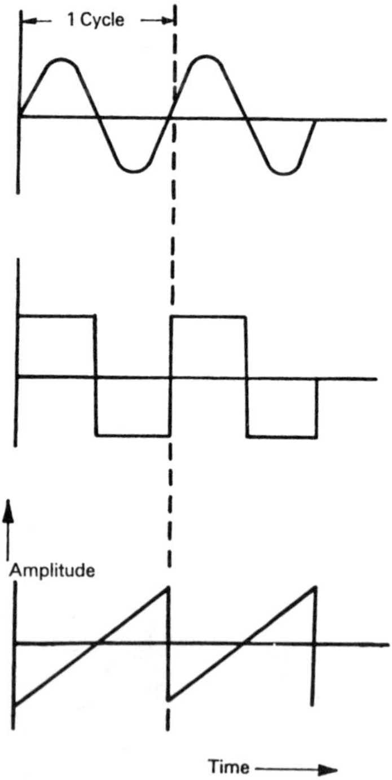

Typical waveforms are:

- sinusoidal
- square
- sawtooth.

The diagrams show two cycles of each waveform.

The amplitude of a waveform is measured in volts (V), millivolts (mV) or microvolts (μV).

Amplitude is measured in terms of peak (p) or peak-to-peak (p.p.) values.

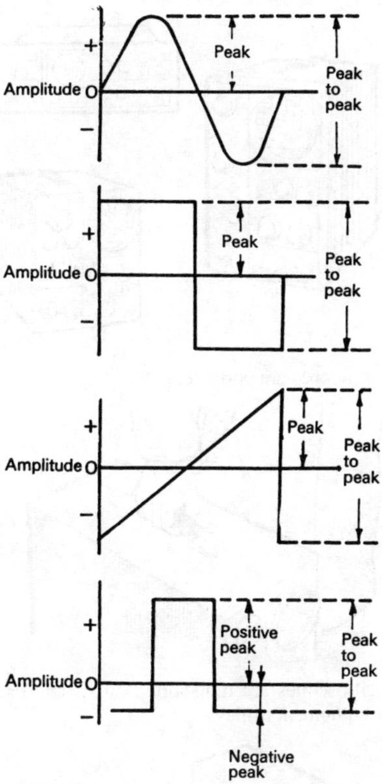

This period of a waveform is the time taken to complete one cycle.

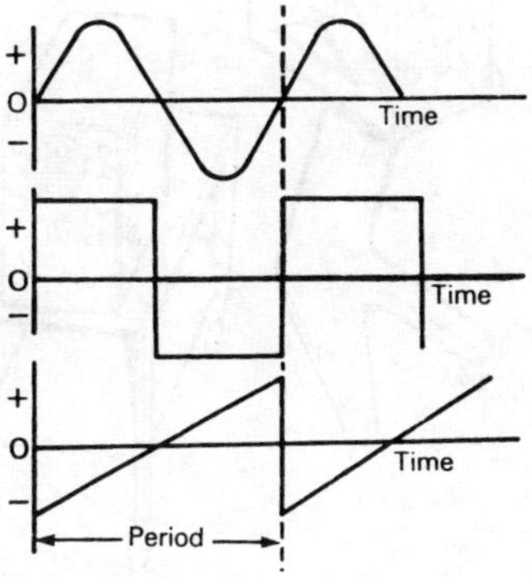

Oscilloscopes range from general purpose to special purpose types.

Oscilloscopes are expensive.

Handle oscilloscopes carefully.

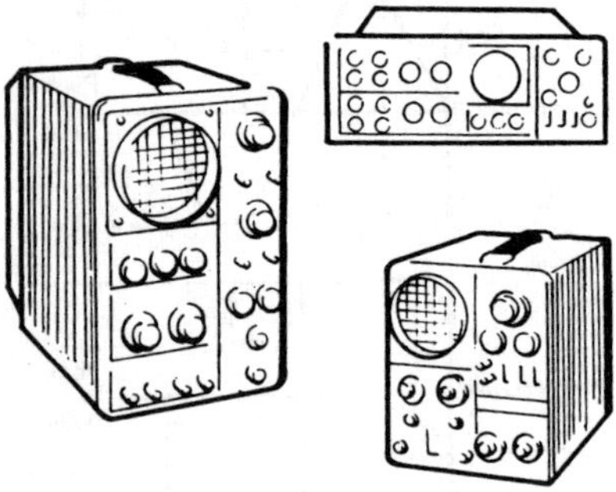

Some oscilloscopes are portable.

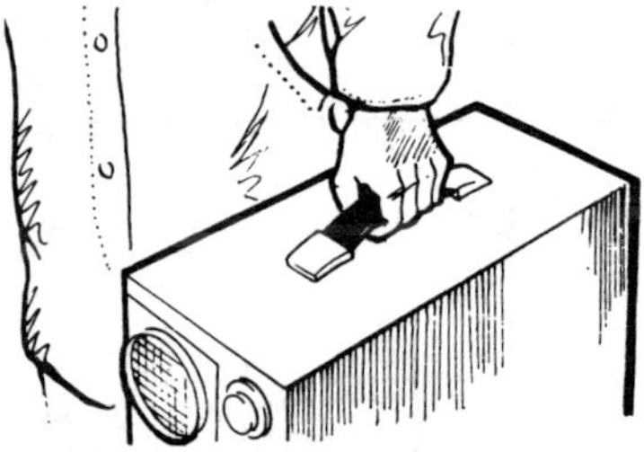

Heavy oscilloscopes are transported on special oscilloscope trolleys or equipment trolleys.

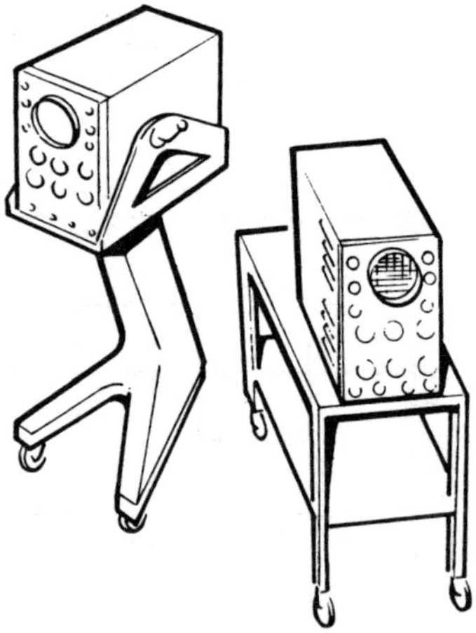

A spot of light is produced on the screen of a cathode-ray tube (c.r.t.) by applying suitable voltages to its electrodes.

The brilliance of the spot is adjusted by a *brightness* control and its sharpness by a *focus* control.

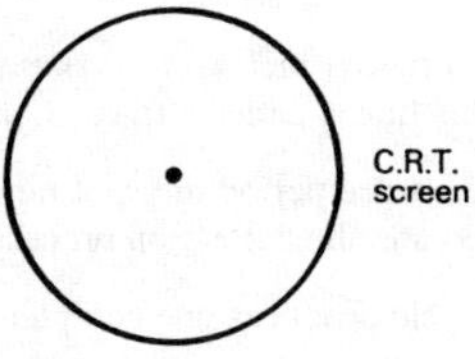

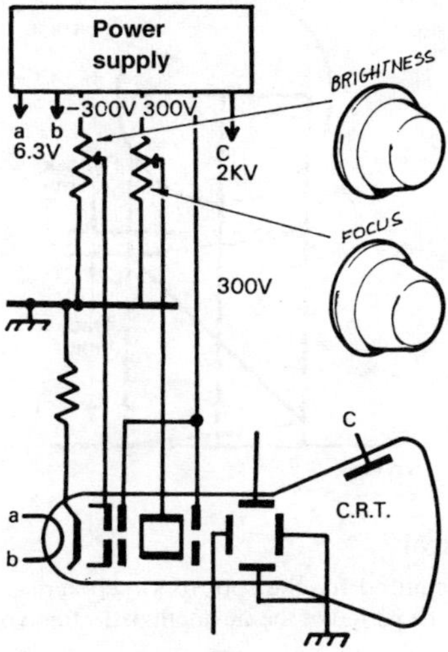

Horizontal deflection

A sawtooth voltage, derived from a built-in time-base generator, is applied to the c.r.t. horizontal deflection plates (X plates).

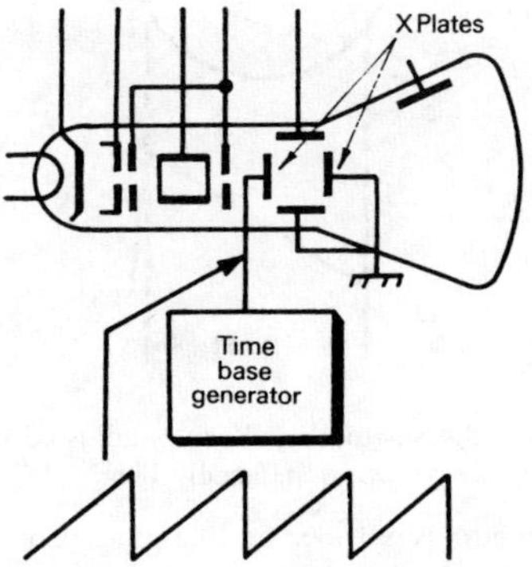

This voltage causes the spot to be deflected (swept) across the screen from left to right at a uniform rate.

If the deflection rate is very slow, the spot movement can be followed by the eye.

If the deflection rate is increased a continuous horizontal line is produced. This line is called a trace.

At the end of the trace period the spot rapidly flies back to the left of the screen and the deflection process is repeated.

One sawtooth cycle produces one complete trace and flyback.

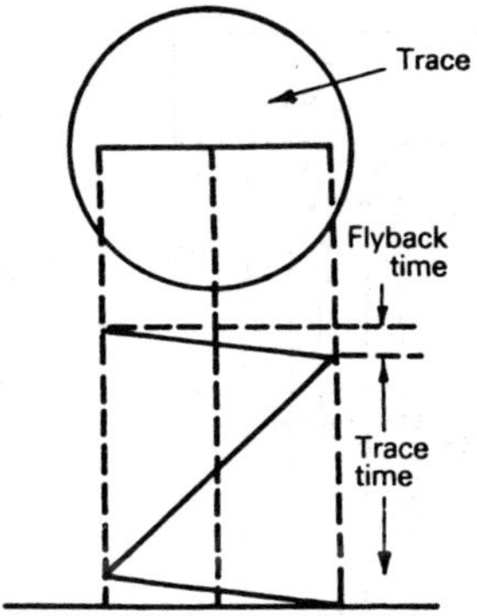

Voltage

Time control

The time required for the spot to sweep across the screen depends on the period of the sawtooth deflecting voltage.

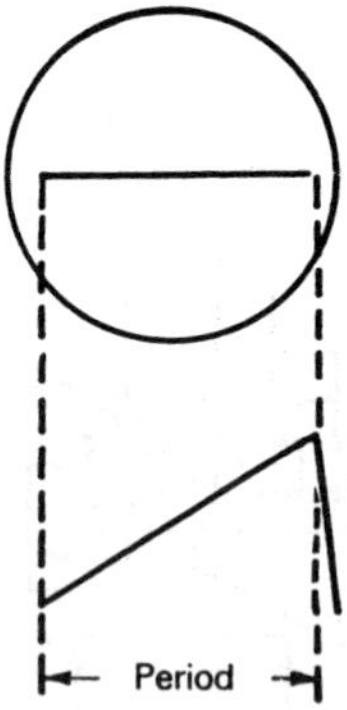

The period of the sawtooth voltage is adjusted by a switch, calibrated in *time per division*, usually 1 cm.

A variable control is provided for fine adjustments.

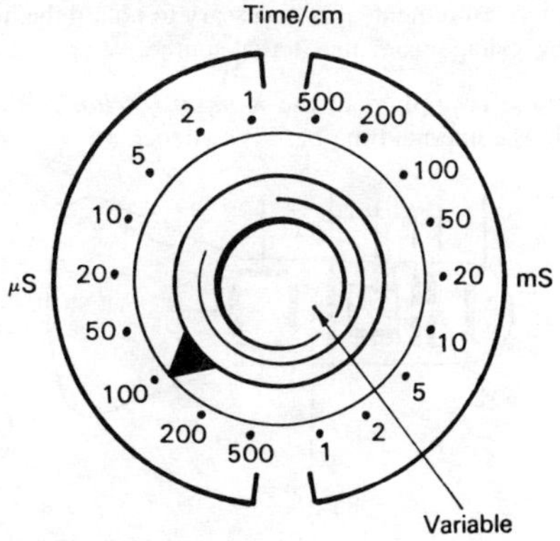

If the switch is set to 100 ms/cm, one centimetre of trace is produced in 100 ms.

Therefore if the length of trace is 10 cm, the spot is deflected across the screen in:

10 x 100 ms
= 1000 ms
= 1 second.

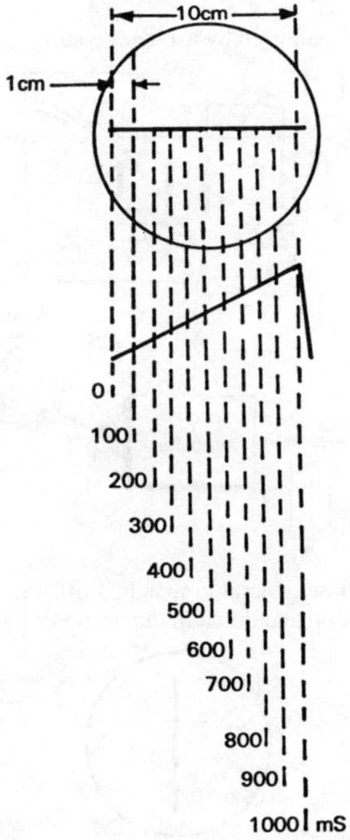

For some measurements it is necessary to obtain the horizontal deflecting voltage from an external source.

The voltage is applied to the *X-input* (or *horizontal input*) terminal. The internal time-base is switched off.

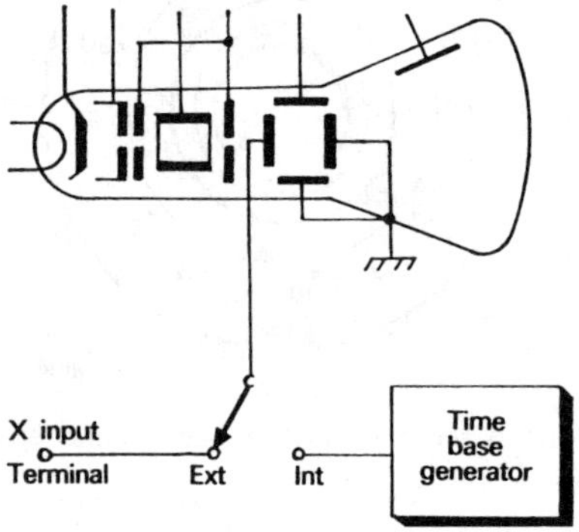

Vertical deflection

The waveform to be examined is applied to the *Y-input* terminal, amplified, and coupled to the c.r.t. vertical deflection plates (Y-plates).

Amplifier gain is adjusted by a *Y-gain* control.

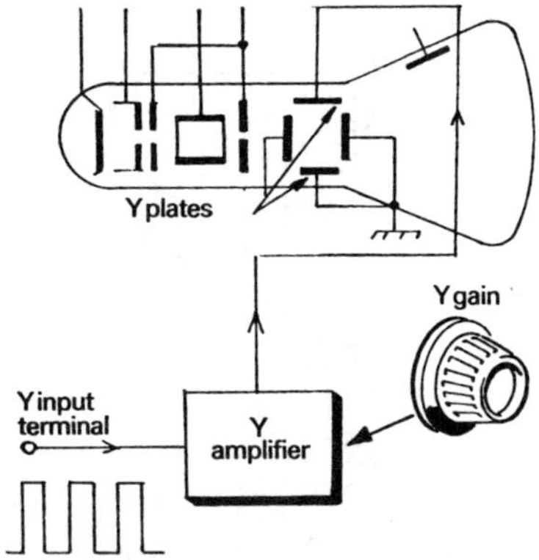

With the time-base generator switched off, the Y-input signal deflects the spot up and down in the centre of the screen.

Voltage control

The vertical deflection is proportional to the amplitude of the Y-input signal.

Signal amplitude at the input to the Y amplifier is controlled by an attenuator.

The attenuation required to produce a given deflection is selected by a switch, calibrated in *volts per division*, usually 1 cm.

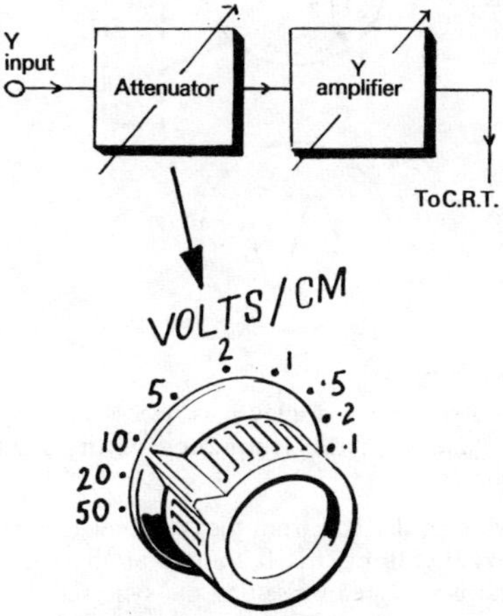

If the switch is set to 10 V/cm one centimetre of deflection is produced for every 10 V applied to the Y-input terminal.

Therefore a 20 V signal causes two centimetres of deflection.

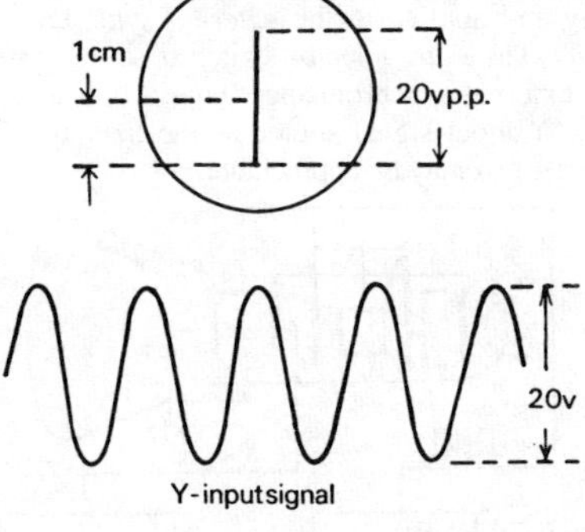

In normal operation the spot will be deflected simultaneously in two directions:

- horizontally due to the time-base
- vertically due to the Y-input signal voltage.

If the period of the time-base is adjusted to correspond with the period of the Y-input signal the form of the Y signal will appear as a stationary trace. By increasing the time-base period to a multiple of the Y-input signal period more than one cycle of the input signal will be displayed.

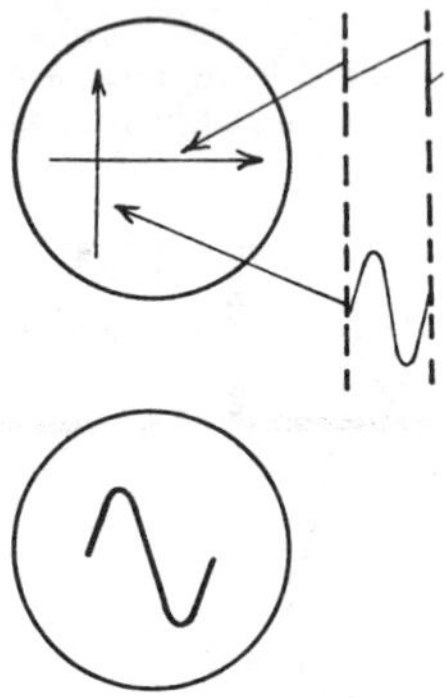

Synchronising

To ensure a stationary display it is necessary to synchronise the time-base with the frequency of the signal under examination.

A small voltage derived from the Y signal is applied to the time-base generator. This is known as the synchronising voltage and is obtained by setting the sync selector switch to internal.

Provided the time-base is adjusted to operate at approximately the correct frequency the synchronising voltage controls the time-base, locking it to the frequency of the Y-input signal. Alternatively, the synchronising signal may be obtained externally and applied to the *external sync.* terminal on the front panel. The *sync. selector* switch is set to *external.* When using an external synchronising signal it is usually better not to use the Y-input signal source as the loading effect on the circuit under test may be appreciable.

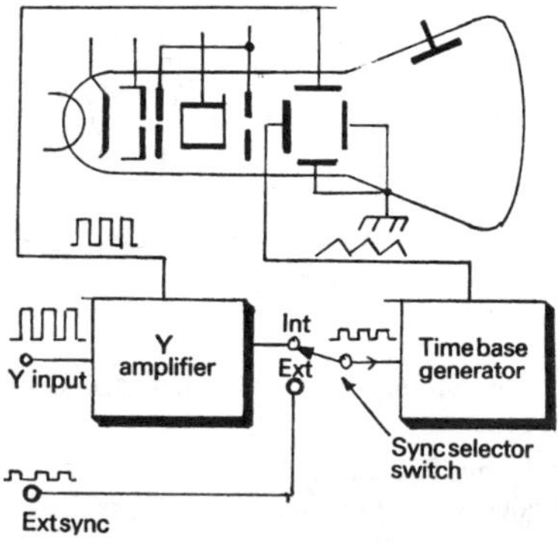

Safety

Earthing

The case of the oscilloscope is usually earthed either:

● by the earthed lead in the mains cable, or;
● by a separate connection from the earth terminal on the case (or front panel) to true earth.

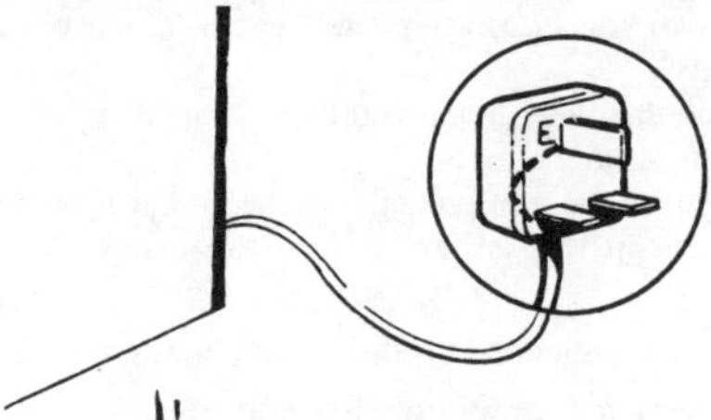

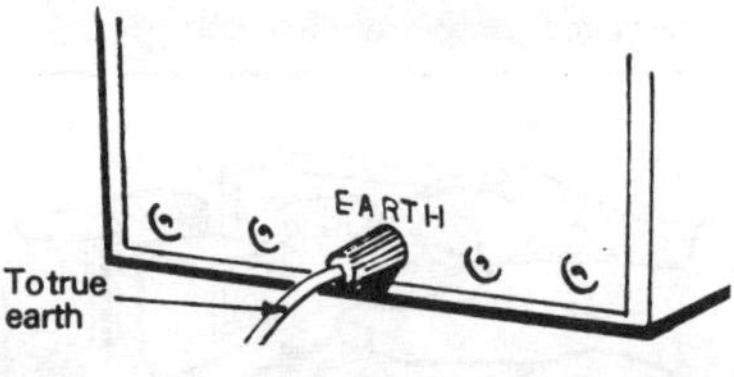

Safety

If the case is not earthed properly there is a risk of:

● electric shock, if the case becomes live
● spurious displays and readings due to ineffective screening.

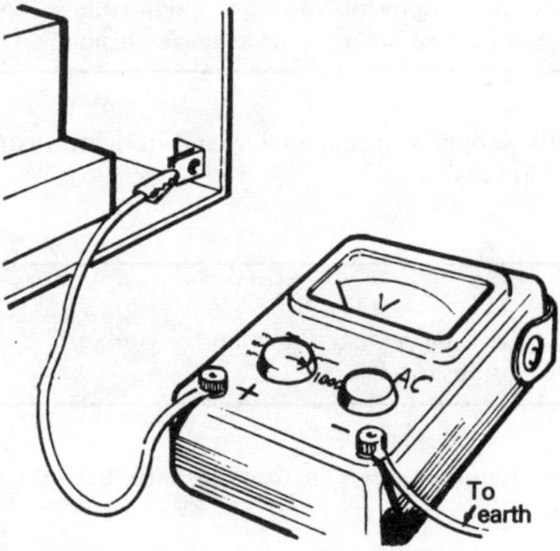

Safety

Mains-operated equipment

Before connecting instruments to mains-operated equipment, check that the chassis of the equipment is not live.

To do this, proceed as follows:

- disconnect the equipment from the supply
- connect a high impedance a.c. voltmeter between the chassis of the equipment and earth as shown in the diagram
- switch the meter to the 1000 V a.c. range
- reconnect to the mains
- switch on the equipment and check that no reading is obtained on the meter (adjust the range switch as necessary).

Use the recommended insulated test clips and prods.

Keep your hands away from live components.

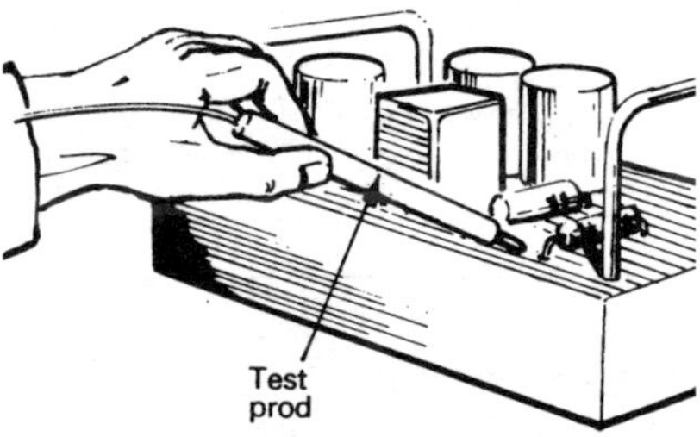

Safety

Do not remove any test equipment from its case.

Many of its components can store a high voltage for some time after the instrument is disconnected from the mains.

An oscilloscope is an expensive and delicate instrument. Handle carefully.

Safety

Never lift a heavy oscilloscope by yourself — use a trolley to transport it.

Read the manufacturer's instruction book before using the instrument.

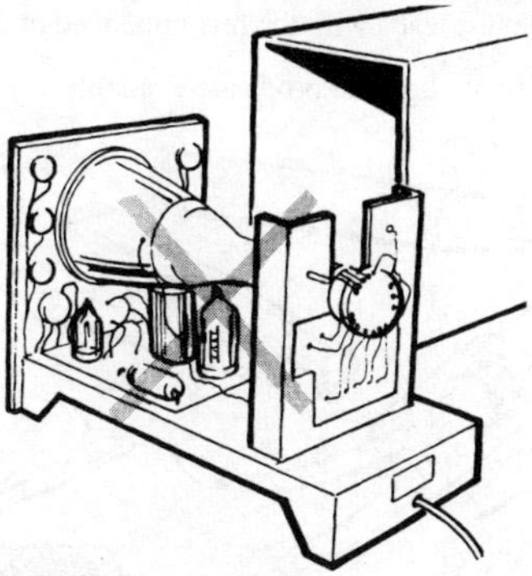

Test leads, clips and prods

Use the recommended test leads.

Screened cable is recommended for most connections and a co-axial socket is provided on the instrument panel for this purpose.

The cable's screening braid forms the earth connection to the equipment being tested.

In addition to a co-axial socket, some oscilloscopes provide separate terminals for the connection of test leads. One terminal is marked *Y-input* or *vert. input*, the other is marked *earth* or with the earth symbol.

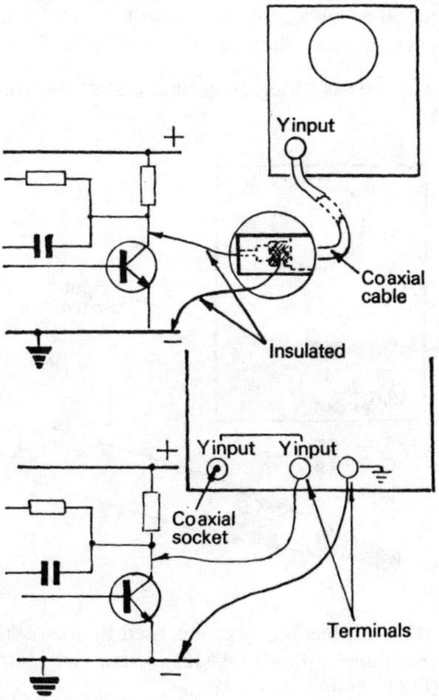

Terminate the test leads with the recommended clips or prods.

A wide variety of clips and prods are available.

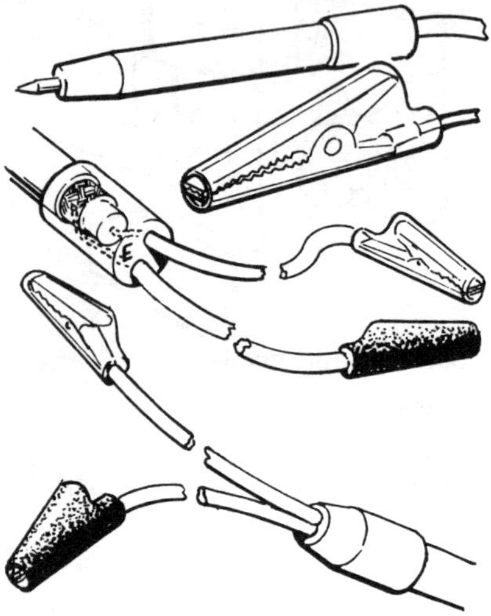

Circuit loading

When an oscilloscope is connected to a high frequency circuit its Y-input capacitance, together with the capacitance of the connecting cable, loads the circuit.

The effect of this is to attenuate and distort the signal being examined.

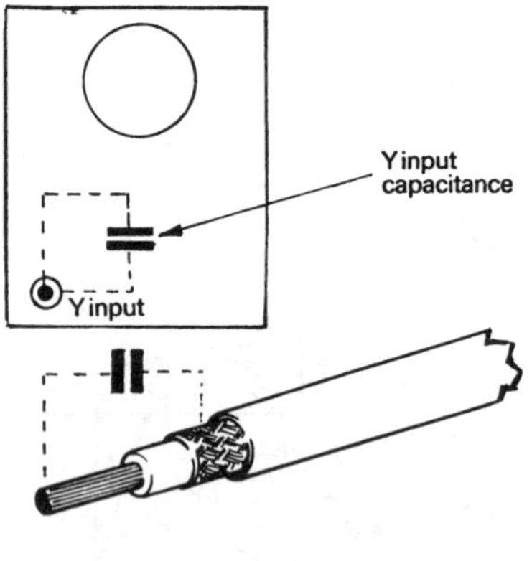

Significant circuit loading is also caused by the oscilloscope's input resistance (typically 1 MΩ) when the instrument is connected to high impedance circuits.

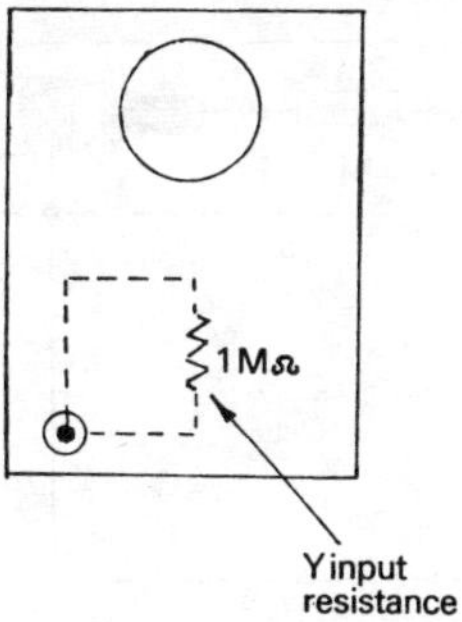

Oscilloscope probes

To reduce circuit loading use an attenuator probe.

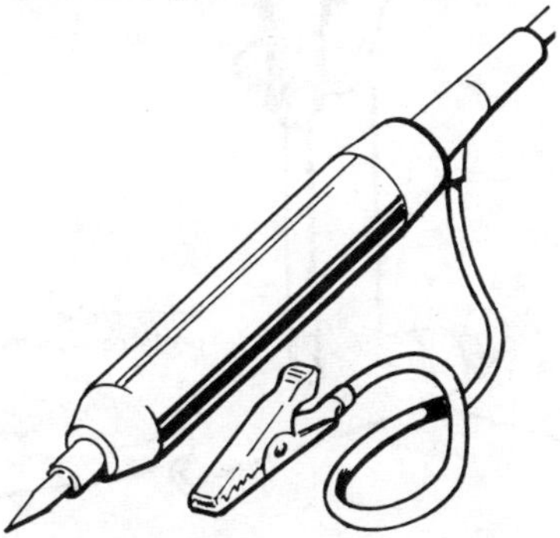

A 10:1 probe contains a resistance capacitance network which decreases input capacitance and gives a 10:1 increase in input resistance.

It also reduces the oscilloscope's sensitivity by 10 times. This attenuation enables signal voltages outside the normal range of the oscilloscope to be measured.

Do not exceed the maximum voltage rating of the probe.

Multiply the *volts/cm* control calibrations by 10 when using a 10:1 probe, and by 100 when using a 100:1 probe.

Before using the probe, adjust it as specified in the oscilloscope handbook.

A typical procedure is as follows:

- apply a square-wave signal to the probe input.

This signal may be obtained from the oscilloscope's *calibration* socket, or from an external source — such as a square-wave signal generator.

Suitable frequencies are 50 Hz and 1 kHz.

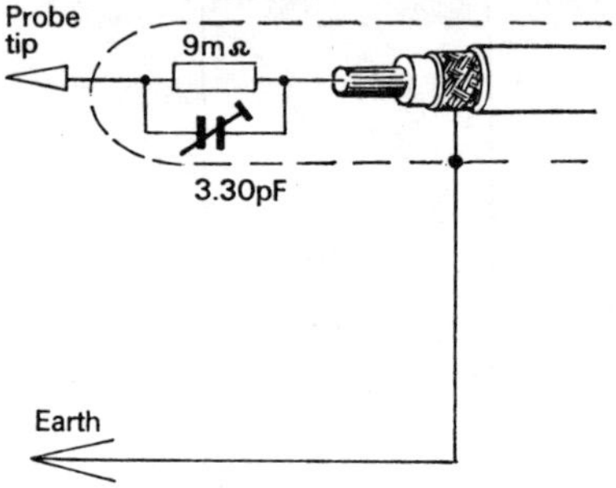

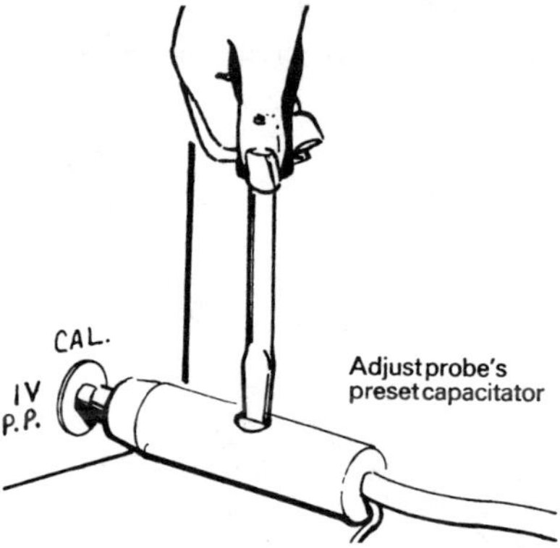

● adjust the probe's pre-set capacitor to obtain flat tops on the waveform displayed on the screen.

Use a *cathode-follower probe* to reduce input capacitance without reducing oscilloscope sensitivity.

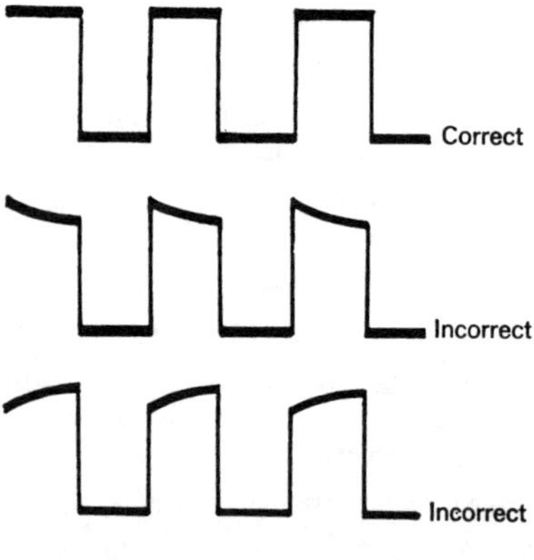

Oscilloscope specification

Sensitivity, bandwidth, rise-time and *input impedance* are important performance characteristics shown in the oscilloscope's specifications.

Sensitivity

This is the Y-input voltage required to obtain one centimetre of vertical deflection (*volts* per *centimetre*), when the Y-amplifier is adjusted to operate at maximum sensitivity.

A typical sensitivity is 50 mV/cm.

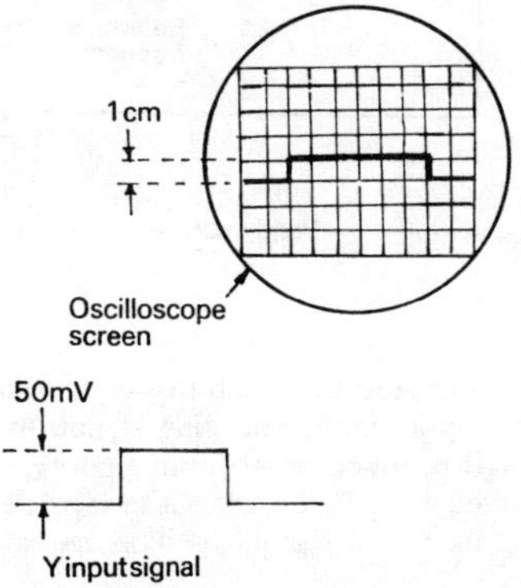

A 400 mV signal will result in a display 8 cm high.

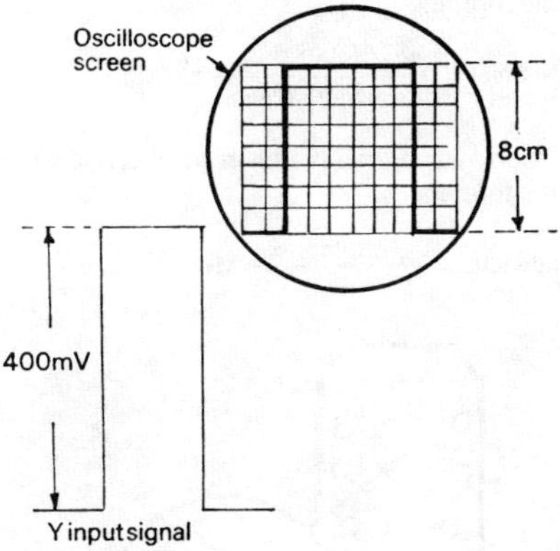

Bandwidth

This is the frequency range over which the gain of the Y-amplifier does not fall below the gain at a specified reference frequency by more than a given value, usually 3 dB.

General-purpose laboratory oscilloscopes have bandwidths within the range d.c.–10 MHz.

High performance oscilloscopes have bandwidths within the range d.c.–150 MHz.

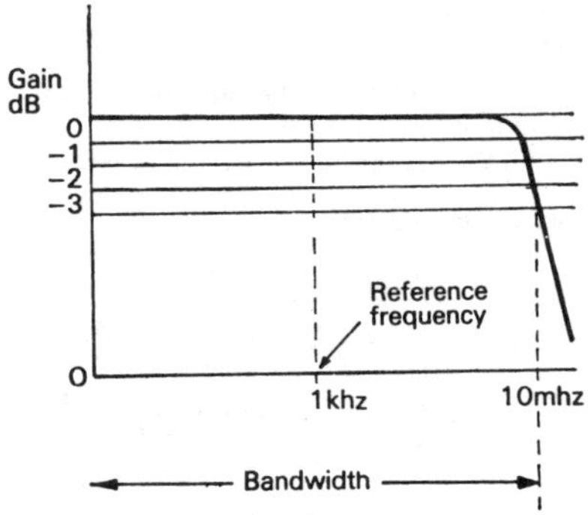

Rise time

Rise time is a measure of the ability of the instrument to accurately display a rapidly changing signal. In general the instrument will produce an accurate representation of a waveform in which the fastest change in input level takes at least five times the rise time of the oscilloscope.

Rise time is related to bandwidth.

To calculate the approximate rise time from the bandwidth specification use the formula:

$$\text{Rise time} = \frac{350}{\text{Bandwidth (MHz)}}\ \text{ns}$$

To calculate the approximate bandwidth from the rise time specification use the formula:

$$\text{Bandwidth} = \frac{350}{\text{Rise time (ns)}}\ \text{MHz}$$

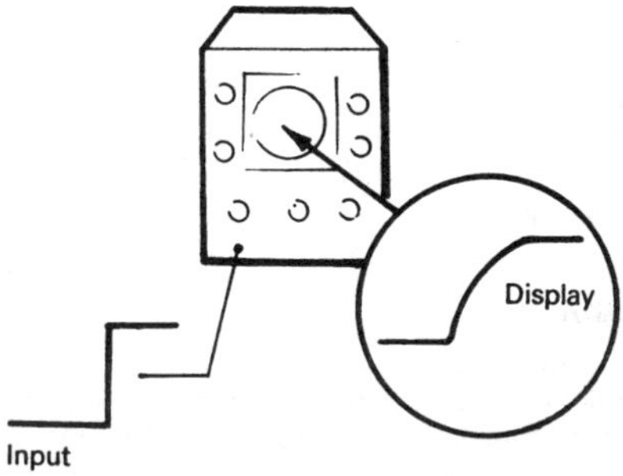

Input impedance

This is usually expressed in terms of resistance in parallel with capacitance.

A typical input impedance is 1 MΩ shunted by 30 pF.

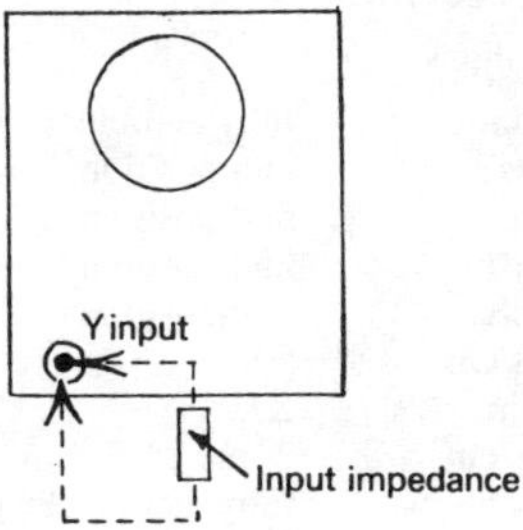

Operating an oscilloscope

The control panel of a typical general purpose oscilloscope is shown below.

Although there is some variation between instruments, operation of the basic controls is similar.

Before using any oscilloscope:

- always read the manufacturer's handbook
- check that the instrument is adjusted to suit the mains supply.

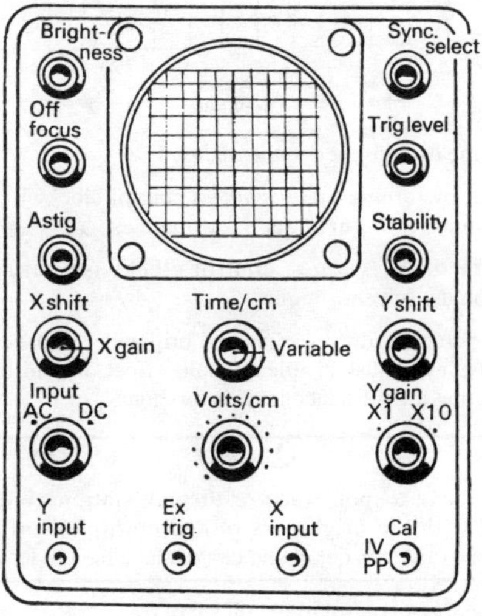

A *graticule* is fitted to the c.r.t. screen to enable voltage and time measurements to be made.

Time is measured on the horizontal (X) axis. Voltage is measured on the vertical (Y) axis.

To obtain a trace

Set control as follows:

Brilliance	fully anti-clockwise
Focus	mid-position
Astig.	mid-position
X shift	mid-position
Y shift	mid-position
Volts/cm	50
Y gain	X1
Time/cm	20 ms
Variable	fully anti-clockwise
X gain	fully anti-clockwise
Input	d.c.
Sync. select	+ Int.
Sync. amp	fully anti-clockwise

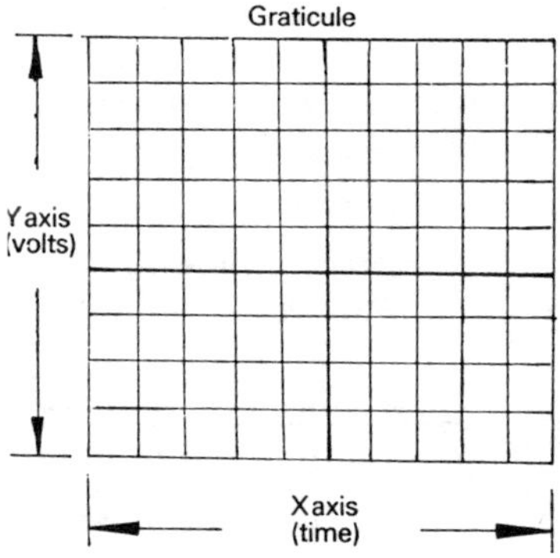

Connect the oscilloscope to the mains.

Switch on by turning the *brightness* control clockwise. Allow the instrument to warm up for a few minutes.

Then turn the *brightness* control clockwise until a trace appears on the screen.

During operation, always adjust the brilliance so that all detail in the display is just visible. A halo effect around the trace indicates that the brilliance level is too high.

> **Quality**
>
> Never leave a spot, trace or display stationary on the screen with the brightness turned up high. Too much brilliance obscures detail and can damage the CRT.

Adjust the *focus* control to obtain sharp trace.

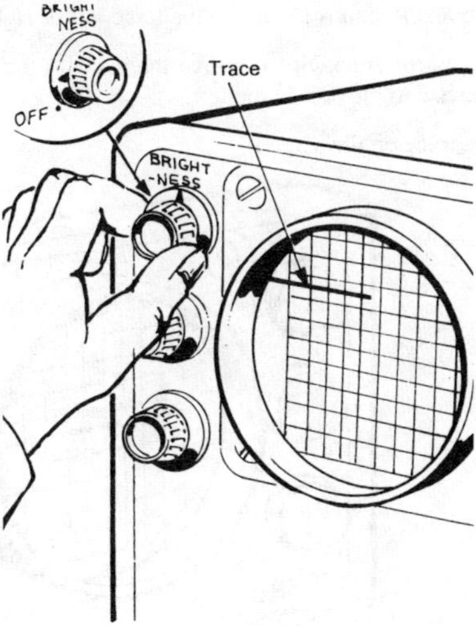

Adjust the *Y shift* control to move the trace upwards or downwards.

Turn the control clockwise to move the trace up and anti-clockwise to move it down.

Centre the trace on the screen.

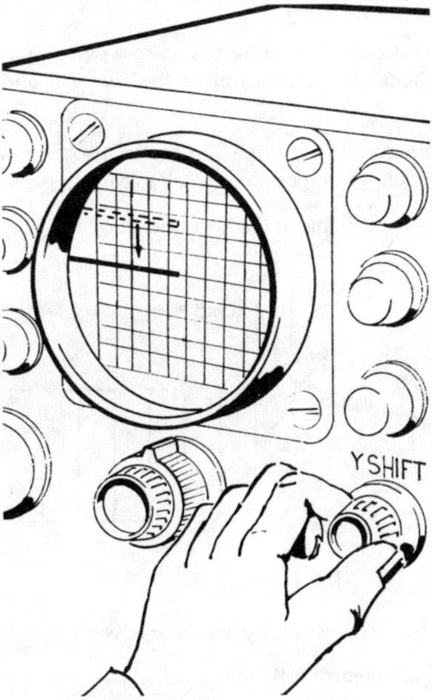

Adjust the *X shift* control to move the trace left or right.

Turn the control clockwise to move the trace to the right and anti-clockwise to move it to the left.

Centre the trace on the screen.

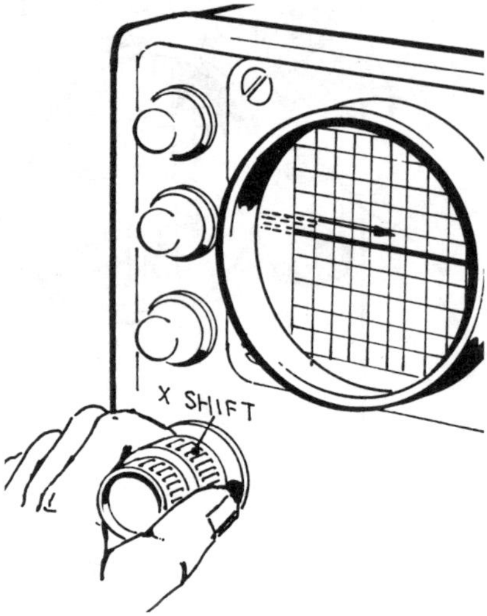

Functional checking

Most oscilloscopes generate a square-wave signal which is used to check the calibration of the *time/cm* and *volts/cm* controls.

The amplitude of the signal is usually 1 V *peak-to-peak* (1 Vp.p.). The duration (period) of each cycle is typically 20 ms.

This signal is available at the *CAL* socket on the front panel.

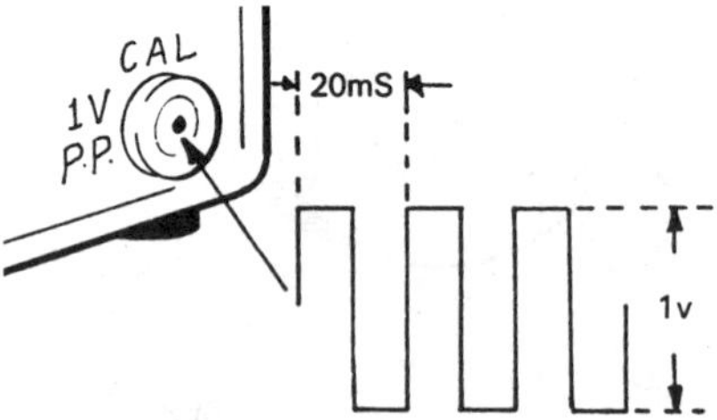

Set the *time/cm* switch to 20 ms.

Set the *variable* control fully anti-clockwise.

Set the *X gain* control fully anti-clockwise.

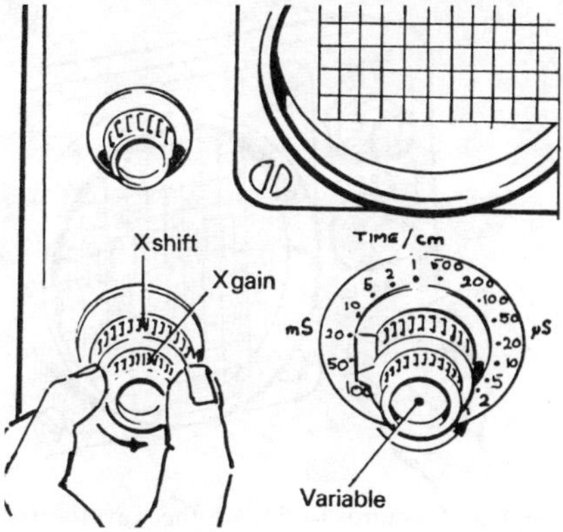

Set the *volts/cm* switch to 0.5 (0.5 V/cm).

Set the *Y gain* switch to X1.

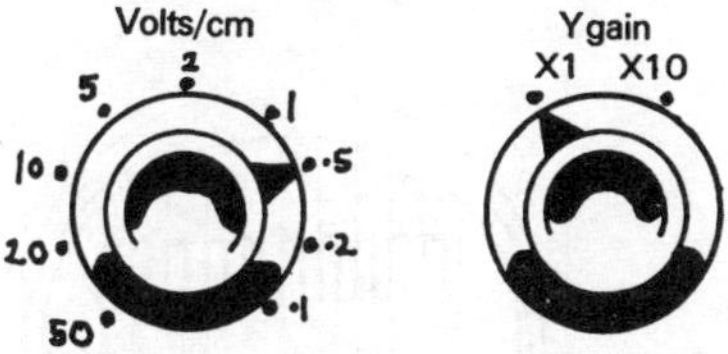

Connect a test lead to the *Y-input* connector.

Connect the test prod to the *CAL* socket.

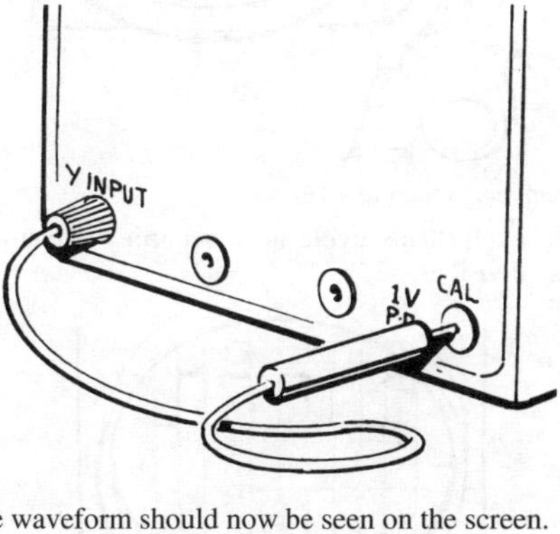

A square waveform should now be seen on the screen.

Adjust the *astigmatism (Astig.)* control in conjunction with the *brightness* and *focus* controls to obtain a sharp display.

The display should be sharp and well defined on both the vertical and horizontal axis.

Once set, the *astig.* control should not require frequent re-adjustment.

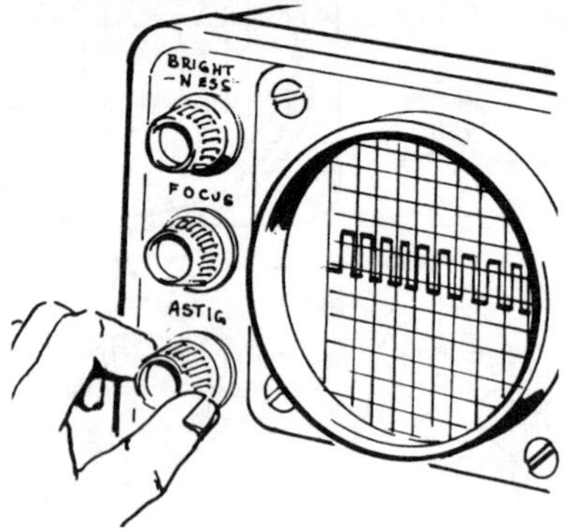

Adjust the *X shift* control to line up the waveform with the vertical lines on the graticule.

Each 20 ms cycle of waveform should occupy one horizontal centimetre division since its period is 20 ms and the *time/cm* switch is set to 20 ms/cm.

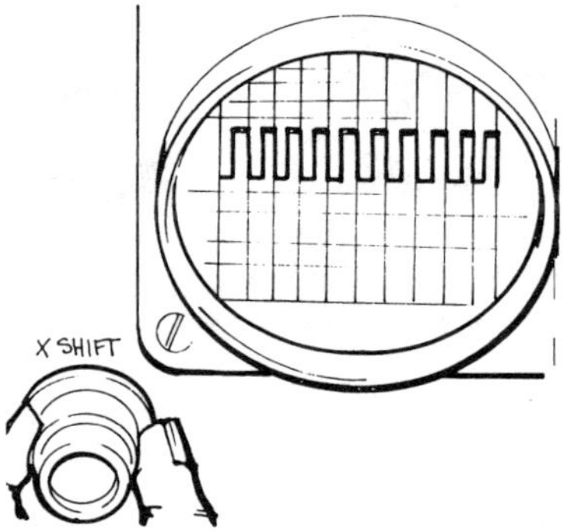

Set the *time/cm* switch to 10 ms/cm.

Note that each 20 ms cycle now occupies two horizontal centimetre divisions.

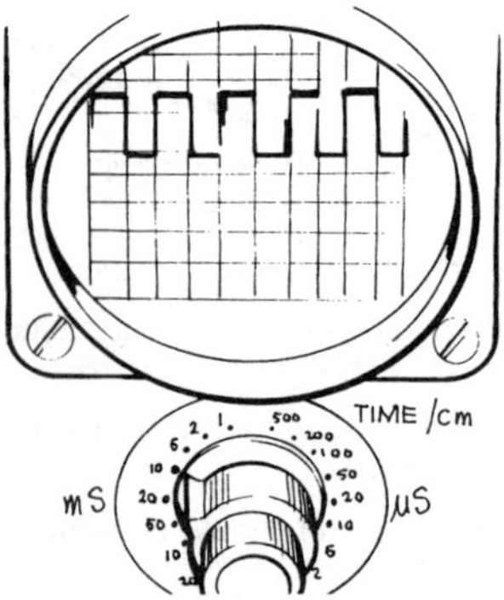

Set the *time/cm* switch to 5 ms/cm.

Each 20 ms cycle should occupy four horizontal centimetre divisions.

In this way, with the *time/cm* switch suitably set, the period of any signal applied to the *Y-input* connector may be measured.

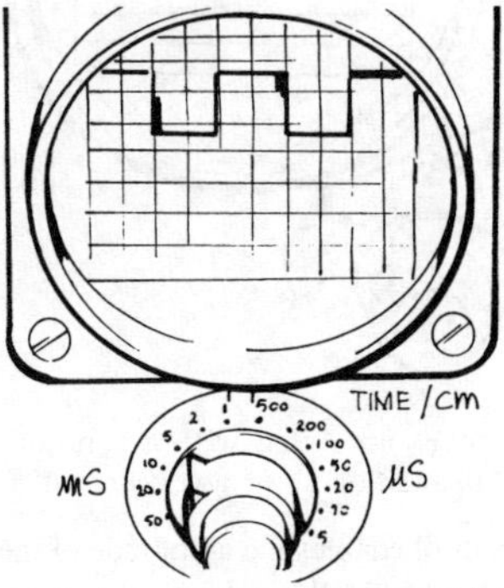

Check that the *volts/cm* switch is still set to 0.5 (0.5 V/cm) and the *Y gain* switch to X1.

Reset the *time/cm* switch to 10 ms.

Re-adjust the *brightness* and *focus*.

Adjust the *Y shift* control until the waveform lines up with one horizontal line on the graticule.

The amplitude (1 Vp.p.) of the waveform should occupy exactly two vertical centimetre divisions since the *volts/cm* switch is set to 0.5 V/cm.

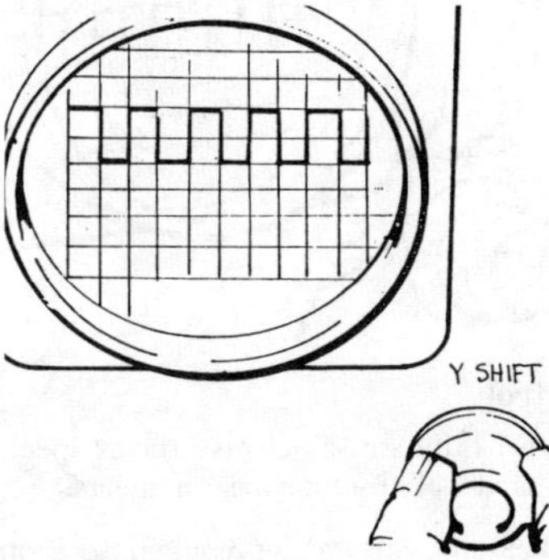

Set the *volts/cm* control to 1 V/cm.

The amplitude (1 Vp.p.) of the waveform should now occupy one vertical centimetre division.

Therefore, with the *volts/cm* control suitably set, the amplitude of any signal within the voltage range of the oscilloscope may be measured.

Disconnect the *y-input* test prod from the *CAL* socket.

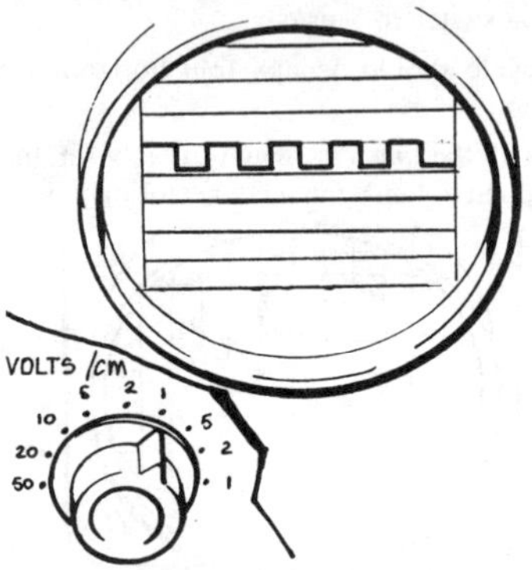

Many oscilloscopes are fitted with a continuously variable voltage control in addition to the *volts/cm* switch.

The *variable* control enables the amplitude of the display to be adjusted to a convenient value.

Calibrations on the *volts/cm* control only apply when the variable control is set to the position specified in the oscilloscope handbook.

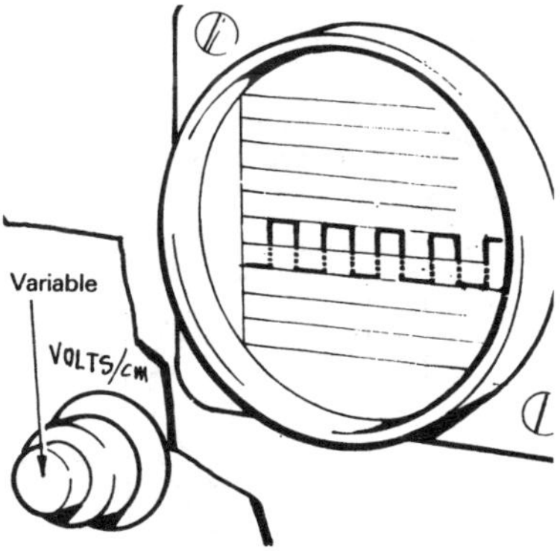

X gain control

Set this control fully anti-clockwise (or as specified in the oscilloscope handbook) for time measurements.

When time measurements are not required the control may be adjusted to expand the trace in a horizontal direction. Maximum expansion is usually between 5 and 10 times the screen diameter.

This facility is useful when examining a small part of a complicated waveform.

Reset the *brightness, focus, shift* and *time/cm* controls as necessary.

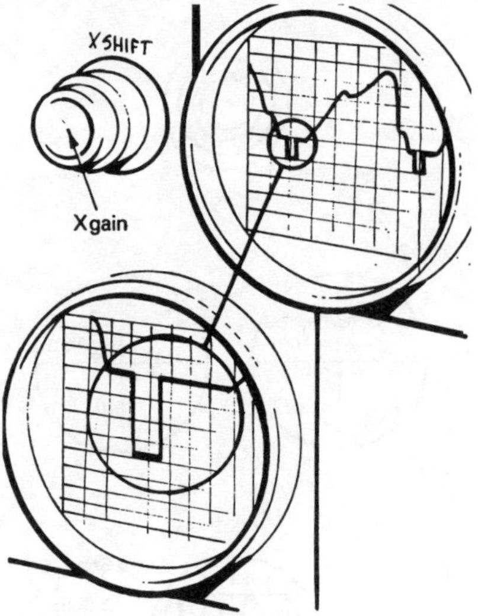

Y gain control

Set this control to X1 to read the values indicated on the *volts/cm* control directly. Set the *Y gain* control to X10 to increase the vertical sensitivity of the oscilloscope.

Divide all values on the volts/cm control by 10. For example, read the lowest setting (0.1) on the *volts/cm* switch as 0.01 and the highest setting (50) as 5.

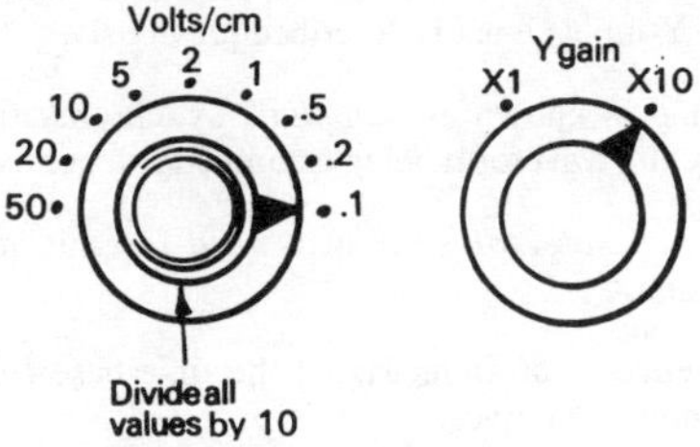

A.C./D.C. switch

Direct voltage.

To measure the direct voltage or alternating voltage with a d.c. component:

- the oscilloscope must be switched to d.c., and adequate warm-up time allowed to prevent *drift* between measurements
- the d.c. component must be measured first, reading to the centre line of the a.c. waveform
- set the a.c./d.c. switch to a.c.
- use the Y-shift and Y-gain controls to obtain a convenient display of the a.c. component.

1.

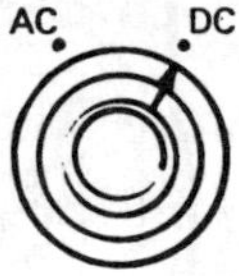

2.

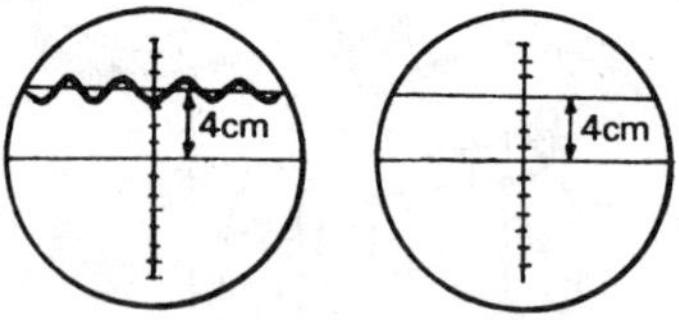

3.

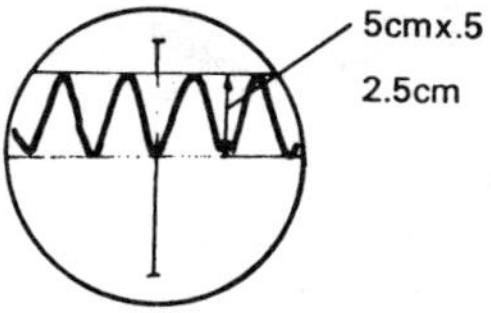

Automatic synchronising

The method used to synchronise the time-base frequency with that of the Y-input signal is described previously.

This method is known as *automatic* synchronisation and is used where the waveform being examined is repetitive.

Set the *sync. selector* switch to *auto* for this method of synchronising.

In the absence of an input signal the time-base free-runs to provide a horizontal trace.

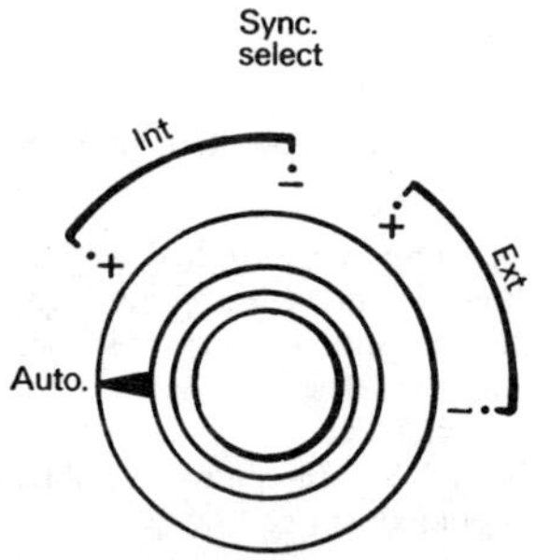

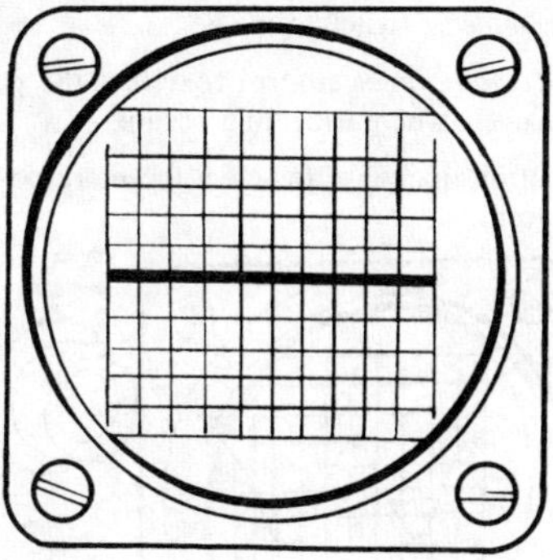

Triggered operation

A second method of synchronising, known as *triggered* operation, is particularly useful when the waveform being observed is a single pulse occurring at random intervals.

Unlike automatic operation, no horizontal trace is displayed in the absence of an input signal.

This method allows the time-base to start the horizontal sweep at a selected point on the input waveform. For example, points 1, 2 or 3 on the positive half-cycle and points 4, 5 and 6 on the negative half-cycle.

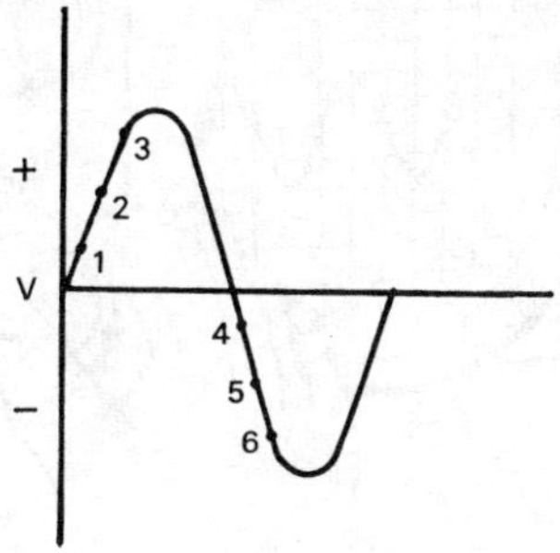

To trigger the time-base from the positive half-cycle.

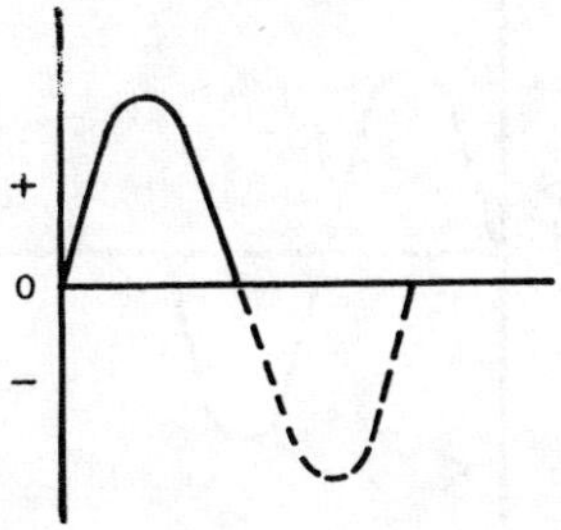

Set the *sync. selector* switch to *Int.* +.

Adjust the *trigger level* control to select the point on the positive half-cycle where triggering occurs.

Turn the control clockwise to select the more positive levels on the waveform.

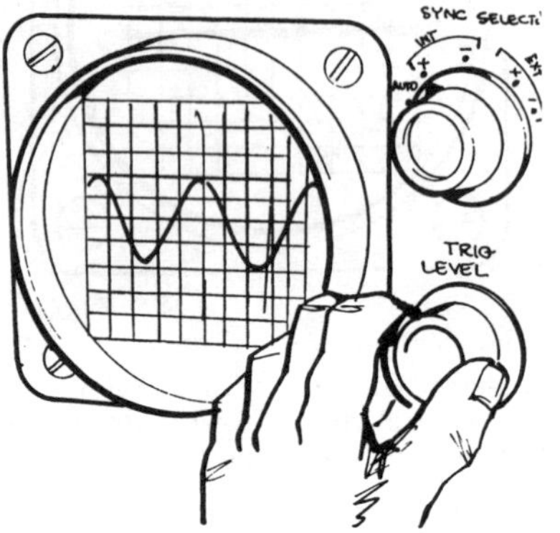

Turn the control anti-clockwise to select more negative levels.

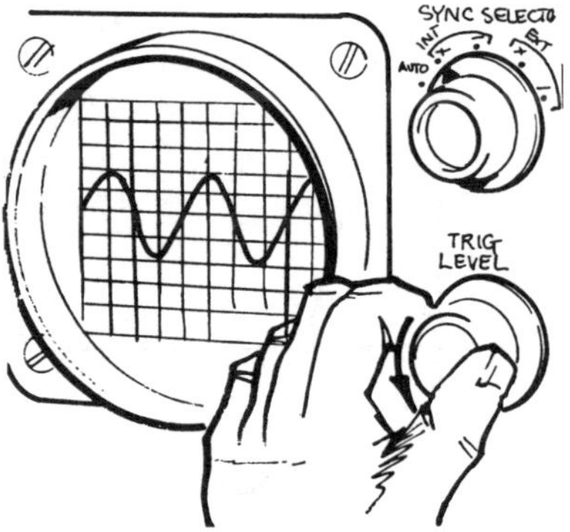

To trigger the time-base from the negative half-cycle of the input waveform.

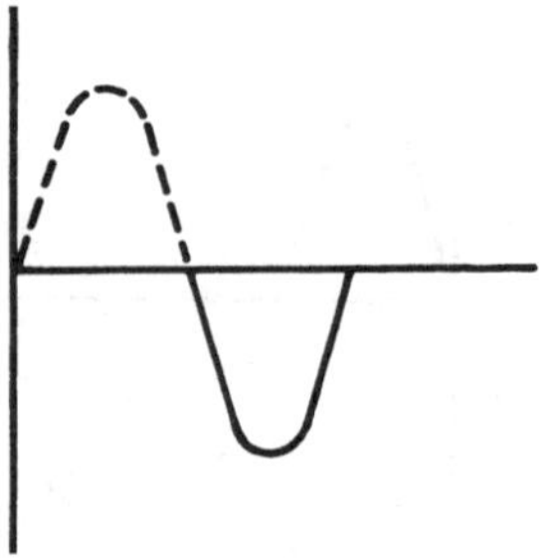

Set the *Sync. selector* switch to *Int.* –.

Adjust the *trigger level* control to select the point on the negative half-cycle where triggering occurs.

Turn the control clockwise to select the more positive levels on the waveform.

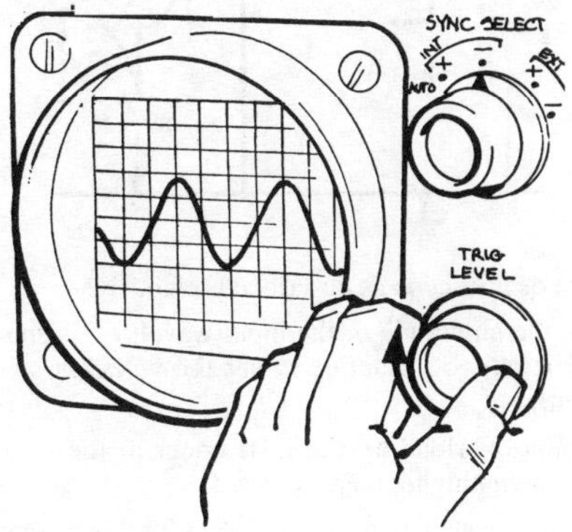

Turn the control anti-clockwise to select more negative levels.

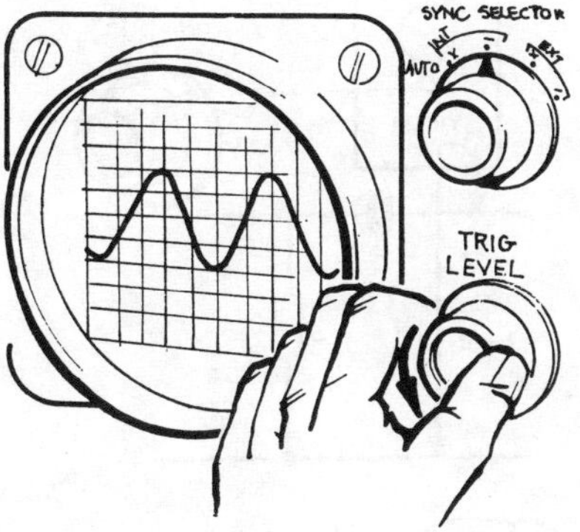

Measuring waveforms in a differentiating circuit

A differentiating circuit is a resistor — capacitor (RC) network used to re-shape waveforms.

Typical input and output waveforms together with their period and amplitude, are indicated in the diagram.

The shape of the output waveform depends on the period (t) of the input waveform and the values of R and C.

Display and measure the characteristics of the input waveform as follows:

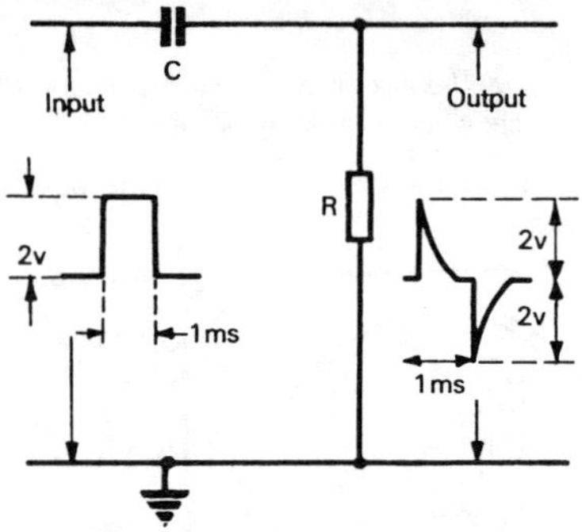

Set up the oscilloscope as described previously.

Although the amplitude of the input waveform is shown to be only 2 V it is good practice to set the *volts/cm* switch to a higher setting.

This avoids overloading the instrument in the event of the amplitude being higher than expected.

Set the *volts/cm* switch to, say, 10 V/cm for this example.

Connect the oscilloscope *Y-input* to the circuit as indicated.

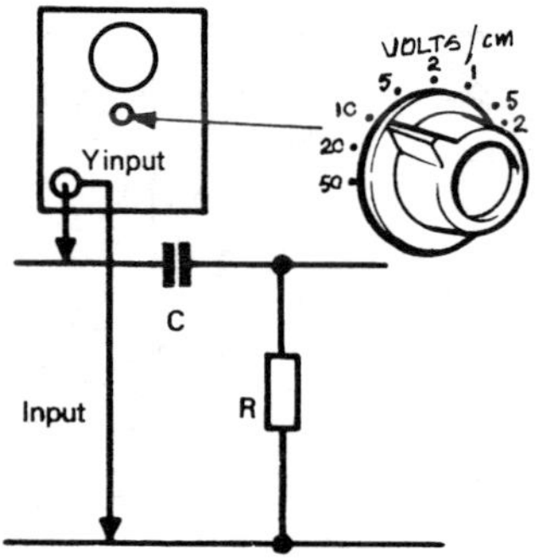

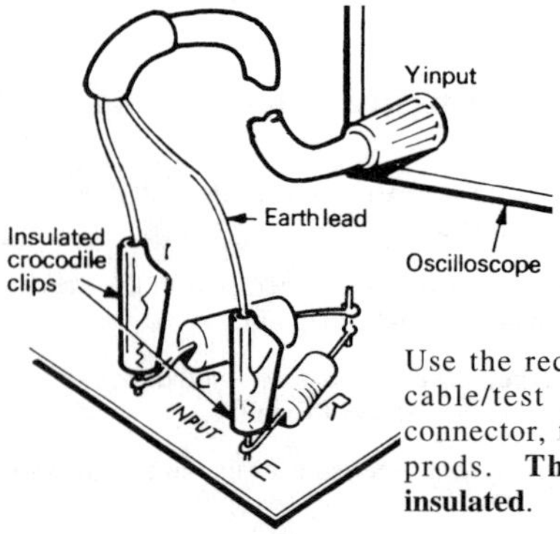

Use the recommended test cable/test leads, *Y-input* connector, test clips and/or prods. **These must be insulated**.

Set the *volts/cm* switch to 1 (1 V/cm).

Since the amplitude of the waveform being examined is 2 V p.p. the display should now occupy two vertical centimetres on the graticule.

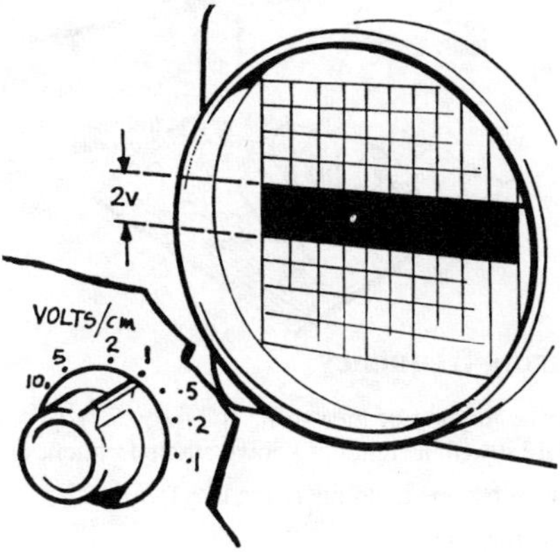

Set the *time/cm* switch to 1 mS to display one cycle of waveform in every horizontal centimetre.

If the width of the graticule is 10 cm, ten cycles will be displayed.

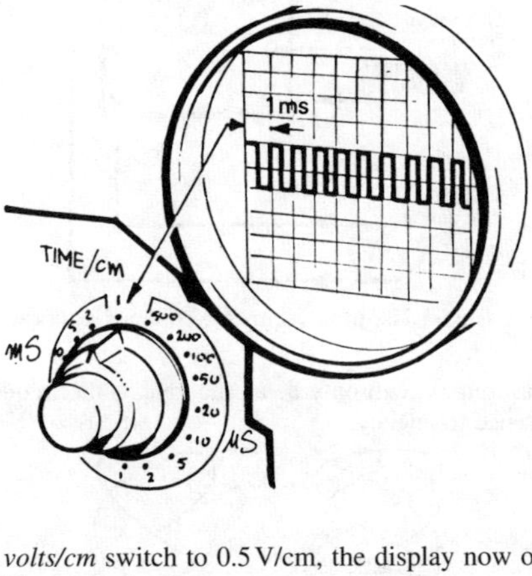

Set the *volts/cm* switch to 0.5 V/cm, the display now occupies 4 cm on the vertical axis.

Set the *time/cm* switch to 500 μS/cm. Each cycle now occupies 2 cm along the horizontal axis. Note that this is only the case when the variable *time/cm* control is fully anti-clockwise (off).

Connect the oscilloscope to the output of the differentiation circuit to obtain the display of the output waveform.

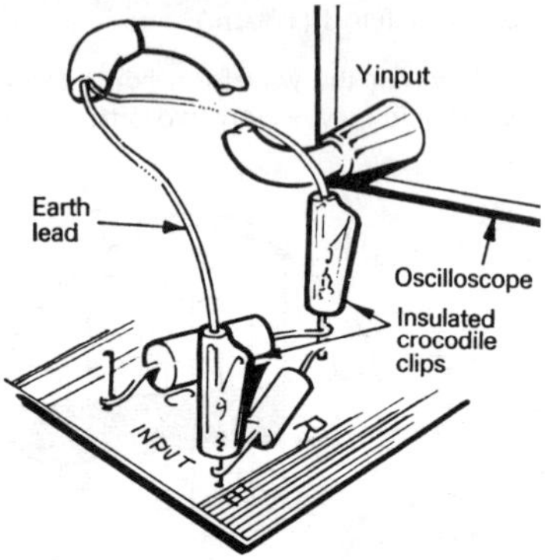

Measuring frequency

An accurate method of measuring the frequency of a signal is to compare it with a signal of known stable frequency.

Use an oscilloscope to compare the two frequencies.

Switch off the time-base.

Connect one signal to the Y-input and the other signal to the external X-input.

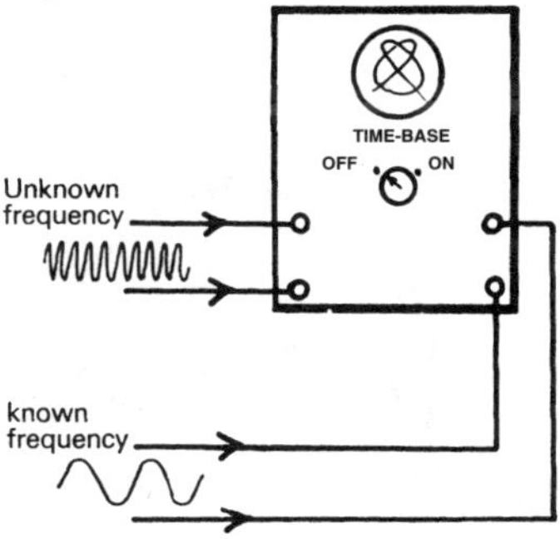

Patterns called Lissajous figures will be displayed on the screen.

The measurement will only be as accurate as the frequency of the reference frequency.

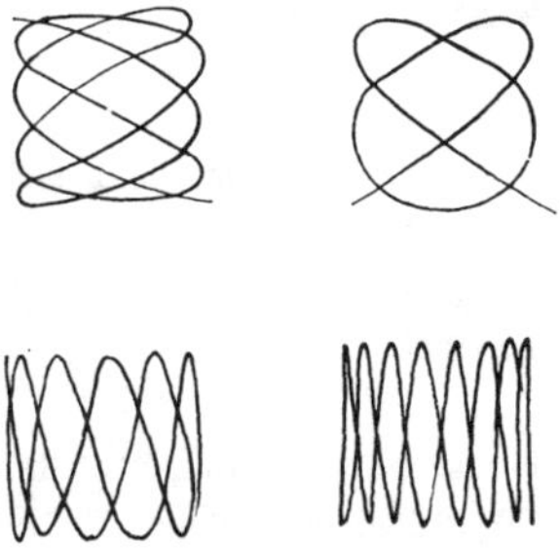

The diagram shows the equipment required to calibrate an oscillator.

A 50 Hz reference frequency of suitable amplitude V_R is obtained from the mains via a step-down transformer T1.

Set the oscillator to its lowest frequency range and adjust the attenuator to obtain an output V_O of the same amplitude as V_R.

Connect the secondary winding of T1 to the oscilloscope X-input terminals and the oscillator output to the Y-input terminals.

Set the oscilloscope controls to display a signal of amplitude V_R. Switch off the time-base.

Switch on the equipment.

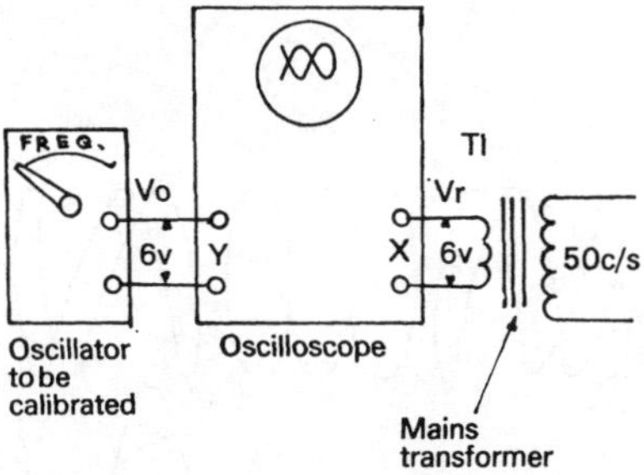

Increase the oscillator frequency until Lissajous figures similar to those shown in the diagrams are displayed.

Carefully adjust the oscillator and oscilloscope controls until a stationary pattern is obtained.

The ratio between the reference and oscillator frequencies is now 1:1 — the oscillator frequency is 50 Hz.

1:1

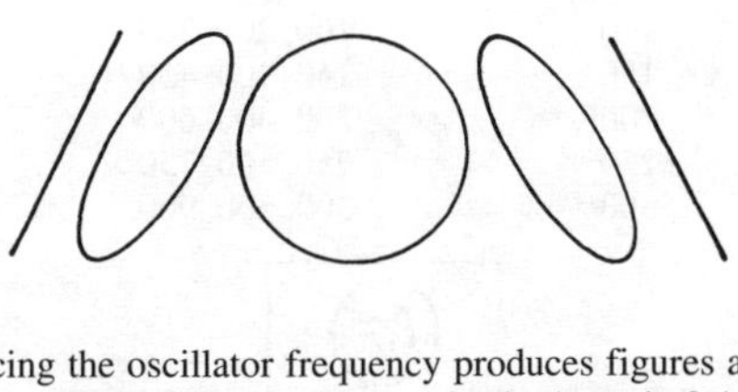

Advancing the oscillator frequency produces figures as shown overleaf, corresponding to frequencies in the ratio 2:1 and 3:1. These correspond to oscillator frequencies of 100 Hz and 150 Hz respectively. In this way the oscillator may be calibrated up to 150 Hz. Examples of Lissajous figures for various frequency ratios may be obtained from reference books.

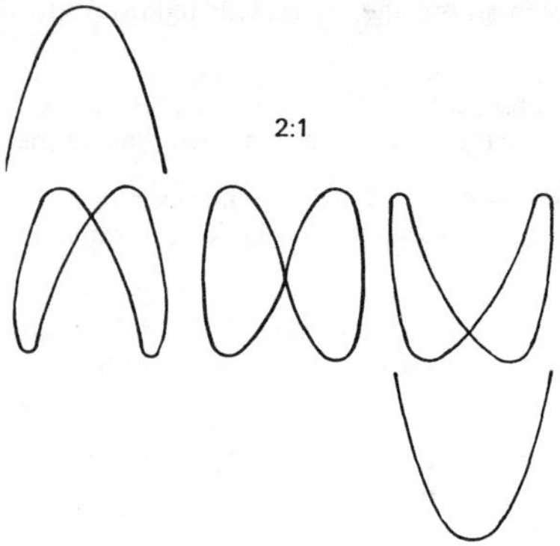

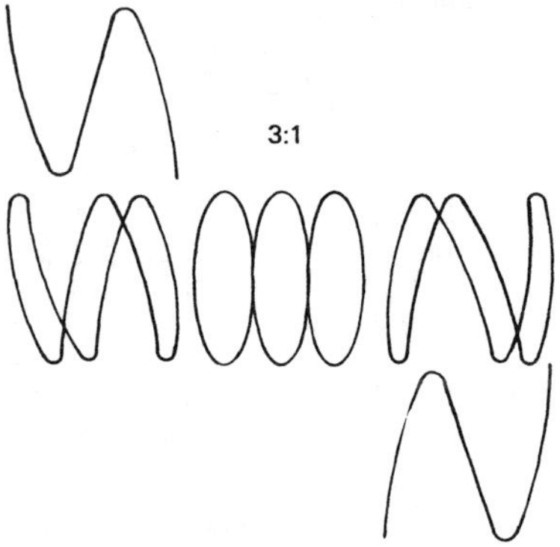

To calibrate the oscillator at higher frequencies use a stable A.F. signal generator to provide the reference frequency.

Examples of reference frequencies and corresponding calibration frequencies are shown in the table below. Use the 1:1, 2:1 and 3:1 Lissajous figures.

Reference frequency (Hz)	Calibration frequencies (Hz)
100	100, 200, 300
150	150, 300, 450
200	200, 400, 600
250	250, 500, 750
300	300, 600, 900

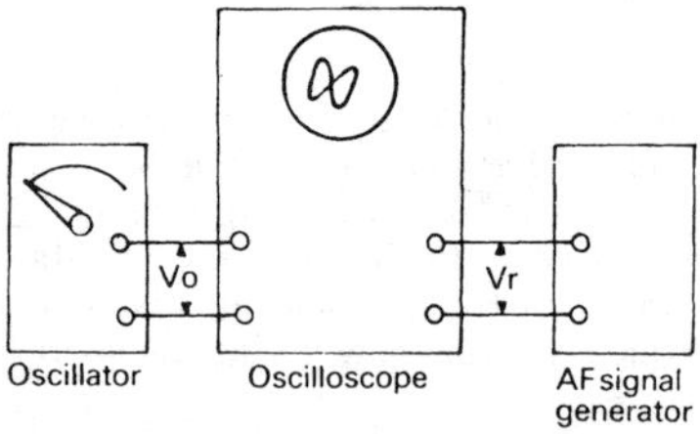

Intensity-modulation (Z-modulation)

Use intensity modulation to measure frequency and time.

Intensity modulation results in the displayed waveform appearing as a series of light and dark segments.

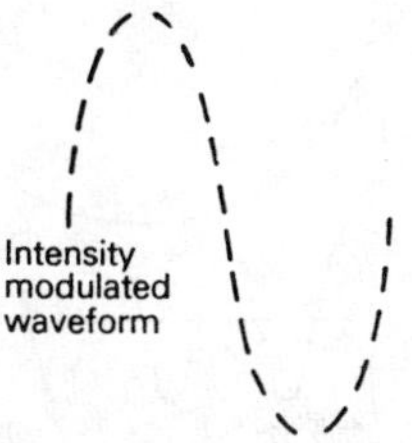

Intensity modulation is obtained by feeding a signal of suitable frequency and of specified waveform and amplitude to the control grid of the c.r.t.

The signal (f_Z) is known as the Z-modulation signal and is applied to the *Z-mod.* terminals on the oscilloscope.

The frequency of the signal must be much higher than that of the Y-input signal. Use a signal of waveform and amplitude as specified in the oscilloscope's instruction book.

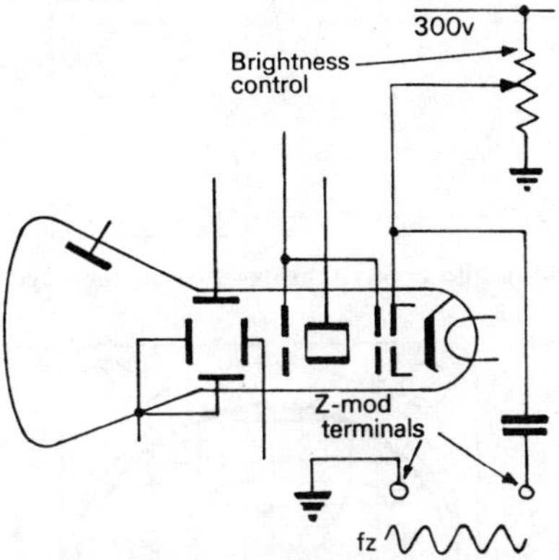

The diagram overleaf shows the effect of a sinewave Z-modulating signal upon the oscilloscope trace.

The c.r.t. bias is reduced during the positive half-cycles of the modulating signal, and increased during the negative half-cycles. The mean level of bias is set by the brightness control.

By suitably adjusting the brightness and focus controls the trace appears as a series of light and dark segments.

Note that one light and one dark segment corresponds to the period T of the Z-modulating signal.

The method of using intensity modulation for measuring purposes is described in the following sections.

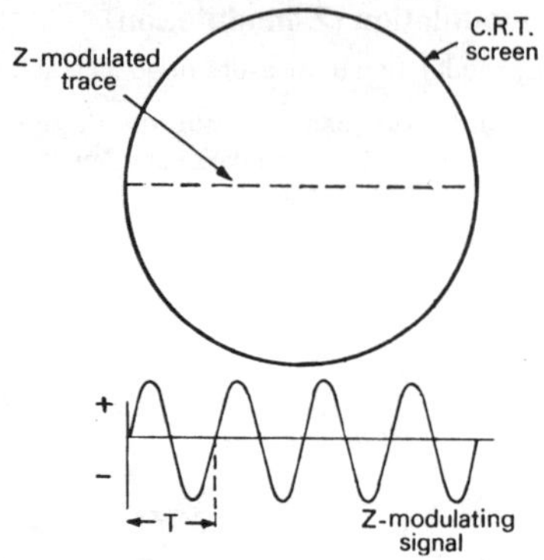

Measuring audio frequency

One method of measuring audio frequency using intensity-modulation is as follows:

Apply the signal of unknown frequency (f_x) to the oscilloscope's Y-input.

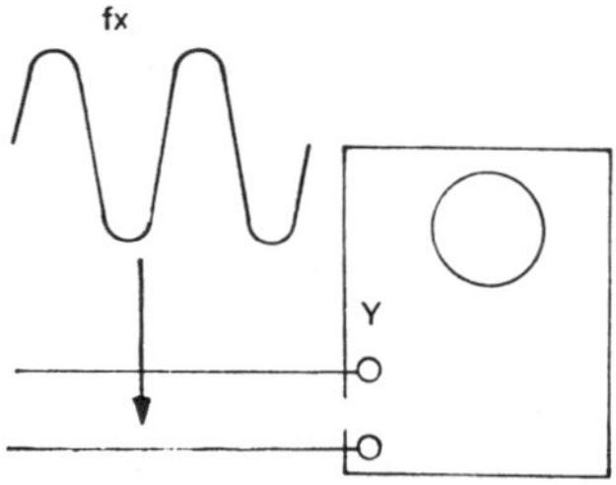

Adjust the oscilloscope to display two or three cycles of the signal.

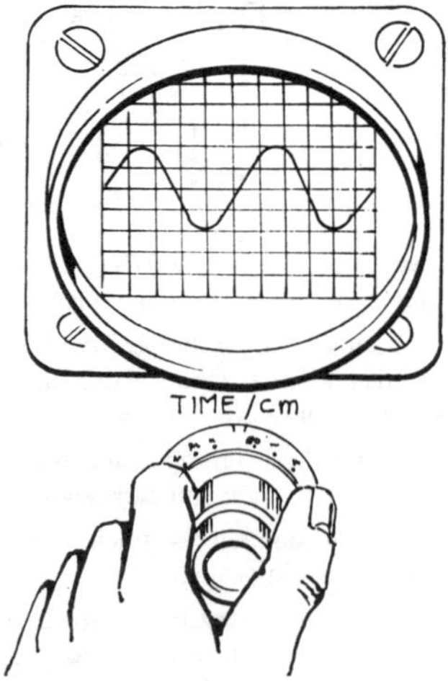

Use a sine or square-wave AF generator to supply the modulating signal (f_Z).

Set the generator to a frequency much higher than the oscilloscope sweep frequency.

Adjust the output control to provide the *Z-mod.* amplitude specified in the oscilloscope's instruction book.

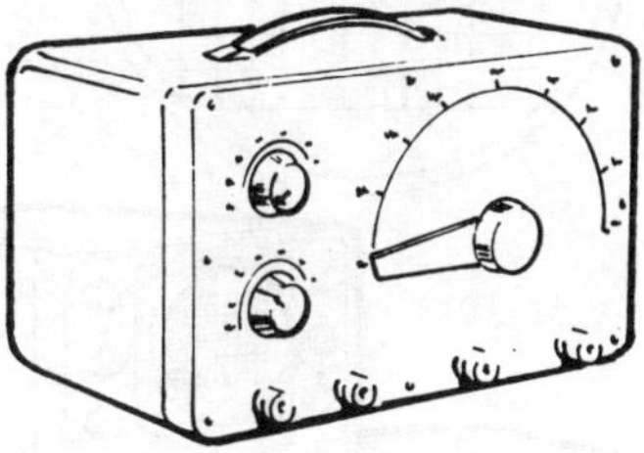

Connect the generator to the *Z-mod.* terminals on the oscilloscope.

These are usually located at the back of the instrument.

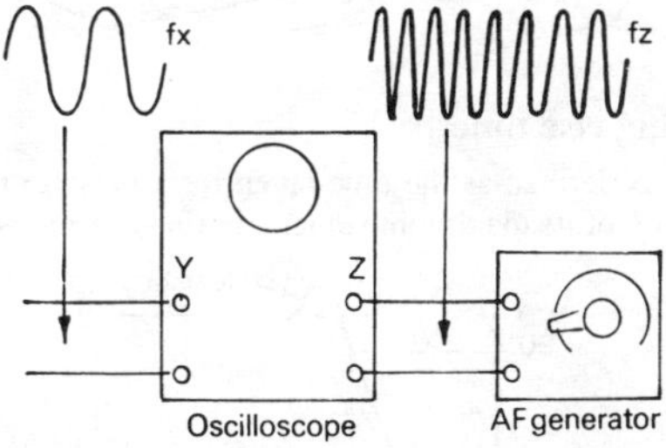

Adjust the generator frequency to obtain a number of segments on the displayed waveform.

Count the number of light or dark segments for one complete cycle.

The frequency of the unknown signal (f_x) is determined from the formula:

$$f_x = \frac{1}{\text{number of segments}} \times f_Z$$

If, for example:

- the generator frequency is 100 Hz
- 10 segments are counted

$$\text{Then } f_x = \frac{1}{10} \times 100\text{ Hz}$$
$$= 10\text{ Hz}$$

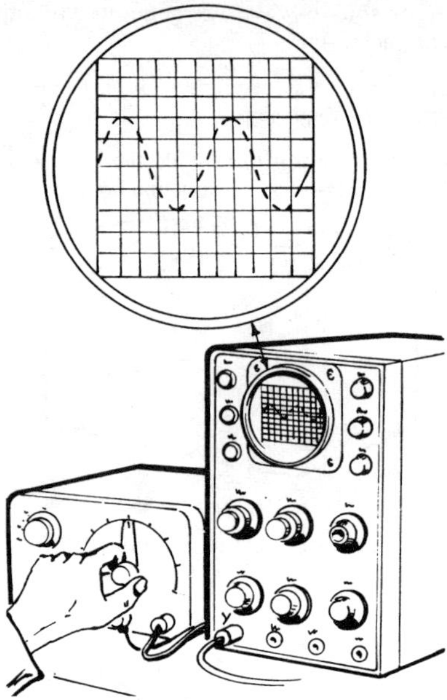

Measuring rise time

Rise time is defined as the time taken for a pulse to rise from 10% to 90% of its maximum value, ignoring any overshoot.

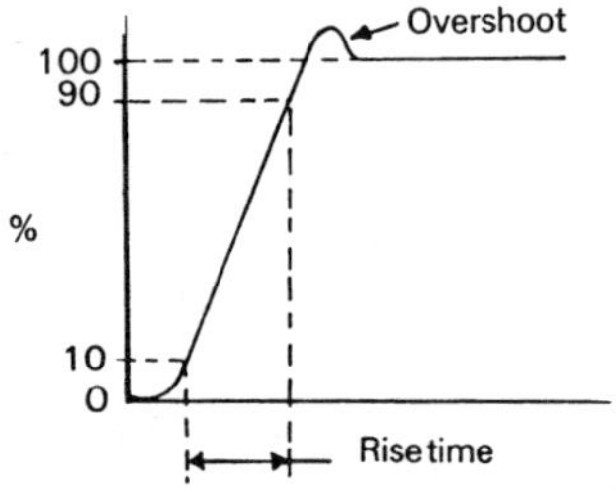

To measure rise time using intensity modulation

Apply the pulses to be measured to the oscillator's Y-input connector.

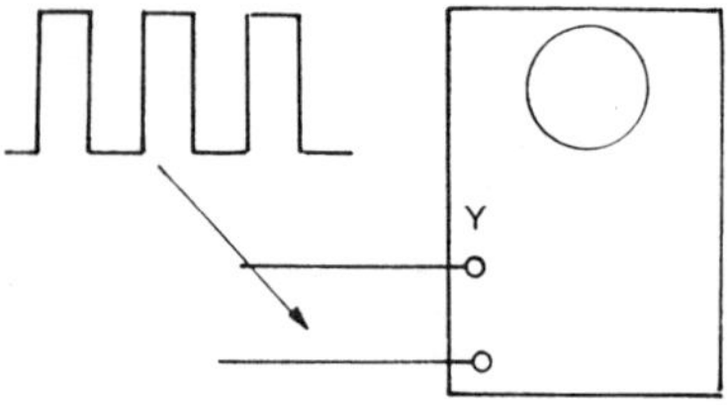

Adjust the oscilloscope to display a single pulse, using the *X- gain* control to expand the display.

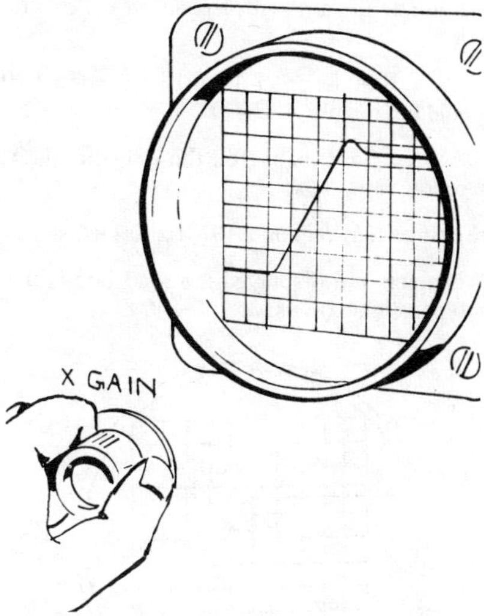

Use an R.F. signal generator (unmodulated output) or an R.F. pulse generator to intensity-modulate the pulse.

Connect the generator to the oscilloscope's *Z-mod.* terminals.

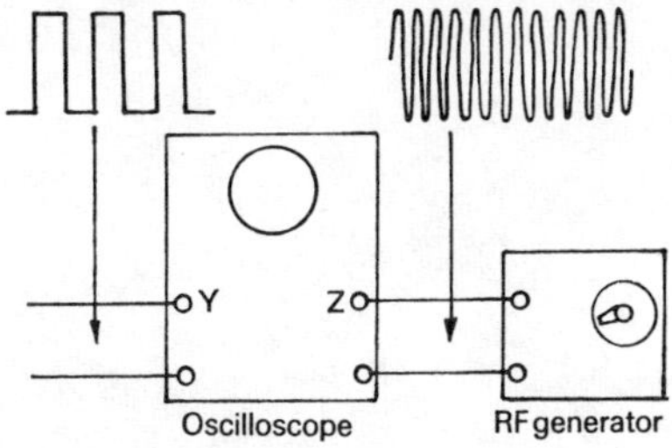

Adjust the frequency of the generator to obtain a number of segments on the slope of the displayed pulse.

The generator frequency will depend on the rise time being measured.

If the rise time is in the order of micro-seconds a 1 MHz (1 μs) signal would be suitable.

If four segments are displayed the rise time is 4 μs, each segment representing 1 μs.

Faster rise times may require a 10 MHz (0.1 μs) signal.

If four segments are displayed the rise time is 0.4 μs, each segment representing 0.1 μs.

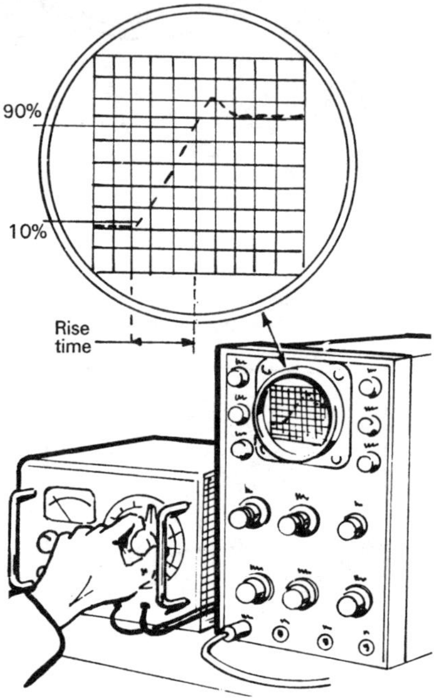

Spectrum analysers

The spectrum analyser is a variation of the oscilloscope principle, displaying waveforms on a cathode ray tube screen to take various measurements. The difference lies in how the waveforms are presented. An oscilloscope displays waveforms in the time domain — that is, a simple representation of signal amplitude against time. As long as signals are related, with a periodic element, a stable display results. However, if signals are unrelated, with an unperiodic element, the oscilloscope cannot produce a stable display. What is needed is a method of displaying signals in the frequency domain — that is, a representation of amplitude against frequency. Difference between time domain and frequency domain analysis is shown on the following page.

A collection of harmonically-related sinusoidal components is shown as would be displayed on both an oscilloscope and a spectrum analyser. Components at frequencies f, 3f and 5f are illustrated in a three-dimensional form, where amplitude, time and frequency are measured along the three axes. The components have amplitudes in the ratio 1.273:0.4244:0.2546. Sinusoidal components of these amplitude and frequency relationships happen to combine (as can be shown mathematically by Fourier analysis) to form a single waveform approximating a squarewave.

When viewed in direction X the three components combine to create a composite periodic waveform of approximately squarewave form, with axes of amplitude (Y-axis) and time (X-axis). This is a time domain display of the waveform, as would be displayed by an oscilloscope.

Viewed in direction Y, on the other hand, the three individual components are seen as functions of amplitude (Y-axis) and frequency (X-axis). This is a frequency domain display as would be displayed by a spectrum analyser.

Principle of spectrum analysers

All spectrum analysers function using the same principle, although how individual spectrum analysers produce their effects varies, by passing the incoming signal through a bank of bandpass filters.

Each bandpass filter is aligned to produce a response which ensures its upper corner frequency intersects closely with the lower corner frequency of the next bandpass filter. Output from each filter is scanned in sequence electronically by a

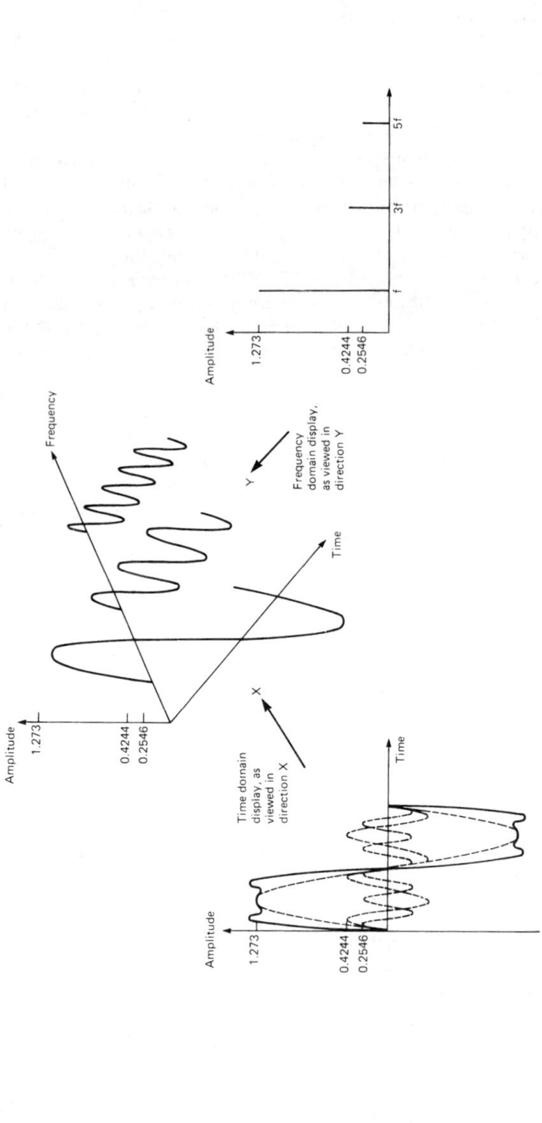
Amplitude
1.273
0.4244
0.2546
Frequency
Time
X
Y
Time domain display, as viewed in direction X
Frequency domain display, as viewed in direction Y
Amplitude
1.273
0.4244
0.2546
Time
Amplitude
1.273
0.4244
0.2546
f
3f
5f

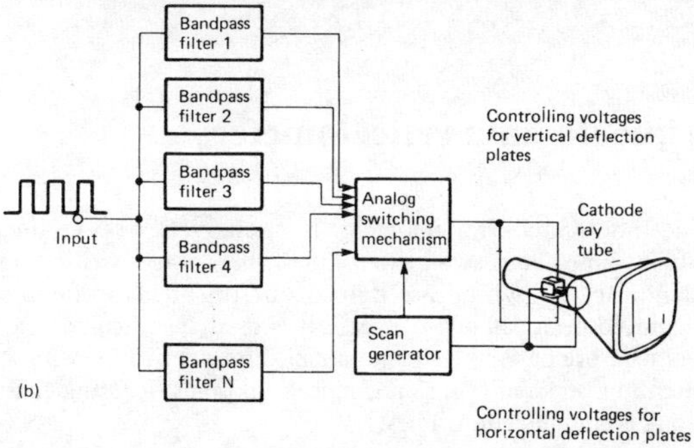

(b)

switching mechanism, such that each filter output in turn is used to define the controlling voltages applied to the vertical (Y) deflection plates of the cathode ray tube. The scan generator which provides the horizontal (X) deflection plate controlling voltage also generates the switching mechanism's controlling pulses.

As each filter is scanned, in turn, a corresponding vertical beam deflection results while the scan generator produces a corresponding left-to-right horizontal beam deflection. The resultant waveform displayed across the screen is a collection of signal amplitudes, one for each bandpass filter.

Time domain reflectometers

The time domain reflectometer is a relatively new product, which came about as an extension of the storage oscilloscope principle, in which a single occurring measurement is displayed on a cathode ray tube. It is used specifically to test and measure parameters in communications networks, with the intention of identifying and, more important, locating faults along communications lines.

It does this by transmitting a defined signal into the network and displaying the returned signal.

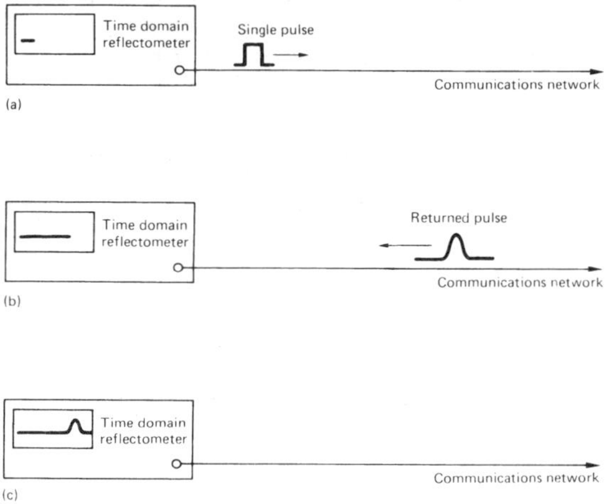

In (a) a single pulse is illustrated as it is transmitted by the time domain reflectometer. In (b) the pulse has reached the end of the network and is returned after a time. In (c) the received signal is displayed on the time domain reflectometer's cathod ray tube display. The shape of the display is sometimes said to be the network's signature.

There are two main types of time domain reflectometer:

- metallic time domain reflectometers (MTDRs) — used in metallic cable communications systems: co-axial, twisted pair and so on
- optical time domain reflectometer (OTDRs) — used in optical cable communications systems: usually optical fibre networks, giving rise to the alternative name fibre optical time domain reflectometers (FOTDRs).

Metallic time domain reflectometers

In a metallic time domain reflectometer, the reflection coefficient ρ (represented by the Greek letter rho) is used to relate reflected signal to transmitted signal according to the expression:

$$\rho = \frac{\text{reflected signal amplitude}}{\text{transmitted signal amplitude}}$$

such that total reflection with no polarity reversal is represented when $\rho = +1$, and total reflection with polarity reversal is represented when $\rho = -1$. If ρ is zero the cable impedance is equal to the characteristic impedance, that is the cable is correctly terminated with no faults.

Two signal types are used in time domain reflectometers:

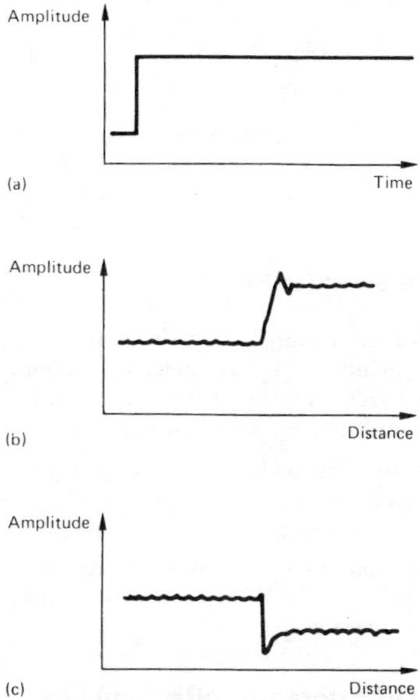

- a stepped change in applied voltage. The signal is transmitted along the cable (a). The reflected signal of an inductive fault or of one which is of higher resistance than the cable impedance is step-up, too (b). A fault which is capacitive or of a lower resistance than the cable impedance creates a step-down returned signal (c). Height of the reflected signal is a measure of the impedance change at the fault
- a pulsed signal. A single (usually one half of a sine cycle) is transmitted (a). The reflected signal is as in (b) if an inductive or high resistance fault occurs, or as in (c) if a capacitive or low resistance fault occurs. Height of the returned signal depends on the impedance change at the fault.

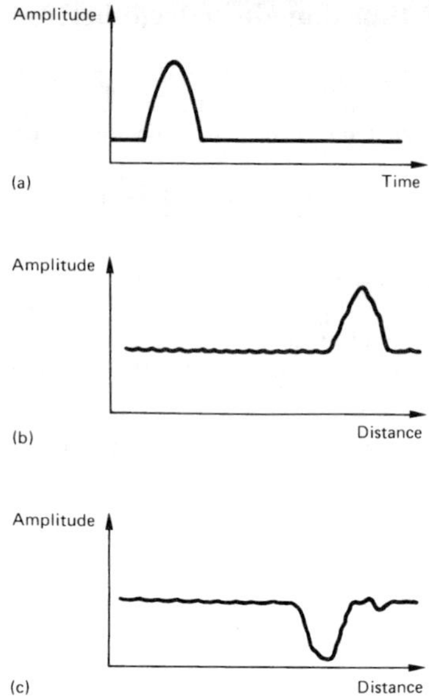

Textbook signatures

Various common faults can be identified by referring to textbook signatures (that is, specific signatures which are known to be caused by specific faults).Some common textbook signatures are shown opposite.

(a) and (b) inductive fault

(c) and (d) capacitive fault

(e) and (f) resistive fault

(g) and (h) connection to an incorrect impedance line — the length of the line which is of the wrong impedance can be measured on the signature.

Optical time domain reflectometers

The optical time domain reflectometer transmits a pulse of light along the optical fibre under test and measures the signal returning after back-scattering (a sort of refraction).

Display of an optical time domain reflectometer is a slope which shows power of the returned signal as shown opposite. As the power diminishes the trace gets lower on the screen, so the slope is a more-or-less continuous line decreasing from left to right.

Faults are displayed as inconsistencies in the decreasing slope. Shown is an inconsistency displayed typically by a poor connection in the fibre network.

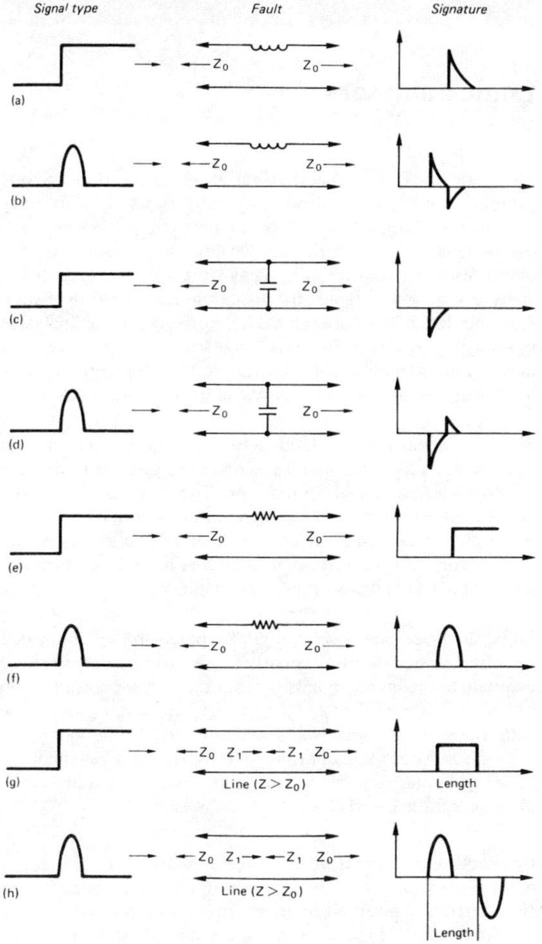

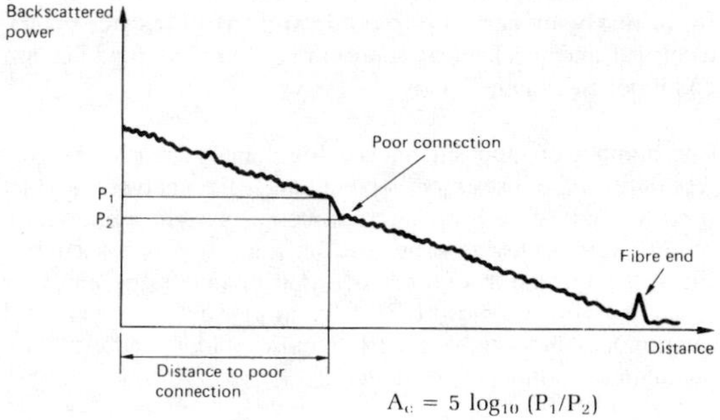

$$A_c = 5 \log_{10} (P_1/P_2)$$

Logic analysers

The logic analyser is another variation on the oscilloscope theme, in which a cathode ray tube is used to display waveforms of signals which occur in a circuit. Where most oscilloscopes, however, display waveforms of signals as they occur (that is, they display real-time waveforms), logic analysers store the signals for display in a non-real-time way. A further difference between oscilloscopes and logic analysers is that oscilloscopes are only used to display one, two, or maybe four waveforms at once. Logic analysers, on the other hand, can display many tens of waveforms simultaneously.

They are specifically intended for use in analysing digital circuits, such as those found in computers and many modern microprocessor-based appliances. These, of course, have developed with time into sophisticated products which do much more than just produce signals at various points within their circuits. As a result, logic analysers have developed also into sophisticated items of test equipment.

Logic analysers are always digitally-based, being controlled by microprocessors themselves. A logic analyser can manipulate incoming signals, storing them, reformatting them into alphanumeric data for display and can even present the data in a way which shows how the observed microprocessor system's software actually runs, step-by-step. A block diagram of a logic analyser shows how the device is essentially a microprocessor-based computer system itself.

Connecting a general-purpose logic analyser to a specific microprocessor system is a simple matter of interfacing the two with a special-purpose interface module which synchronises the logic analyser to the microprocessor system, formats data from the microprocessor system in a way the logic analyser can understand, and provides necessary electrical and mechanical connections. Interface modules are commonly nicknamed pods.

The number of input channels a logic analyser has more-or-less defines the instrument's potential. If the analyser is to be used to analyse an 8-bit microprocessor system with, say, a 16-bit address bus, an 8-bit data bus, and five or six control lines, the minimum number of input channels the analyser ideally needs is about 30. To analyse fully a 16-bit microprocessor system similarly requires a logic analyser with about 46 or so input channels.

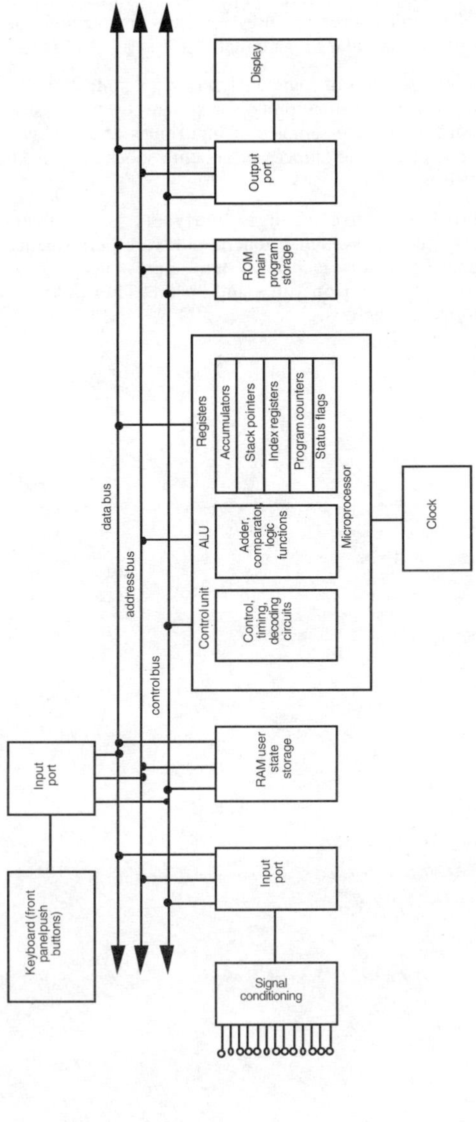

data bus
address bus
control bus
Display
Output port
ROM main program storage
Registers
Accumulators
Stack pointers
Index registers
Program counters
Status flags
ALU
Adder, comparator, logic functions
Control unit
Control, timing, decoding circuits
Microprocessor
Clock
RAM user state storage
Input port
Input port
Keyboard (front panelpush buttons)
Signal conditioning

Main modes of use

Logical analysers have evolved from digital storage oscilloscopes and, like them, can be used to display signal inputs as strict functions of time. This mode is known as timing analysis, and is used to observe signals immediately before and after a particular trigger point.

Another main use of logic analysers is to study overall logic performance of a microprocessor system, by taking straightforward logic state samples of input lines at every cycle or multiple of the microprocessor system's clock. This is known as state analysis.

A third main use of logic analysers is to observe a microprocessor system's general software performance, by comparing factors such as system address usage, program statement execution lines and so on. This is known as performance analysis.

Inspection procedures

Preparation and procedure

Keep your inspection bench tidy. Keep all equipment clean.

Obtain and read all drawings and specifications required to inspect the assembly.

Inspect to the inspection schedule. Be systematic and thorough. Mistakes are more costly to correct later.

If in doubt → Ask!

When the equipment has been inspected and accepted, route to the next operation.

Rejecting work

Reject for rework all items which contain defects.

List the defects on a reject report. Indicate the precise location of each defect.

Report repetitive defects to your inspection supervisor, and to Quality Control. Where possible investigate the causes of repetitive defects. Inform the production supervisor of your findings.

Re-inspect all reworked items.

Ensure that areas adjacent to the rework have not been damaged.

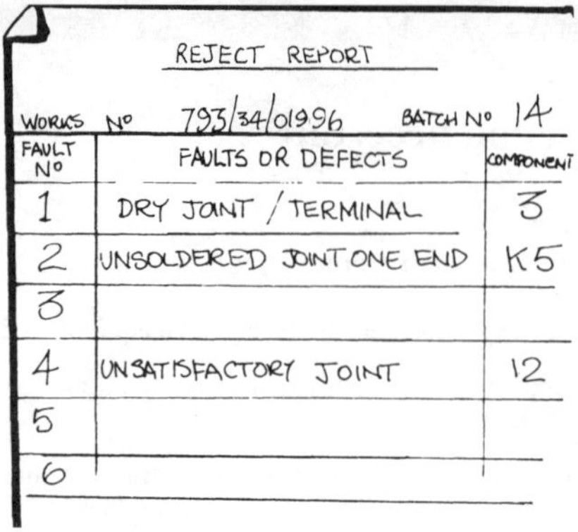

REJECT REPORT

WORKS No 793/34/01996 BATCH No 14

FAULT No	FAULTS OR DEFECTS	COMPONENT
1	DRY JOINT / TERMINAL	3
2	UNSOLDERED JOINT ONE END	K5
3		
4	UNSATISFACTORY JOINT	12
5		
6		

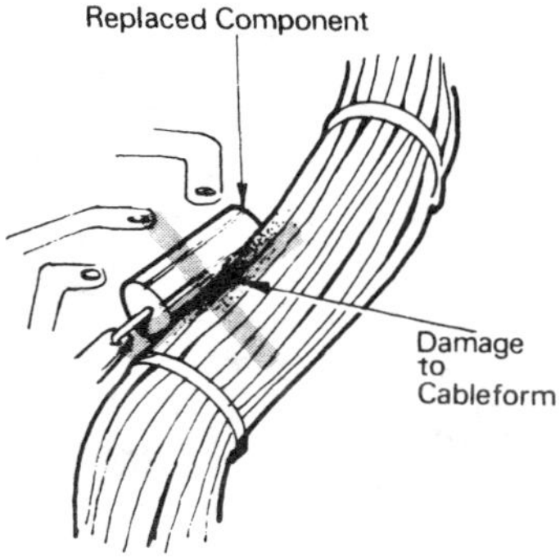

Minor defects

Minor defects do not influence the functioning of the equipment. It is often company practice to accept these to avoid unnecessary rework.

Be guided by relevant specifications, company practice and your own discretion.

Concessions

On Government contract work a *concession* must be raised by the planning engineers, or by inspection supervision, before minor defects are passed.

Ensure that a concession has been granted before allowing a defect to pass.

Defect report forms

Enter all the defect information required by the Quality Control department on defect report forms. Include minor defects.

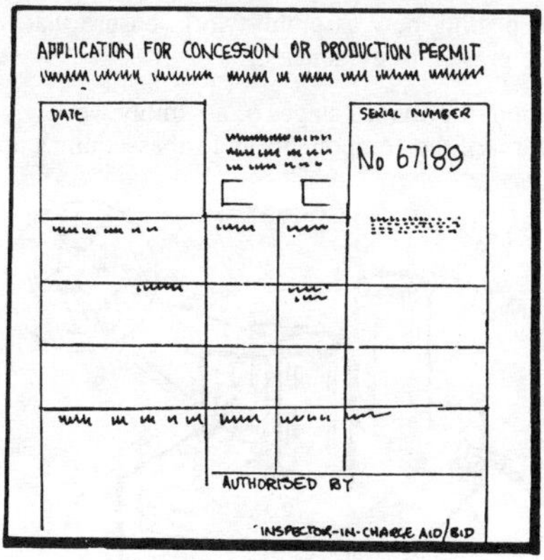

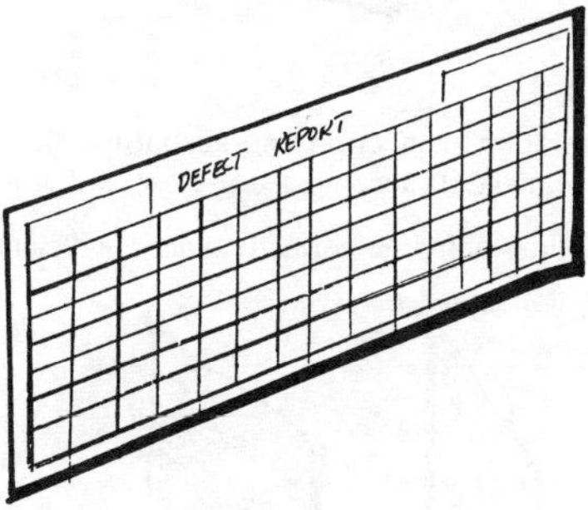

Acceptance

Inspection stamp

Every inspector has a inspection stamp with his personal identification number.

Keep your stamp under your personal control and for your sole use. If lost, report immediately to your supervisor.

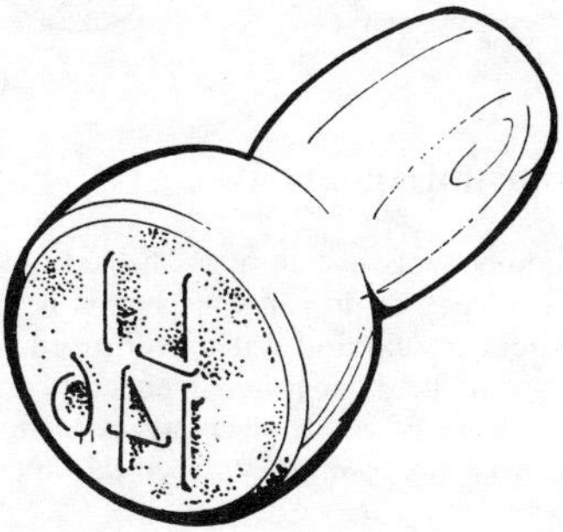

Before inspecting new assembly work, ensure that inspection stamps are present as evidence of:

- inspection of previous stages of assembly work
- inspection of the components and subassemblies used.

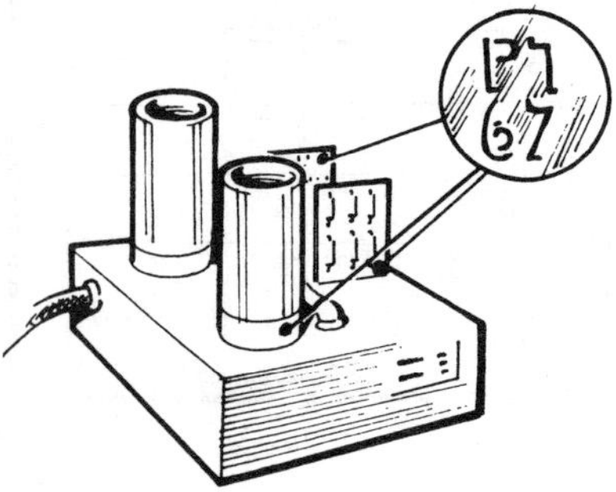

After completing inspection and accepting the work done, stamp the item and its accompanying flow card or log book.

Ensure that the stamps are correctly placed and clearly visible.

Bought-in components

Bought-in components usually pass through goods receiving inspection, evidence of previous inspection is examined, and further inspection carried out as required. You may be required to ensure that *bought-in* components come to the shop floor with evidence of goods receiving inspection approval in the form of inspection stamps, stamped documents, or release note numbers.

Release notes

Some Government contracts require that both the components used and the finished equipment are accepted by Government inspectors. This acceptance depends on satisfaction that all contract conditions have been met.

Components and equipments which receive Government approval are *released*. They are shipped on a release note.

Bonded stores

A condition of Government approval is the maintenance of a bonded store where only *released* components and materials are stored. Components may only enter the store on a release note, and be removed to the shop floor under an approved procedure.

Crown agents and inspection agents

Other customers may employ agents to inspect goods before they leave the factory, e.g. Crown Agents deal with many equipments for export.

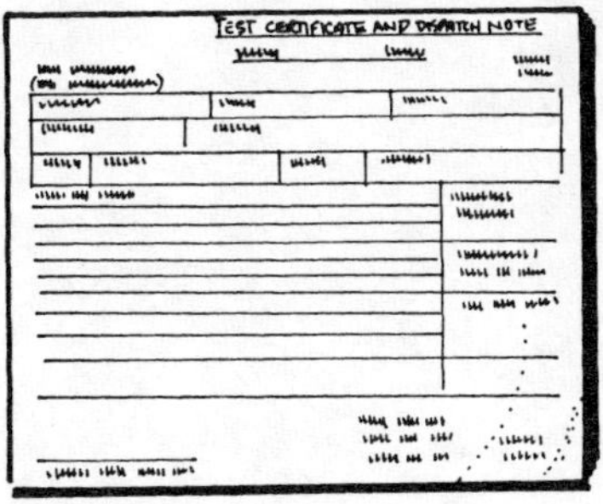

Contracts and works orders

Ensure that you are familiar with the contract inspection conditions of the equipment you are inspecting. These are found most easily on the *Works Order*. In particular:

● ensure that the equipment is built to the customer's requirements, and that modifications he requires are incorporated
● note the correct packaging requirements — ensure that they are stipulated clearly on the despatch of release note before allowing the equipment to be despatched
● note the inspection conditions and pass the equipment through the correct procedure.

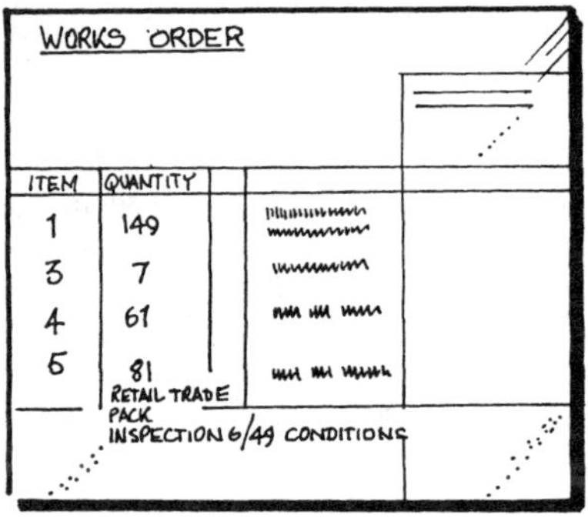

Inspection techniques

100% inspection

This is the inspection of:

- all work done since the last inspection, or
- the whole of a batch, as opposed to a sample.

Electronic equipment is normally subjected to 100% inspection in stages during production.

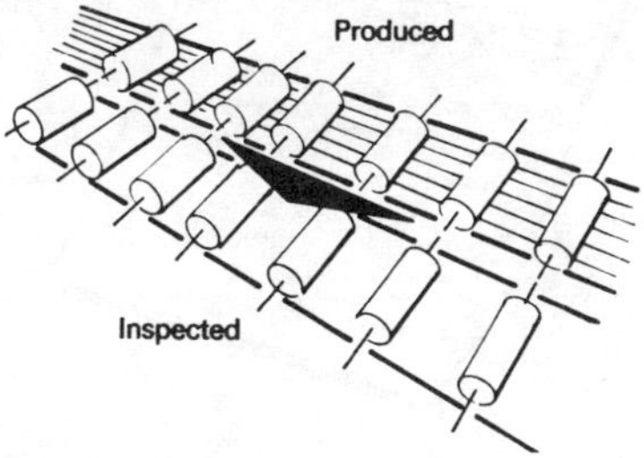

Inspection stages

The sequence of assembly of electronic components into electronic equipment is:

- building components into a number of small sub-assemblies
- building sub-assemblies into larger assemblies.

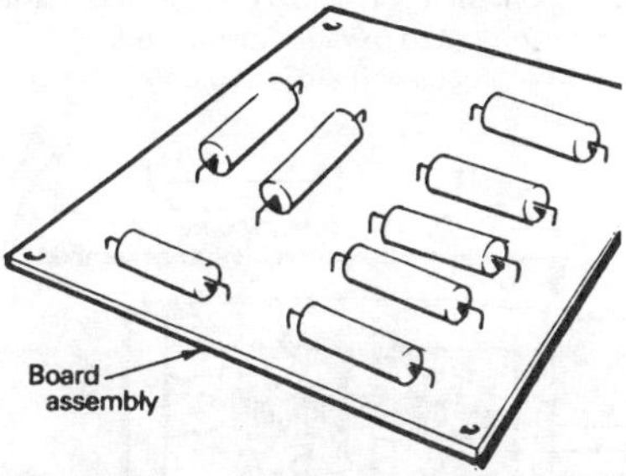

Inspection takes place at the end of each sub-assembly stage when only work carried out since the last stage is inspected.

When several inspected sub-assemblies are assembled, only the work done in assembling them together is inspected.

Inspection is also carried out:

- after difficult or critical operations
- after operations where the defect level is known to be high.

Final inspection

The equipment is given a final inspection after test.

Special attention is paid to:

- general appearance
- areas which have been disturbed by test. These should be listed, by the test department, on a test repair report
- the equipments documentation. This must be correct before the equipment is sent to despatch.

Reason for inspection stages

Inspection is carried out in stages in preference to a single operation on completion of the equipment because:

- work carried out on a defective component is wasted
- some work is obscured by subsequent work
- it is easier to inspect small areas at a time.

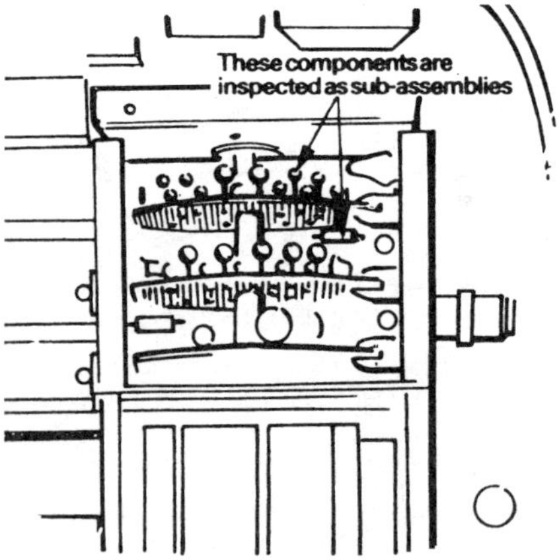

Sample inspection

Sample inspection is the inspection of:

- a sample of the work done since the last inspection, or
- a sample out of a batch of components.

The whole of the work done is accepted or rejected on the result of the sample.

Samples are taken at random.

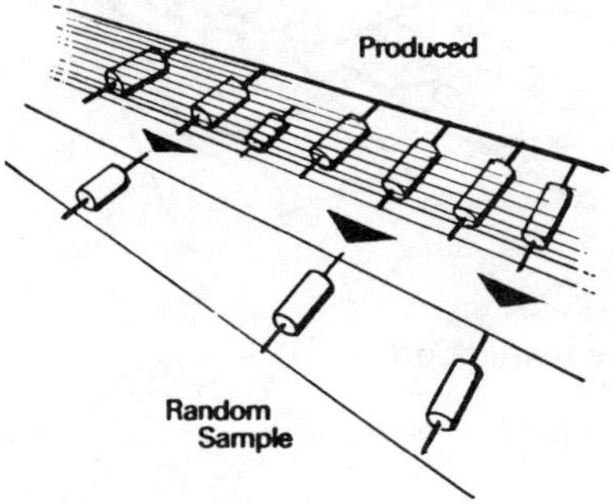

The size of the sample inspected depends on:

- the complexity of the item
- the standard required
- whether the standard usually produced is good or bad.

Double sampling

In double sampling a second sample is taken if the first fails. Rejection follows the failure of the second sample.

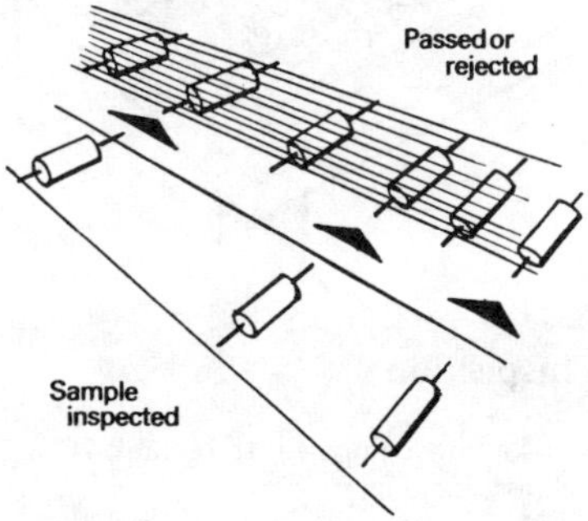

Control sampling

Samples of products from automatic or semi-automatic processes are inspected at specified intervals. The results are plotted on a control chart. Trends towards future defects can be seen and action taken before production becomes outside specification.

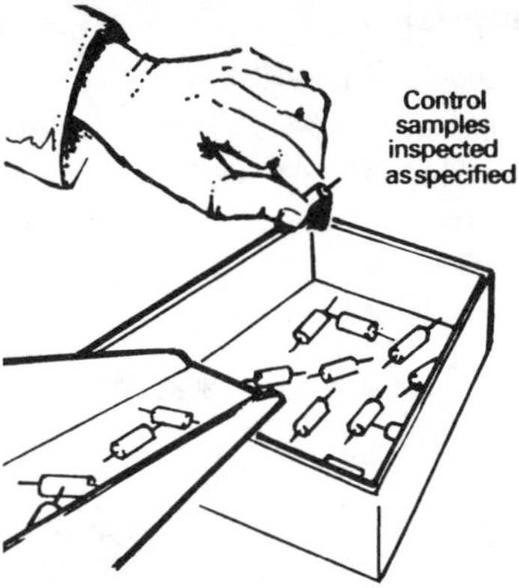

Selective inspection

Only those parts of the assembly operations are inspected which are:

- critical to the equipments functioning, or
- subject to a high defect rate, e.g. it may be necessary to inspect only the soldered joints on an assembly.

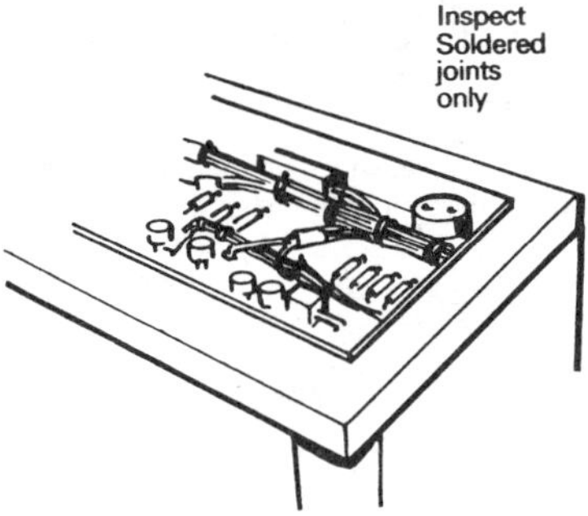

First-off inspection

The first-off of a new equipment to be built is subjected to a critical inspection, including:

- the inspection of each component, including *bought-in* components and metalwork
- checking all drawings for ambiguities and contradictions. The circuit diagram is used as the basis for these checks
- thorough inspection of the assembly
- noting all ambiguities in the assembly instructions
- noting departures from engineering codes of practice.

History sheet

A history sheet or first-off inspection report is drawn up detailing all comments during this inspection. This gives information to bring new designs and drawings into line with production requirements and to remove design defects.

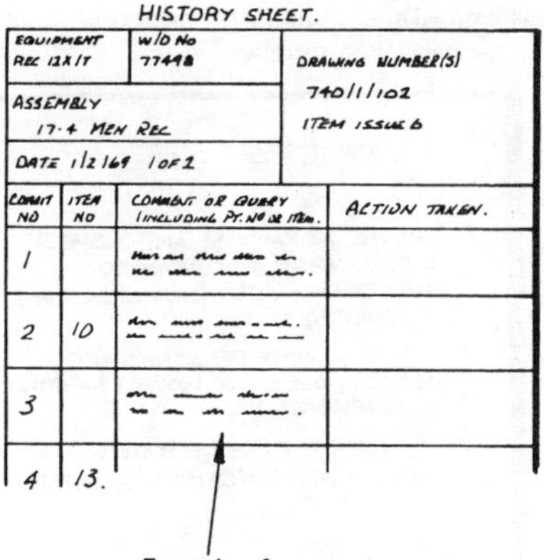

Examples of comments

Cable form X372 - suggest reroute via backplate to avoid edge of bracket 224. Will decrease cable length.

Bracket X416 - JIG & gauge reqd. to comply with dimension.

Screw 24712 - excessive length suggest 100mm.

Item 7 incorrectly coded on assembly drawing as item 4.

Potentiometer S4876 - unable to identify - coding hidden.

Patrol inspection

Patrol inspection may supplement or be an alternative to stage inspection. It places emphasis on continuous control of the manufacturing processes. This is done by:

● inspection by control sampling techniques. Sample inspections are carried out according to a specific, yet flexible pattern. This ensures a continuous monitor of all aspects of production

Control limits are specified. If defects found on sample inspections exceed this limit 100% inspection is temporarily restored on the particular operation.

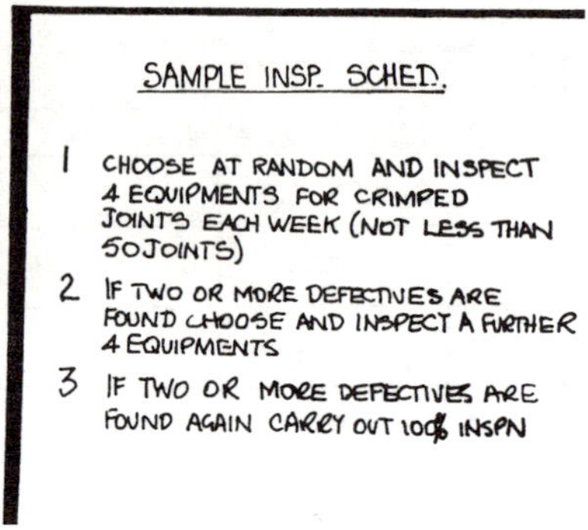

● close monitoring of the methods of production, and the efficiency of tools to ensure a high level of quality. This is sometimes called process control.

Process control

Pay attention to:

● the tidiness of the operator's bench. Files and caustic liquids are common causes of scratching and discoloration of the equipment

● wire clippings and other scrap should be cleared from the bench. The equipment must be kept clean.

Soldering

Check:

- the quality of the soldering iron bit
- the use of the correct soldering iron
- that the correct operating procedure is being followed.

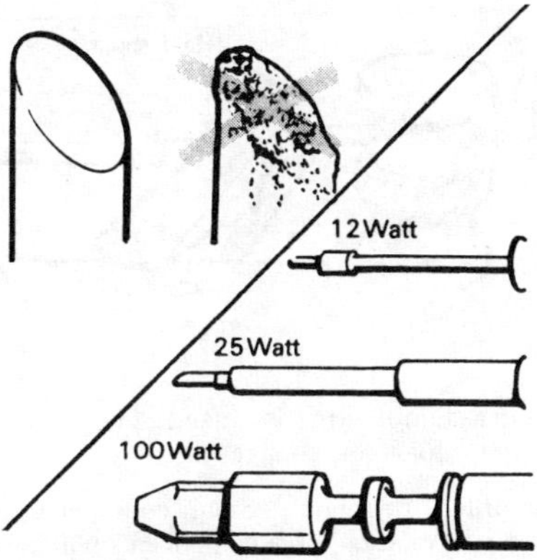

Wrapped and crimped joints

Ensure that wrapping and crimping tools are kept in good condition.

Component mounting

Components are delicate.

Ensure that heat shields and heat shunts are used as required during soldering.

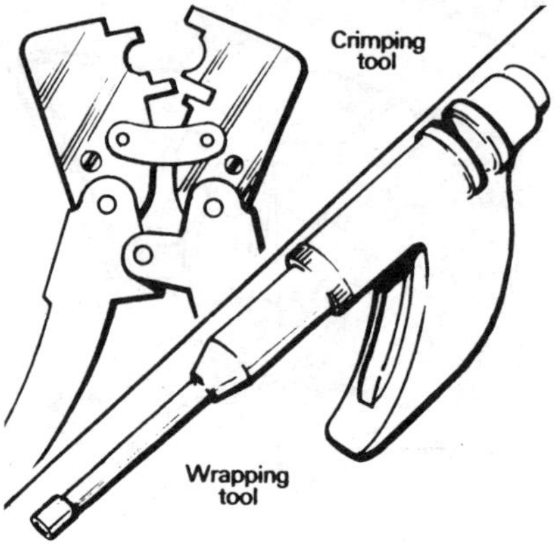

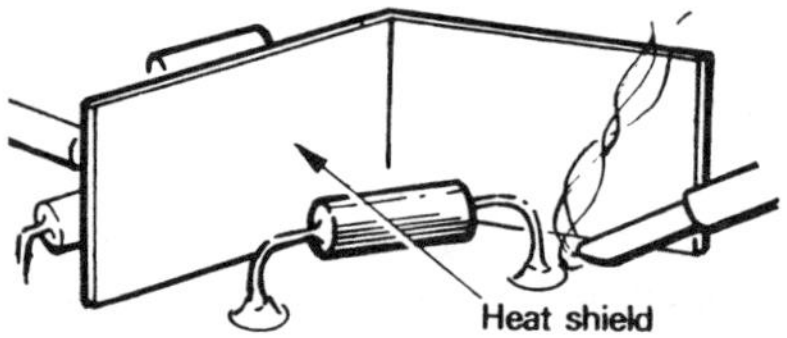

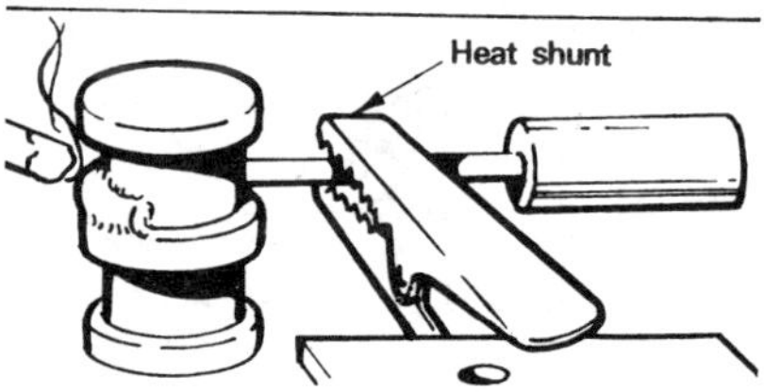

Check that assembly jigs are used. They must provide sufficient protection to the components.

Check the order of assembly. Small components should be assembled first. Adjacent large components will prevent them from being held firmly in place during soldering.

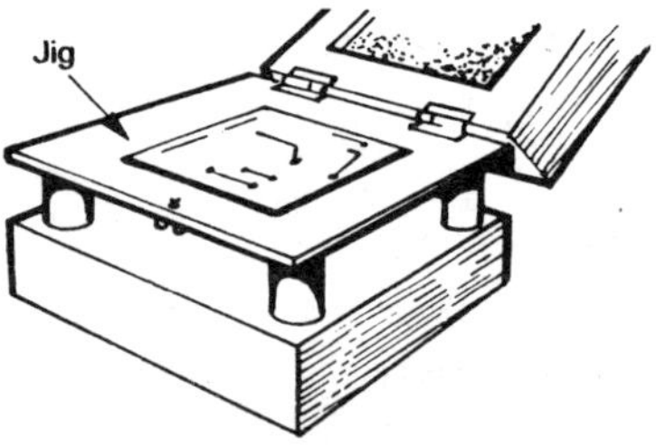

Printed circuit boards and assemblies must be handled correctly. They must be stored in racks, or in polythene bags.

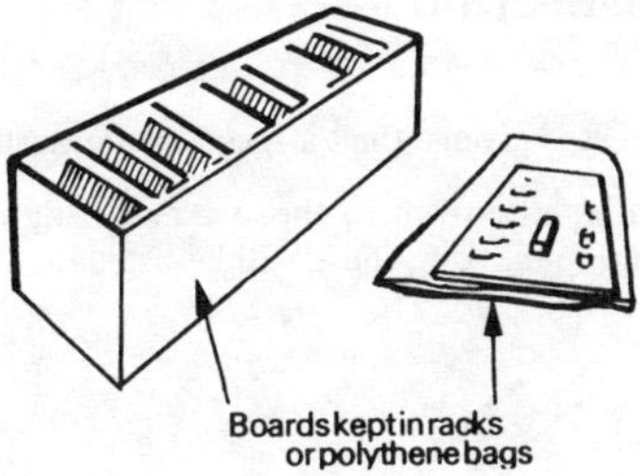

Final inspection

The equipment is given a final inspection after test.

Pay particular attention to those areas which have been disturbed during the test. These will be listed on the test repair report.

Inspect for damage to the area around each repair, e.g. look for:

- burned insulation on leads
- track lifting on printed circuit boards
- loose parts
- burred screws
- bent plug and socket pins.

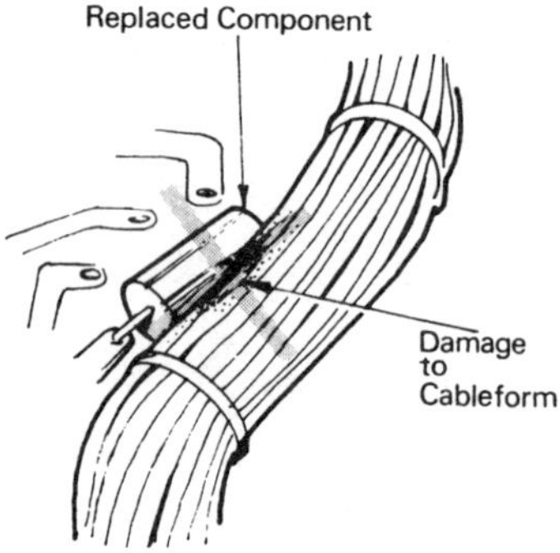

Ensure that all moving parts operate correctly e.g. that control knobs turn smoothly.

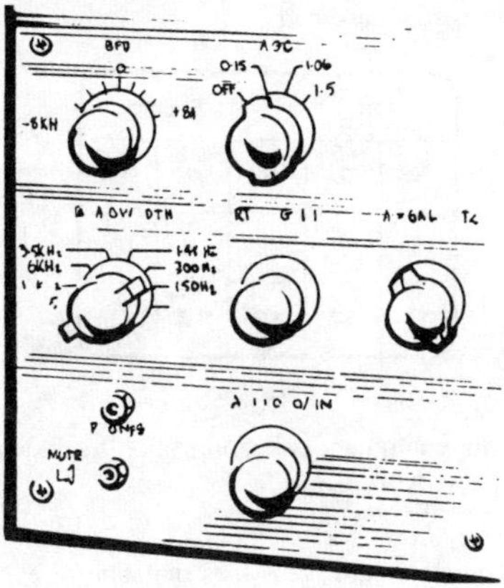

Pay special attention to the appearance of the equipment. This is of great significance to the customer.

Ensure that the general appearance of the front panel is neat and clean. The paint work must be unscratched.

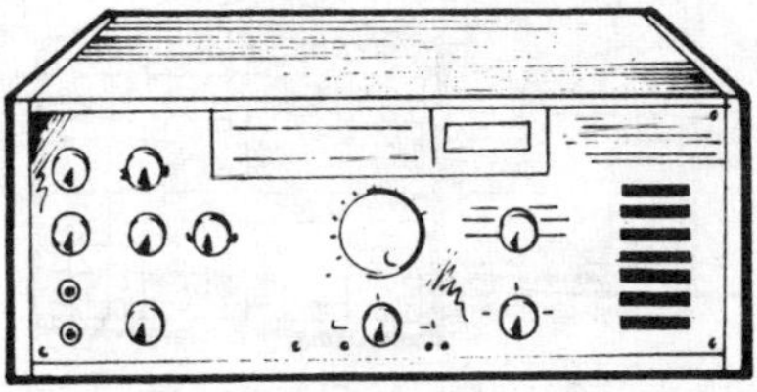

Check that the labelling is correct and undamaged.

Inspect plated surfaces, particularly chromed handles, for scratches and discoloration.

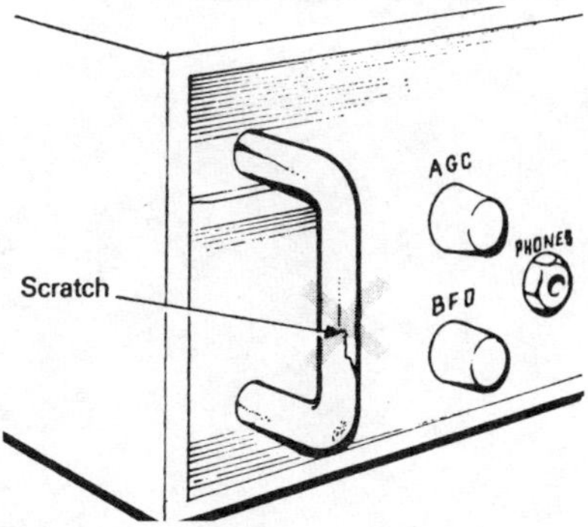

Ensure that modification labels are marked up to date.

These indicate the modifications carried out on that particular unit since it was first designed.

Mod			Record	
1	2	3	4	5
6	7	8	9	10
11	12	13	14	15

Ensure that the equipment is accompanied to despatch by all the relevant paper work, e.g:

- test documentation
- modification state card. This lists the state of each modification label on the equipment
- despatch or release note. This must include the correct information of packaging.

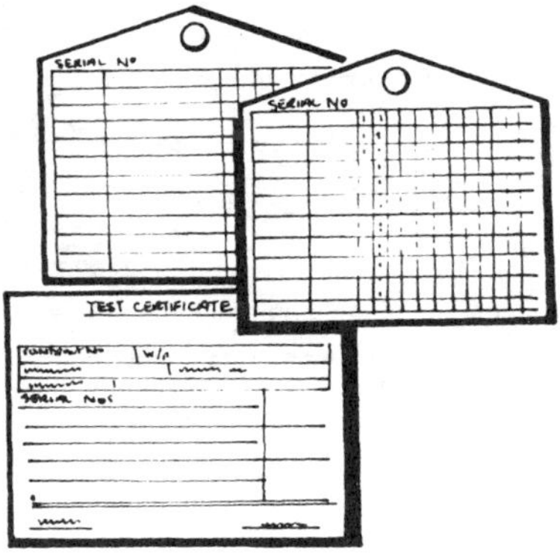

Component identification

Transistors

Transistors are 3-terminal devices. The terminals are called:

- base
- emitter
- collector.

They may be classified as:

- silicon or germanium
- NPN or PNP.

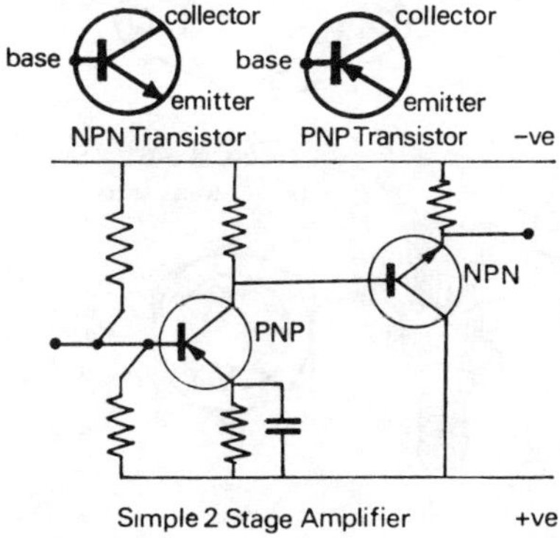

Transistor configurations

Common case configurations are shown:

To-18, To-72 Low power transistors. The To-72 has a fourth lead connected permanently to the case and used to earth the case.

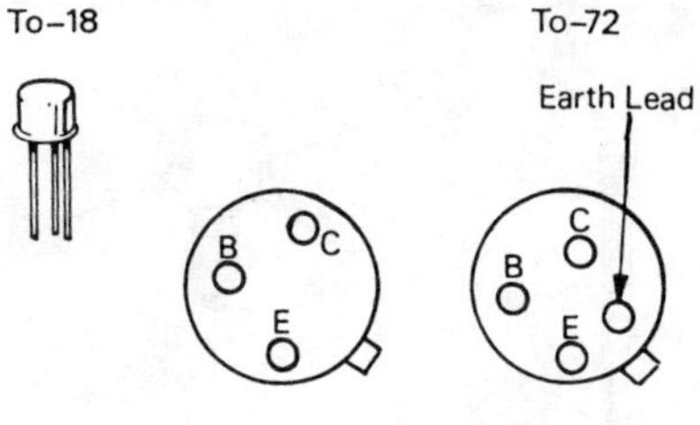

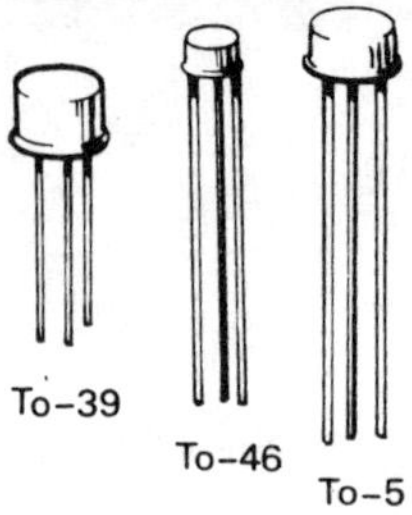

To-5, To-39, To-46 Medium power transistors.

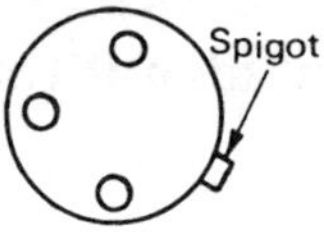

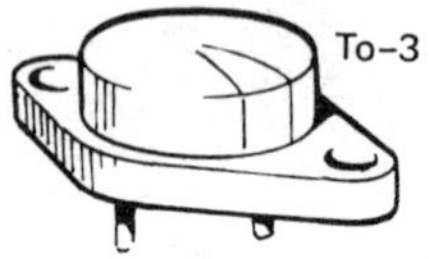

To-3 High power transistors. Note that connection to the collector electrode is the case.

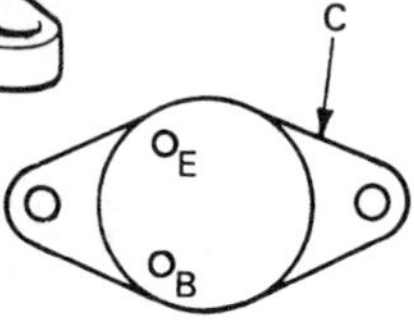

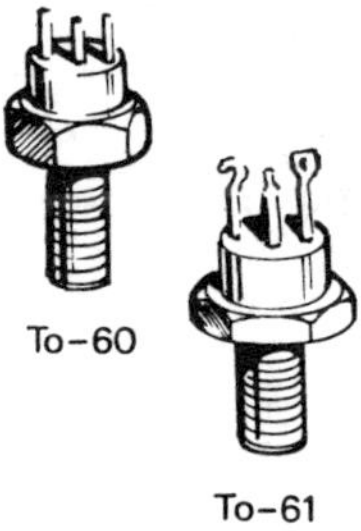

To-60, To-61 Stud mounted power transistors.

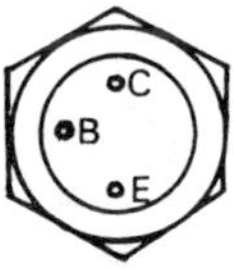

New types of configuration

Transistors encapsulated in plastic are inexpensive to manufacture.

For use in microminiaturised circuitry, special transistors have been produced. In general these are small.

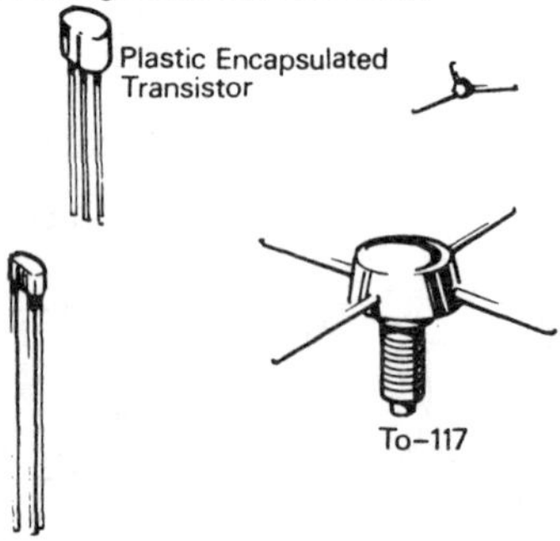

Dual transistors

These have two transistors mounted within one case.

Safety

Never open transistors, some contain dangerous substances.

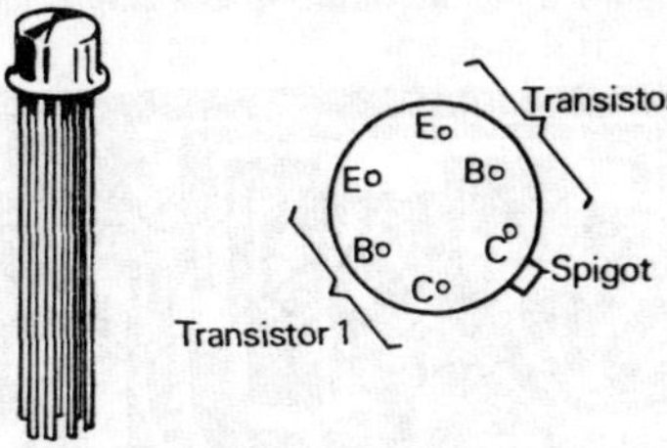

Field effect transistors (F.E.T.s)

These are different from conventional transistors. They have a very high input impedance.

They have three terminals called:

- drain
- gate
- source.

Usually they are in To-18, To-5 or plastic encapsulated.

Because of the high input impedance of F.E.T.s special care must be taken to avoid damage. With some types of F.E.T.s it is necessary to:

- wrap a tinned copper wire around the leads when handling or soldering
- use a low voltage iron for soldering.

If in doubt → Ask!

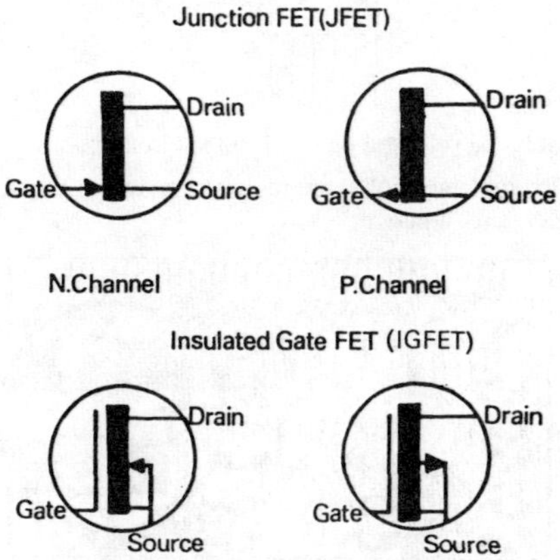

Silicon integrated circuits (S.I.C.s)

A complete electronic circuit containing transistors, diodes, resistors and capacitors, with all the interconnections can be manufactured by a special technique on to a silicon chip of about 1 or 2 millimetres square. The example on the page opposite shows a circuit containing 34 transistors.

A highly magnified picture of the silicon chip embodying the circuit function is shown opposite.

Actual chip size 1 mm. ▪

The chip is mounted in a case, in this example, in a *flat pack*.

Completed S.I.C.

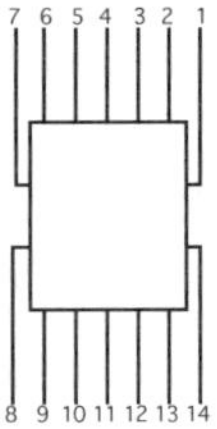

S.I.C.s may be soldered on to a printed circuit board.

In this example a number are mounted on to the board to form a complex circuit network.

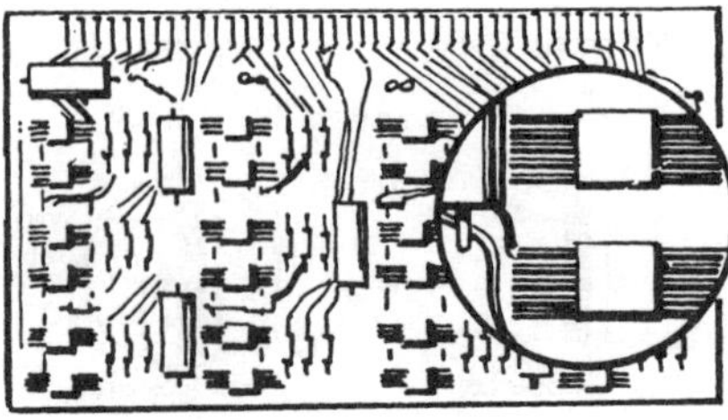

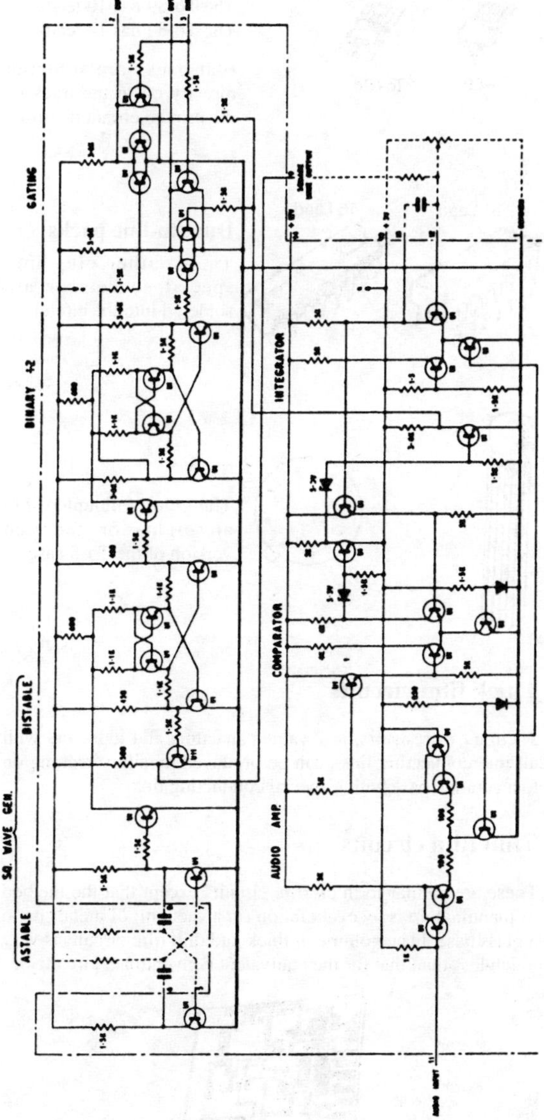
ASTABLE
SQ. WAVE GEN.
BISTABLE
BINARY ÷2
GATING
INTEGRATOR
COMPARATOR
AUDIO AMP.

The main types of S.I.C. packages are:

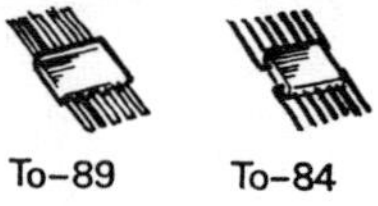

Flat packs

The To-89 has 10 leads.
The To-84 has 14 leads.

Flat packs are soldered directly on to the track of the printed circuit board.

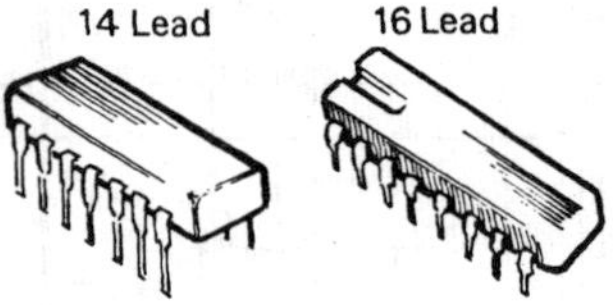

Dual-in-line packs

These either plug into special sockets or are soldered into the circuit.

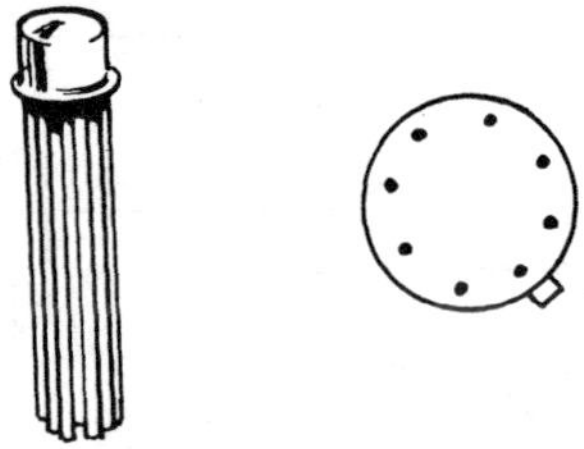

To-5

The chip is mounted into an eight- or ten-lead version of the To-5 case.

Thick film circuits

A circuit of resistors, low value capacitors and inductors with all interconnecting links can be produced by silk screening on to a ceramic base with a special conducting link.

Thin film circuits

These are similar to thick film circuits except that the method of manufacture is by evaporation (in a vacuum) of metals on to a glass base. (The volume of thick and thin film circuits is very much less than that for the equivalent conventional circuitry).

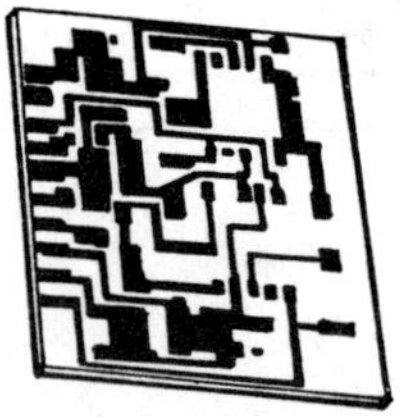

Transistors, diodes, integrated circuits and capacitors are soldered to thick or thin film circuits as required in order to complete the circuit.

Quality

Handle integrated, thick film and thin film circuits with extreme care; they are expensive and easily damaged.

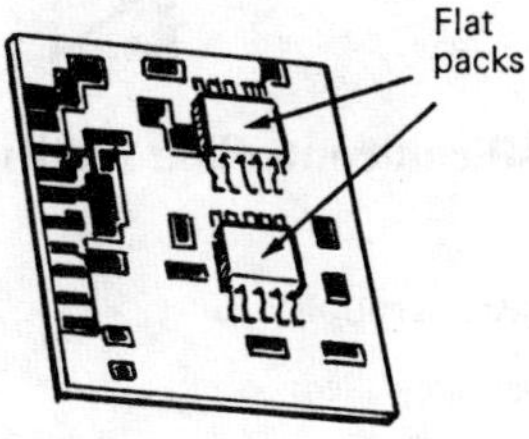

Diodes (Rectifiers)

Diodes are 2-terminal devices; the terminals are called:

- anode
- cathode.

They are made of either silicon or germanium.

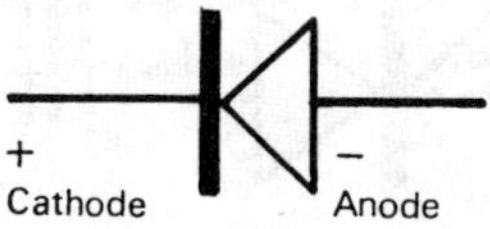

Typical diode case constructions are shown.

The positive and negative terminals are indicated by one of the following:

- the diode symbol on the body
- a red spot or coloured band at the cathode end
- a +ve sign at the cathode end.

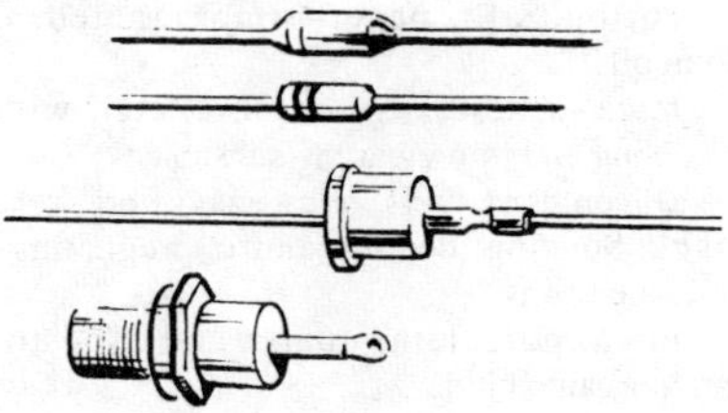

Zener diodes

Zener diodes are used to provide a voltage reference. They are manufactured in nominal voltage ratings between 5 and 200 V.

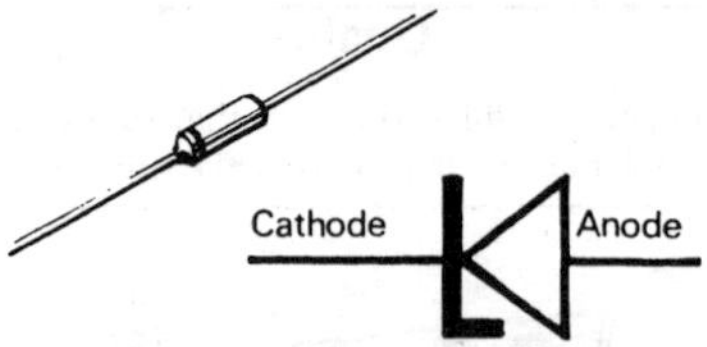

Thyristors (silicon controlled rectifiers) (S.C.R.s)

Thyristors are used in control circuitry.

The three terminals are called:

- anode
- cathode
- gate.

To-18, To-5 and stud mounted are the most usual configurations used.

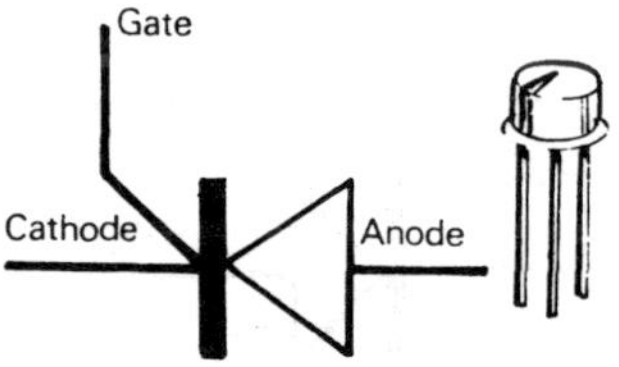

Recognition and application of materials

Insulating materials

Are often classified according to the maximum temperature at which they can operate without harm.

Typical materials in the main classes are:

Class Y cotton, silk, paper—without impregnation, PVC
Class A cotton, silk, paper—impregnated, coated or immersed in oil
Class B mica, asbestos, glass fibre, etc., with suitable bonding, coating and impregnating substances
Class H silicon elastomers, mica, glass fibre, asbestos, etc., with suitable bonding, coating and impregnating materials such as silicone resins
Class C mica, porcelain, quartz and asbestos with or without an inorganic binder.

In practice, Class Y is not widely used because the absence of impregnation means that the materials pick up moisture from the air and, becoming damp, they lose their insulating qualities and deteriorate by chemical action.

Attention must be drawn to the existence of many other specifications issued by various bodies.

S.R.B.P.

Layers of treated paper are compressed to produce a hard smooth board. Many grades and forms are available. Colours are mainly yellow and brown.

Used for terminal boards, barriers.

Mica is a fairly flexible mineral. It withstands temperatures up to red heat.

Phlogophite mica called *amber* is widely used in electric irons, toasters, hair dryers, etc., in which the elements are wound on mica formers to withstand the heat.

Micanite. Flakes of mica are bonded together to form micanite sheet.

Mica paper or foil. Available in sheets up to 0.5 mm (0.02 in) thick, it has a shiny, slippery surface almost colourless apart from darkish spots.

It is usually impregnated for mechanical strength with consequent alteration to colour. It may be cut or punched cleanly and without splitting.

Tubes are wound from foil with or without reinforcement by woven glass cloth layers.

Asbestos

Asbestos is a fibrous mineral. Its essential property is that it will withstand temperatures up to white heat. The fibres appear white when split apart and are very flexible and silky.

Plastics

Polythene, a wax-like, translucent material, is used mainly for high voltage and high frequency (H.F.) applications where temperatures are not excessive, e.g:

- co-axial cables for radio work
- relay cables for radio work.

Irradiated Polythene is used for certain cables and wiring and for sleeves at the termination of conductors.

Polyvinylchloride (PVC), a tough, rubber-like material, widely used for insulating and covering wires and cable for use up to 11,000 V and 90°C.

Used also in tube and sheet forms.

Moulded components are readily made of PVC.

Polytetrafluoroethylene (PTFE), a wax-like material. Available in small sheets and in blocks, rods and cylinders (for machining), tubes and tapes, moulded forms. It has a very low coefficient of friction (gears and bushes, etc.) but it is an expensive material. It is extensively used for co-axial radio frequency cables.

Polyvinylidene fluoride, after exposure to radiation, provides a tough, flexible material used in tubular form to cover conductors, joints, terminations, etc.

Application:

- place over component (A)
- heat from 120° to 175°C according to type (B)
- material shrinks and forms a tight well-fitting sleeve (C).

Available in colours it is widely used in air-insulated switchgear up to 11 kV.

Nylon. As a *synthetic enamel*, it is used for covering wires and as a *moulding* is used for bearings and gear wheels and various LV applications.

Polyurethane. A synthetic resin available in the form of:

- foam for thermal insulation
- powders for mouldings and bearing, gears, electrical components
- extrusions for cable coverings
- film, sheet and tape for electrical insulation.

Rubber

Rubber is sometimes used for cables and wires up to 6600 V but is largely being replaced by synthetics. It is not suitable for use in oily situations or when exposed to sun and weather unless protected by an oil and weather resisting sheath (polychloroprene). Maximum operating temperature is 60°C.

Taping and other insulating materials

Cotton
Varnished to prevent moisture. Used largely for low voltage insulation.

Silk, rayon, nylon
Often used as a backing for other materials, e.g. mica.

Terylene (woven)
Used in tape form for low and medium voltage insulation.

Terylene paper
Usually backed with press-paper.

Glass fibre
It will not stand chafing.

Fibrous Types

Leatheroid
A hard tough material.

Terylene film (Melinex)
Available in tape form or backed with presspaper.

PTFE (Polytetrafluoroethylene)
Good resistance to attack by strong acids, oils, and solvents.

Mica
Used in form of tape, more generally in sheets.

Metals (ferrous)

Cast iron
An alloy of iron with up to 4% carbon. It is a grey, brittle, coarse grained material having low tensile and shear strength, but high compression strength. Mainly used for the quantity production of intricately shaped components not needing great tensile or shear strength, e.g. resistance grids, cases, stands, etc.

Steel
An alloy of iron with up to 2% carbon together with other elements, e.g:

- nickel, chromium, silicon, molybdenum or tungsten — according to the properties required

- mild steel (construction), stainless steel (anti-corrosion, non-magnetic), high tensile steel (bolts and tension members), *electrical* (magnetic) steel, etc.

Silicon-alloy steels
These are used for the magnetic circuits of electrical machines — rotating and static. They have been developed because of their low magnetising losses.

Metals (non-ferrous)

Copper
This is a soft ductile, reddish-coloured metal of low electrical resistivity, second only to silver.

Cadmium-copper and chromium-copper alloys are frequently used because of their mechanical strength and good conductivity.

Brass
This is a pale, yellow alloy of copper (over 50%) and zinc with traces of tin, lead, etc., to give special properties. Less ductile than copper but of higher tensile strength, it is widely used for

screws, nuts and bolts, small terminal stems, brush holders, cable lugs, bearing surfaces, castings, pressings, deep drawing, extrusion.

Manganese-bronze is actually high-tensile brass containing a small proportion of manganese. It is not bronze and contains no tin.

Bronze
This is a yellow alloy of copper and tin with some zinc and lead, *Gunmetal* contains about 90% copper with zinc and a small amount of tin. It is of a richer, warmer colour than brass in general, but the composition of brass and bronze varies so widely that colour alone does not distinguish one from the other. Bronze is used mainly for pumps, valves, castings and pressings, bearings and cams, overhead line (1% tin).

Phosphor-bronze contains about 0.25% of phosphorus and is used mainly for springs, spring washers, castings, etc.

Aluminium
This is a soft, ductile, silvery metal notable for its lightness. It is normally resistant to corrosion except in sea water. Principally used alloyed with zinc, silicon, manganese, magnesium, etc., to increase its strength.

It is used for castings, pressings and die castings, extrusions, conduit work, cast-in rotor bars, overhand H.V. transmission lines (with steel core for strength), underground L.V. distribution cables (cores and sheaths), foil type transformer windings, electro-magnet coil windings, capacitors, fabricated tanks, containers, etc.

Nickel
This is a soft, ductile, silvery material mostly used in alloys.

Pure nickel is used for plates in alkaline cells, nickel-plating, etc.

Copper nickel alloys are used for resistance wires.

Iron nickel alloys are used for resistance to corrosion, particular magnetic properties.

Nickel-silver (copper-nickel-zinc) is used for contacts, resistance wires, basic material for silver-plated goods.

Lead
This is a very soft, ductile, grey metal notable for its heaviness. Mainly used alloyed with small amounts of tin, cadmium and antimony for accumulator plates, covering cables, solder, fuse wire, etc.

Silver
This is a soft, ductile, precious metal having low electrical resistance. Too costly to be used in bulk, it is widely used in the form of silver solder and for the plating of copper, brass, etc.

Inspection of printed circuit boards

Printed circuit boards are usually inspected before delivery to the inspection shop. The criteria for their acceptance are as follows.

Dimensions

The length, width and thickness of the board must be correct to drawing. Twisting and bowing must be within specified limits.

Holes must be in the correct position, of the correct size, and centrally placed in the printed circuit pads. Usually a minimum gap of 0.4 mm between the hole and the edge of the track is specified.

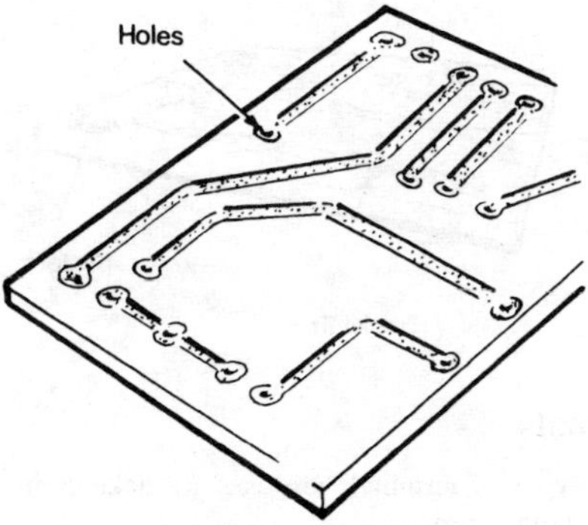

Plating

Plated surfaces must be free of stains, blisters and flaking. The number of pin holes present must be below specified limits, e.g. pin holes must not affect over 25% of the circumference of a plated-through hole, and must not be present in more than 10% of the holes.

Coding

Coding must be correct and legible.

Roller soldering

The track must be tinned evenly to drawing. There must be no visible contamination (discoloration), and no solder where it is not required.

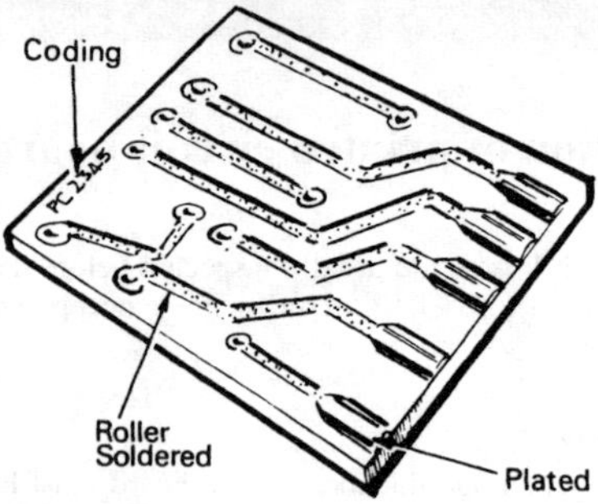

Damage

There must be no damage such as broken, damaged, or lifted track, blistered terminals, or excessive pin holes.

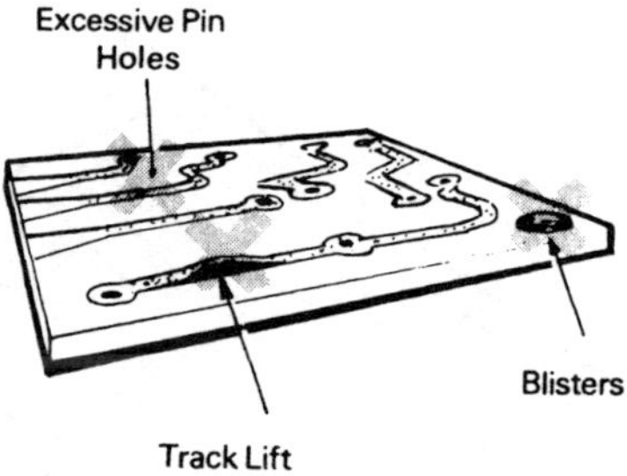

Terminals

Various types of terminals are used to make connections to printed circuit boards.

Terminals are assembled by:

- riveting. The terminal itself is a form of rivet
- a push fit. The terminal is held by a serration or taper on the terminal
- screwing in
- staking.

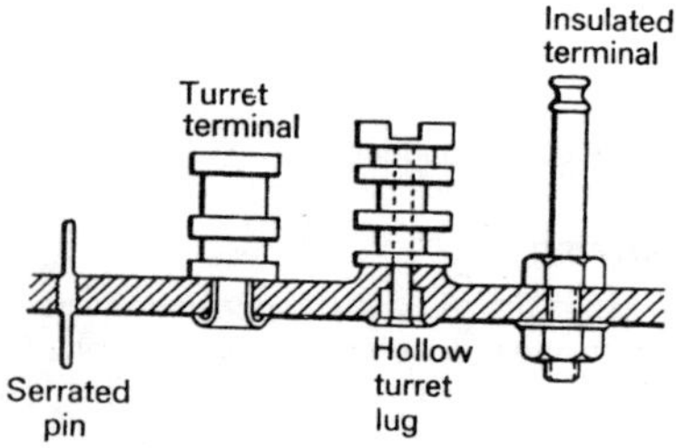

● ensure that the correct terminals have been used
● ensure they are tight
● check for damage to the board or to the terminal, e.g. cracking around the base of the terminal is a common fault
● report all defects.

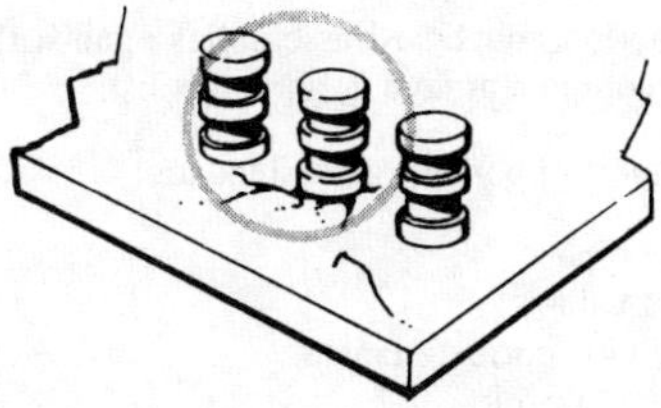

Edge connector sockets

Some printed circuit boards have edge connecting points which mate with sockets.

Ensure that:

● there is no dirt on the edge connecting point, or in the socket
● the connector is not damaged. Look particularly for loose or damaged pins.

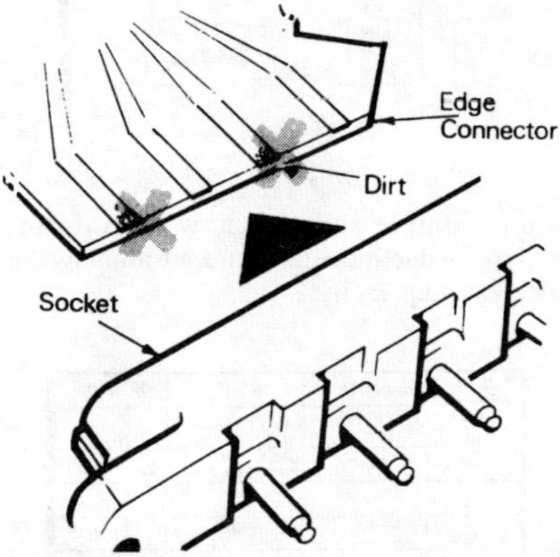

Inspection of component mounting

Inspect printed circuit board assemblies against the assembly drawing (or component layout) and parts list.

Ensure that the components assembled are:

- the correct type
- the correct value
- mounted in the correct manner
- in the correct position.

Ensure that all the required components have been mounted.

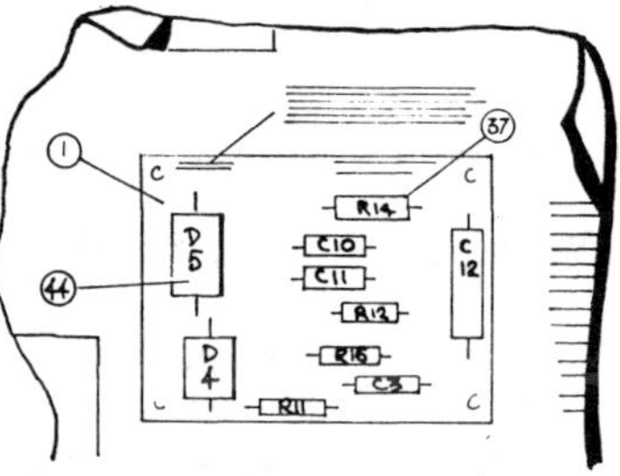

Below is an example showing how the information on component type, value mounting, and positioning is portrayed on the drawings and parts list.

ITEM	DRAWING NO.	NAME	No OFF	REMARKS
34	HK 167480 UAC/01/11	CIRCUIT DIAGRAM		
35	BX 194671/1	CAPACITOR PLASTIC 0.49UF	2	C10-C11
36				
37	DEF 64189 140/[illegible]	RESISTOR 330Ω ± 5%	1	R14
38				
39				
40	No 17/001/4	SEMI-CONDUCTOR	1	D3
41				
42				
43	196/H/119/1	SEMI-CONDUCTOR		D4
44	990 141 1181	SEMI-CONDUCTOR		D5
45				
46				
47				

Leads through printed circuit boards

Inspect the forming of the leads.

● ensure that the component lead is not bent at the point of exit from the component. The gap between the component and the bend must be at least 1.5 mm.

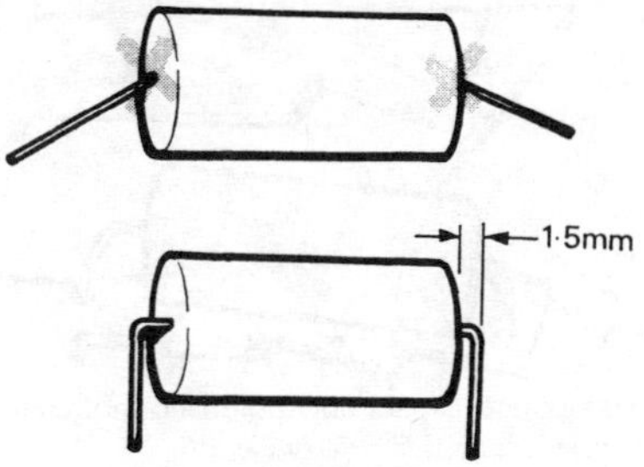

● ensure that the lead is not formed too sharply. The radius of the bend should be at least equal to the diameter of the lead

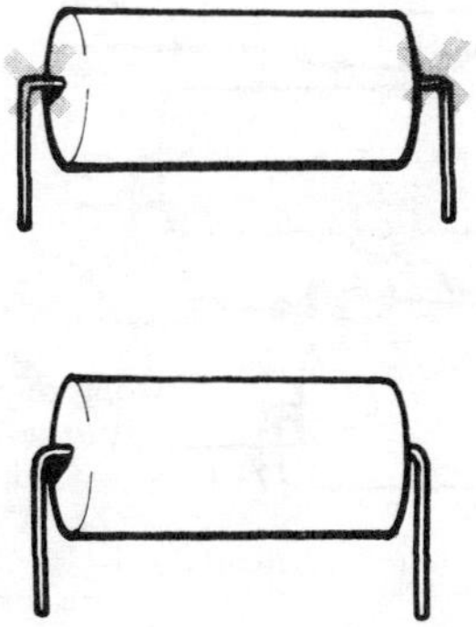

● ensure that the bends are symmetrically positioned with respect to the component

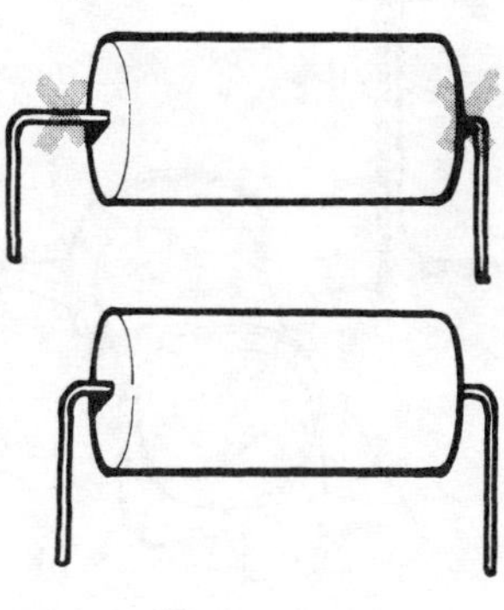

● ensure that the leads are formed to align with the fixing holes in the board.

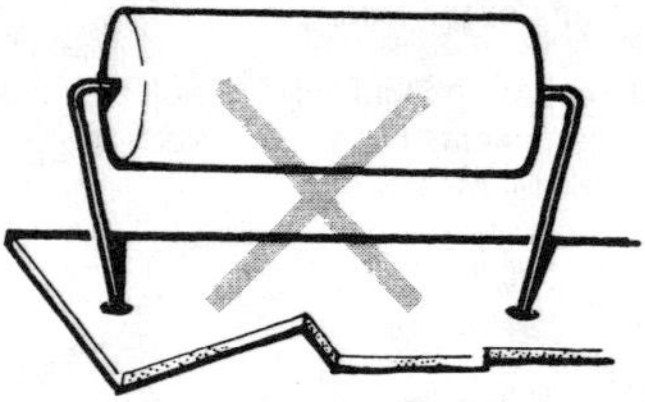

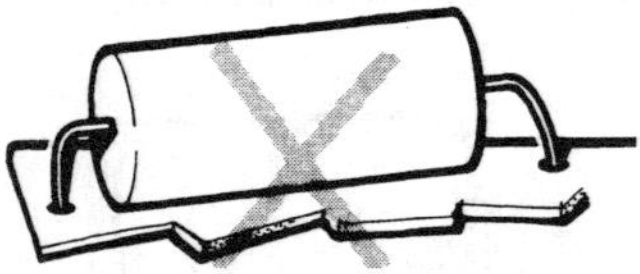

Small resistors and capacitors should be mounted flush with the board unless specified otherwise.

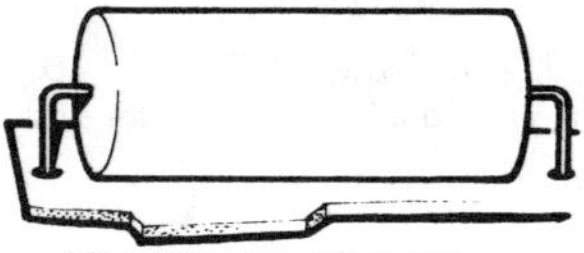

Components not flush with the board must be supported, e.g. with beads.

Components mounted vertically must sit squarely on the board.

The shoulder of printed circuit board components should be flush with the board.

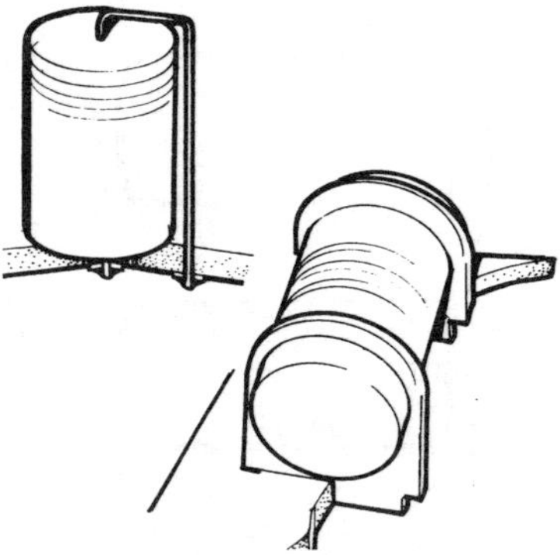

Ensure that, where possible, components are mounted so that:

- component identification is visible
- the identification of components reads from left to right in relation to the drawing
- rows of components form a neat line
- components are in line with the component symbols on the printed circuit board, where these exist.

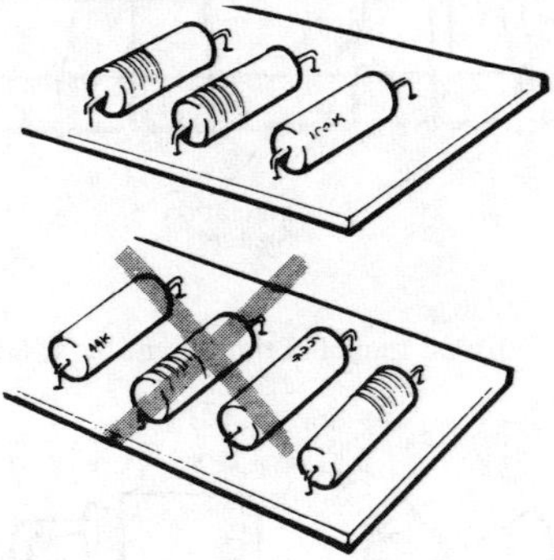

Transistors

Ensure, by checking the position of the location code with the drawing, that transistor leads have been assembled to the correct positions.

Check transistor mountings to ensure that:

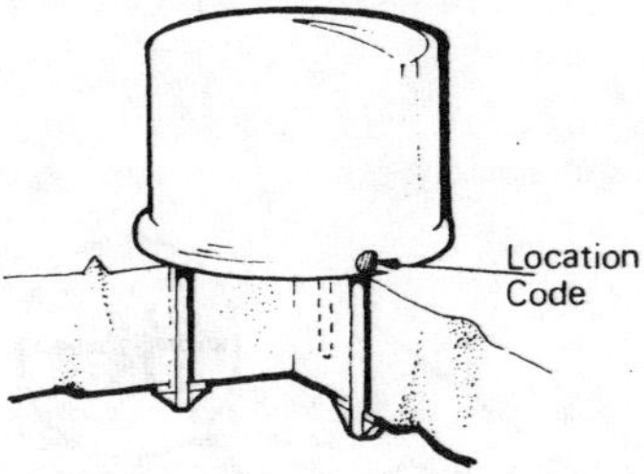

- spacers are assembled where required

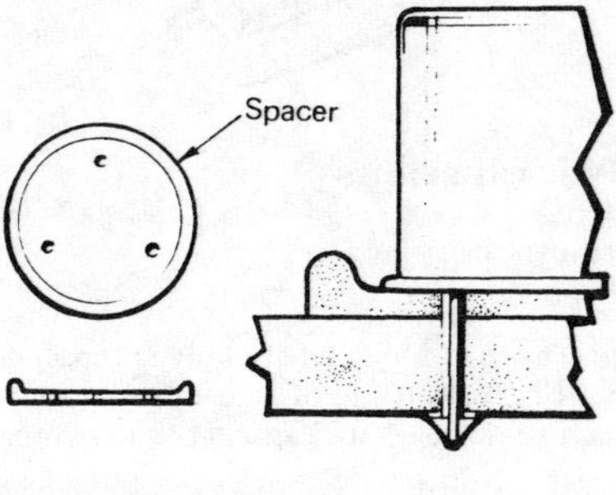

● insulation sleeves are present as required. They must be the specific colour and length

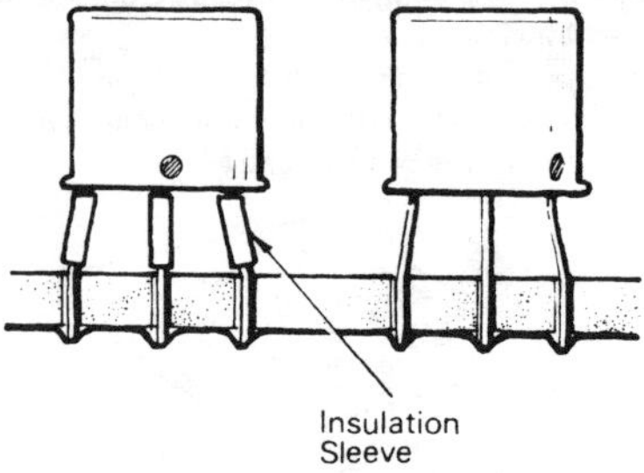

● transistor clips are mounted tightly and hold the transistor firmly.

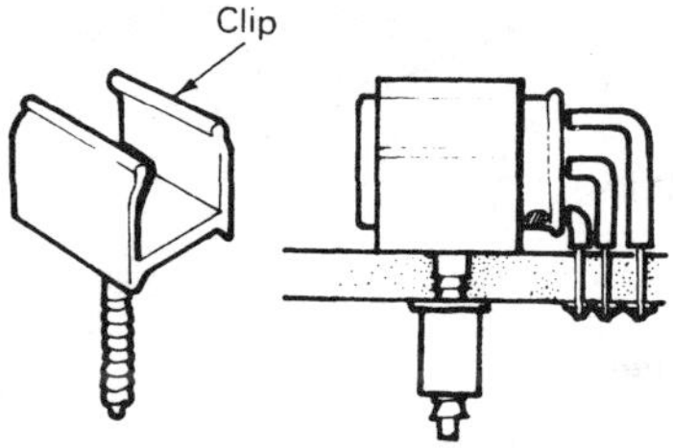

Diodes

Check diodes for polarity.

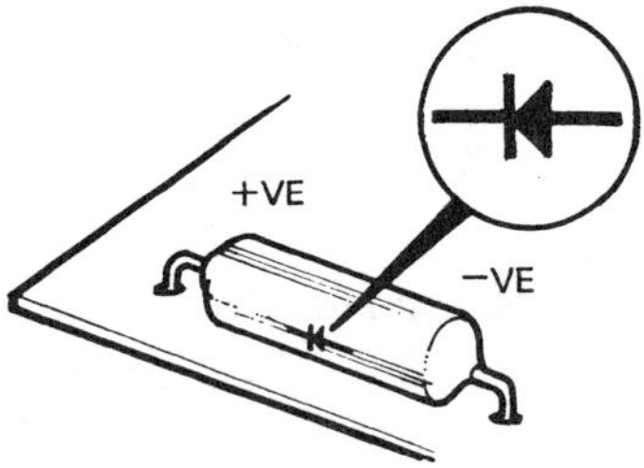

Electrolytic capacitors

Check electrolytic capacitors for:

● polarity
● shelf life. The shelf life of an electrolytic capacitor depends on the type. Check shelf life of actual capacitor used. If the shelf life has expired reject the capacitor to be reformed.

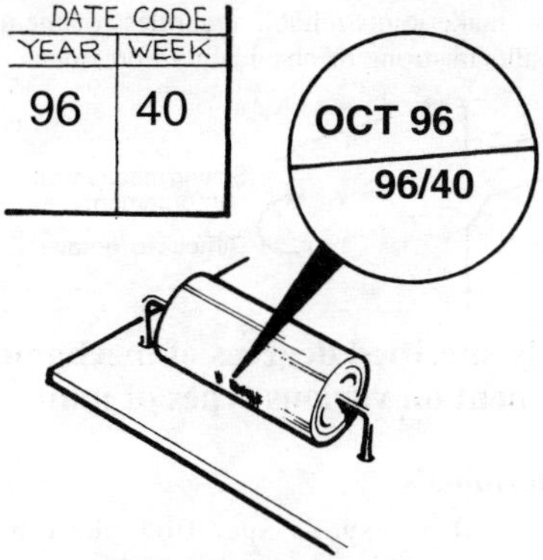

Mechanical reinforcement

Mechanical reinforcement is the physical bond between the component lead and the connecting point prior to soldering.

The degree of mechanical reinforcement required is specified on the drawings, or in general requirements specifications.

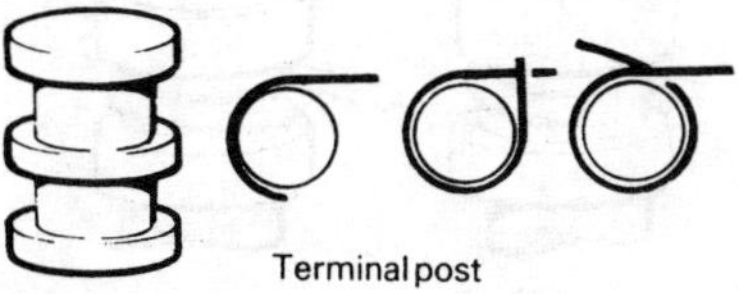

Terminal post

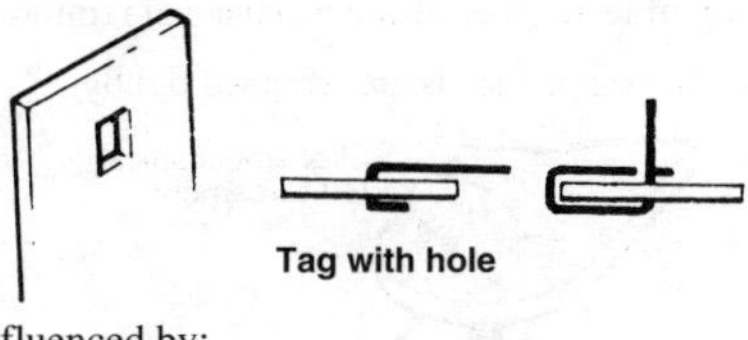

Tag with hole

This is influenced by:

● the need for easy removal of the components. This results in limited mechanical reinforcement

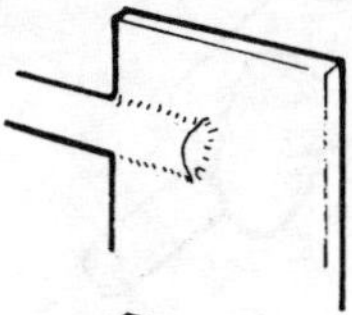

No mechanical reinforcement

Very easy to remove

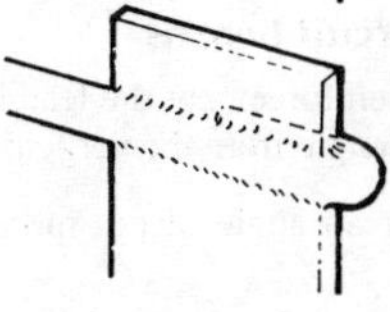

Some mechanical reinforcement

Easy to remove.

● the need to make joints reliable regardless of the quality of soldering results in strong mechanical reinforcement.

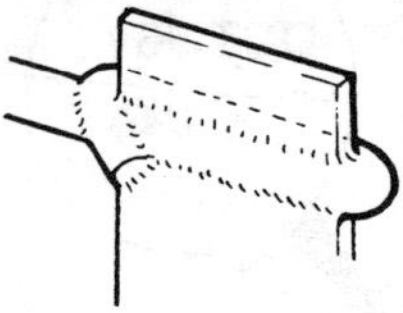

Commonly specified degrees of mechanical reinforcement on various types of joint

Leads on terminals

● a wrap of 180° is usually specified where minimum reinforcement is required
● for a strong mechanical reinforcement a full wrap of 360° plus a tab is common.

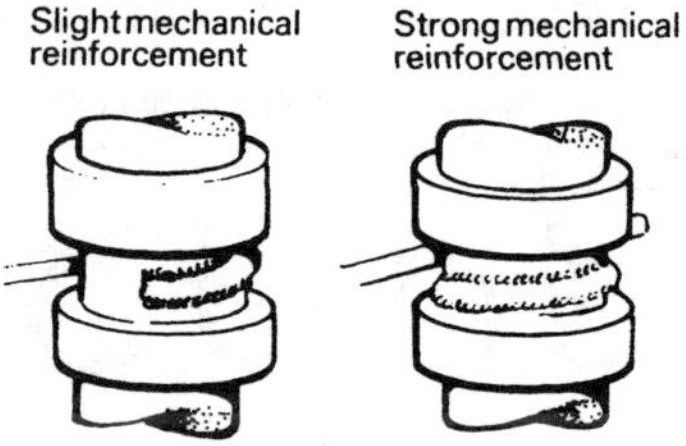

Leads through tag holes

● the lead is usually bent over 90° or less to give easy removal
● a full wrap of lead gives strong mechanical reinforcement.

Ensure that all wrapped leads are wrapped tightly.

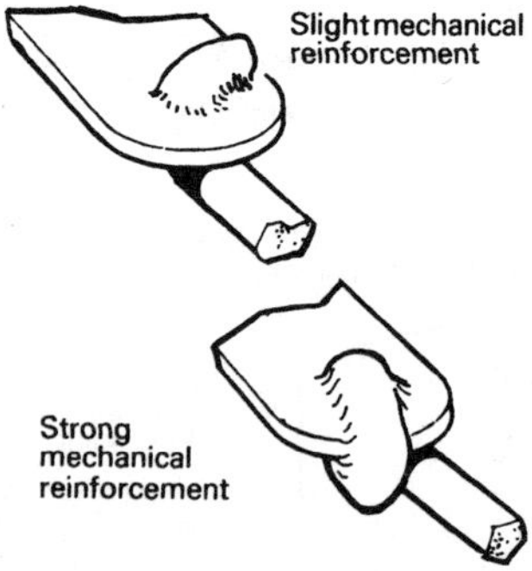

Leads through printed circuit boards

To give minimal mechanical reinforcement the lead is croppe off straight, or bent at a slight angle from the vertical.

Specifications usually call for an angle of not more than 30° from the vertical.

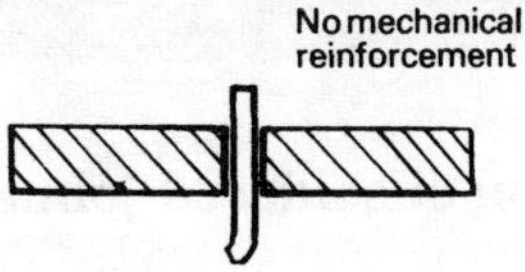

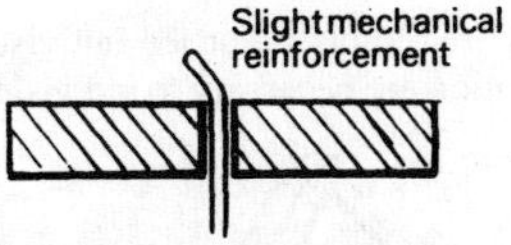

Ensure that the vertical lead clears the pad by at least 0.75 mm.

The lead must be visible through the solder, or its presence cannot be verified.

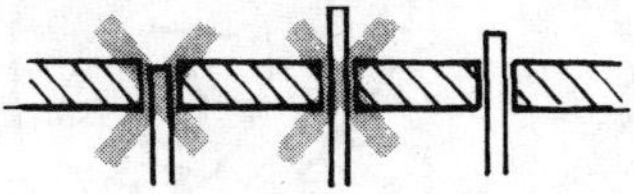

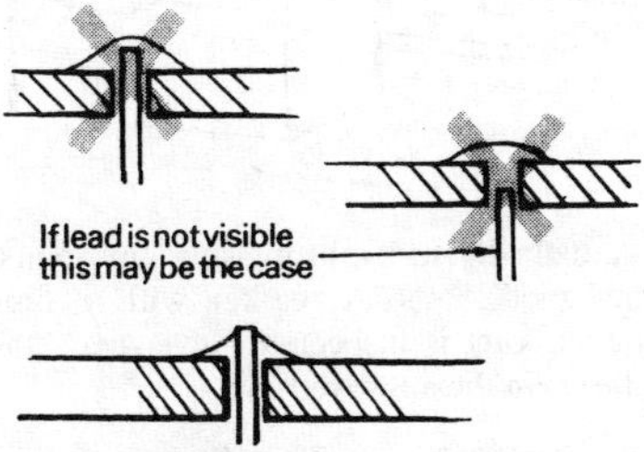

For strong mechanical reinforcement the lead is bent over flat with the board. Ensure that it is:

- flush with the board
- centrally placed along the printed circuit path or pad
- not too long or too short. The length of the lead should not normally be greater than the diameter of the terminal pad, nor shorter than its radius.

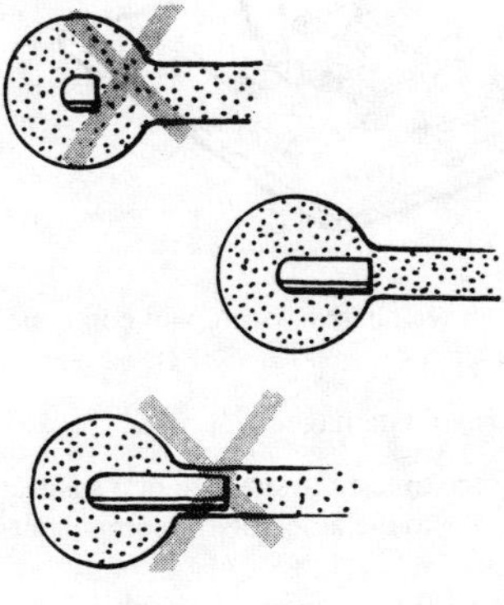

Inspection of soldered joints

The inspection of soldered joints is carried out visually. Good lighting and magnification are necessary to aid vision.

Inspect systematically to avoid missing any joints. It may be specified that good joints are marked with a coloured dye to ensure that each joint is inspected. Never mark the joint with anything other than the approved dye.

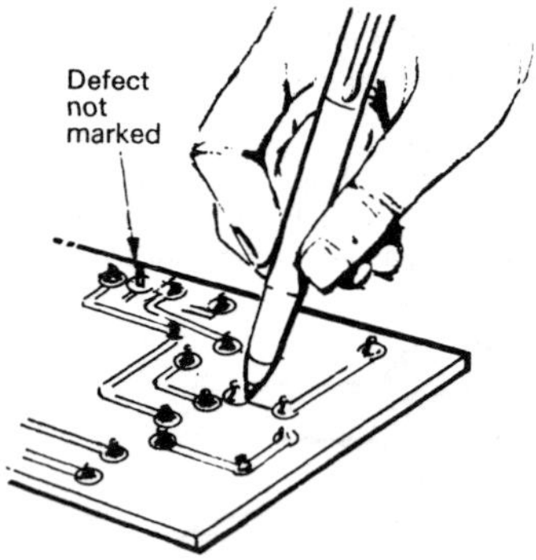

Use a probe to move nearby wiring and components out of the line of view.

Move gently to avoid damage.

Do not apply force to test the strength of a suspect joint, as this may cause damage to the assembly or component.

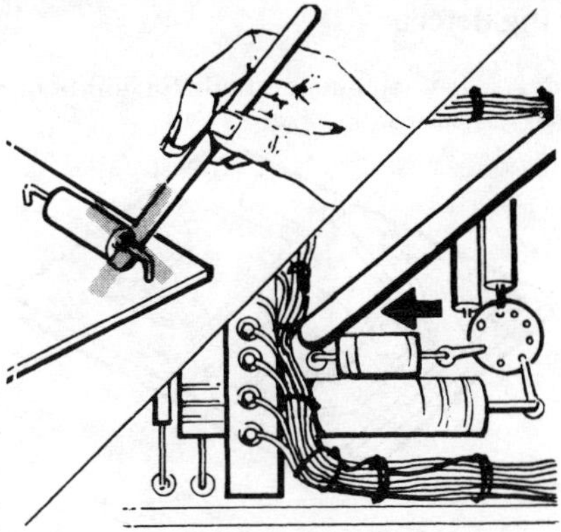

Good soldered joints

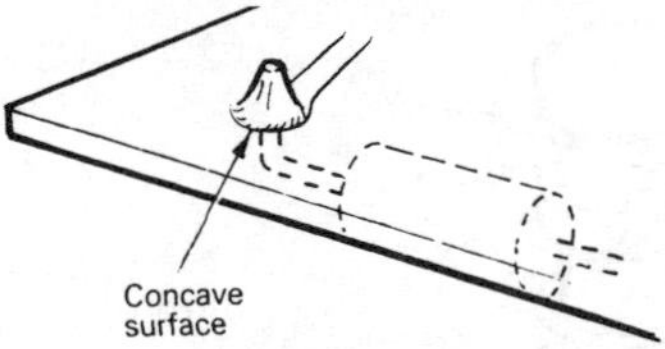

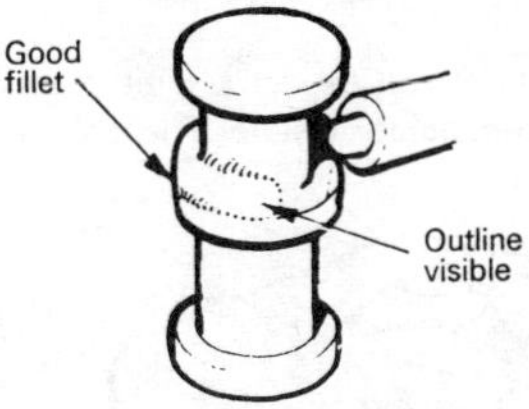

A good soldered joint:

- has a bright shiny appearance, and is free of surface flux
- is free of holes, scars or spikes
- has a good fillet between conductor and terminal
- has a concave surface
- leaves the outline of the conductor clearly visible.

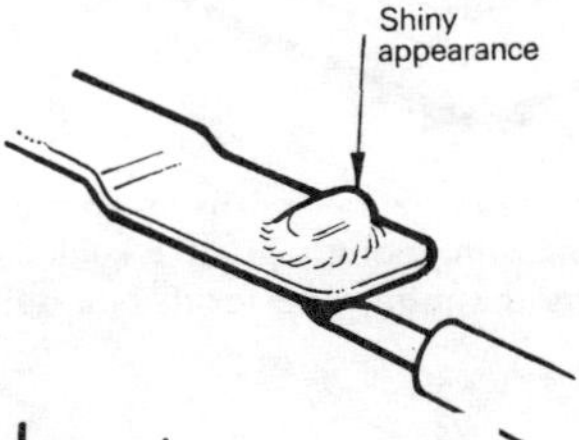

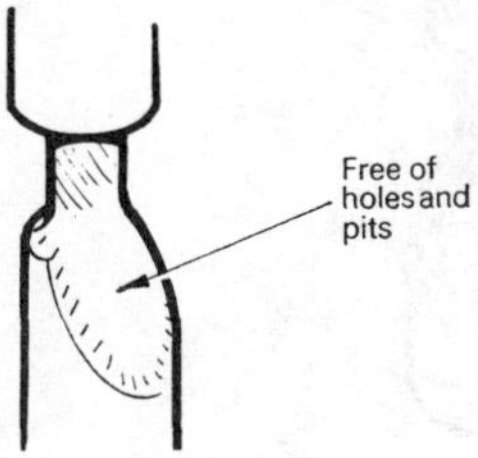

Soldering defects

No solder. It is a common fault for joints to be left completely unsoldered.

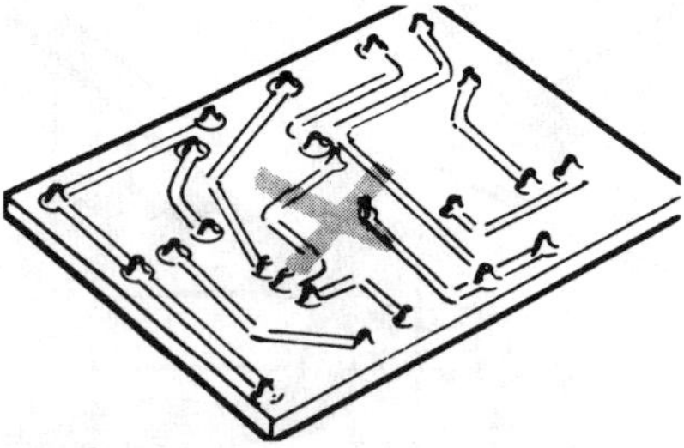

Insufficient solder. There must be enough solder to form a good fillet between the conductor and the connecting point.

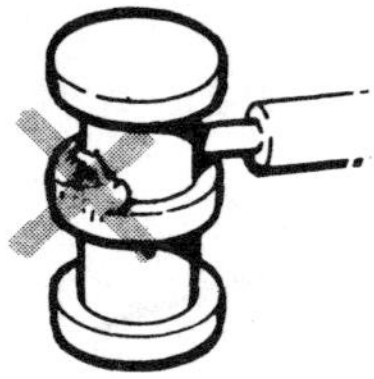

Excess solder. It is impossible to inspect a joint unless the outline of the conductor is completely visible. There should be no blobs of solder.

Crystallised solder. This can be caused by movement between conductor and connecting point while the solder is solidifying, or by reheating the joint. The result is a dull, heavily pitted surface.

Poor wetting. Wetting is a term used to describe the flow of the solder over the surfaces. Poor wetting can be caused by:

- an unclean or oxidised surface
- inadequate flux
- untinned surfaces
- inadequate heat.

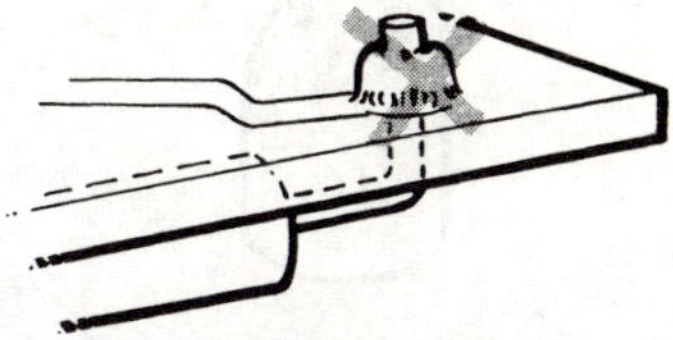

Poor wetting is most easily detected by a high angle between the solder and the connecting point.

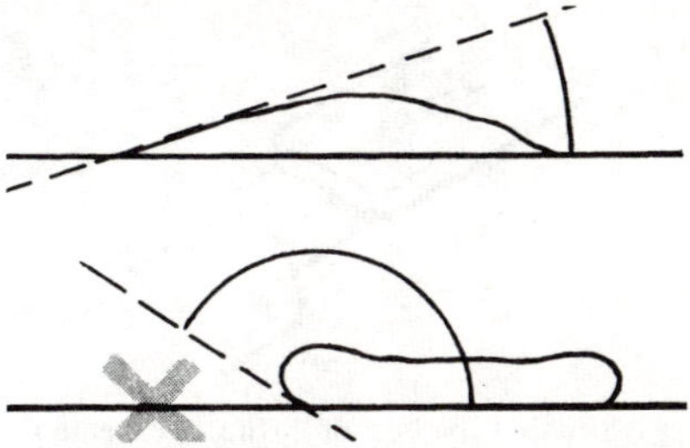

Cold solder. This is very poor wetting due to insufficient heat. The fillet does not blend into the connector, or connecting point, and the solder has a lumpy appearance.

Dry or resin joint. This is the prevention of sound electrical and physical contact by a layer of solidified flux or dirt. The joint is usually poorly wetted. Black areas may be visible, particularly at the junction between the solder and the connecting point, or conductor.

Some dry joints appear wetted. Dark patches in the solder are the only indication of unreliability.

Overheating. This burns the flux and prevents the solder from running evenly. The result is a poorly wetted joint with embedded charred flux.

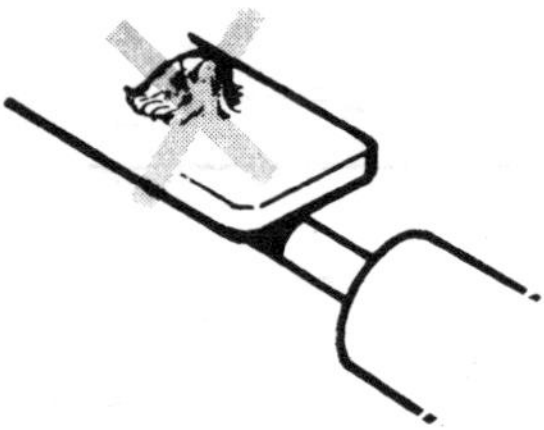

Wicking. Solder travels rapidly up overheated stranded wire, making it brittle, particularly where the solder ends.

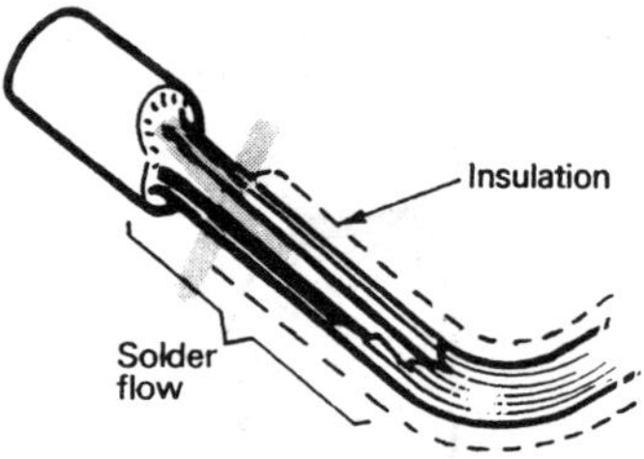

Holes and pits. These are caused when the heat applied is insufficient to drive out all the air from the joint until the solder is solidifying. The number of holes or pits permitted is usually specified.

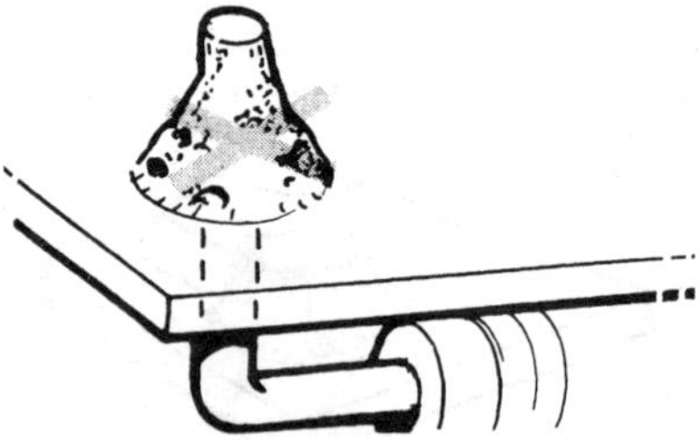

Solder spikes. These are caused by removing the iron before removing the solder, or by overheating the joint. They are a common cause of short circuits.

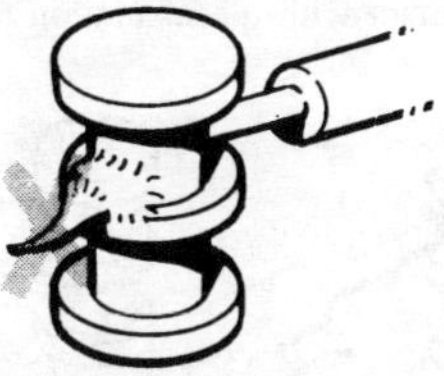

In addition to inspecting the joint itself, check for:

Short circuits. These may be caused by:

- loose strands of wire

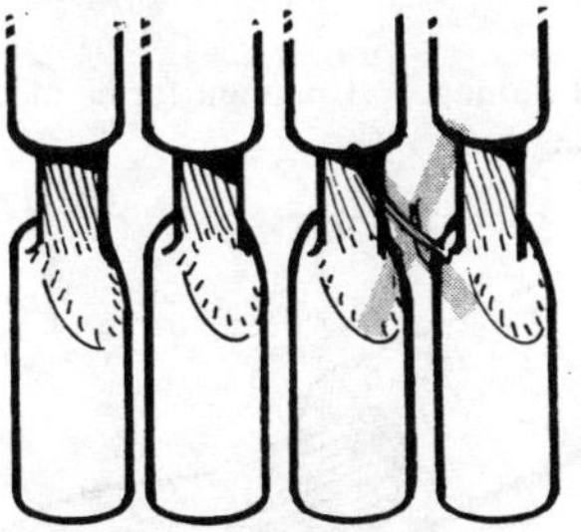

- the solder of two or more joints being in contact
- excessive solder coming into contact with part of the chassis

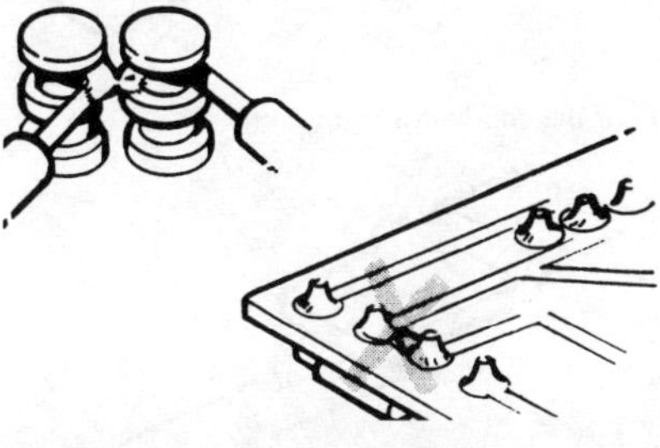

- solder splashes or other foreign bodies in the equipment. These can also cause damage to mechanical parts, e.g. gears.

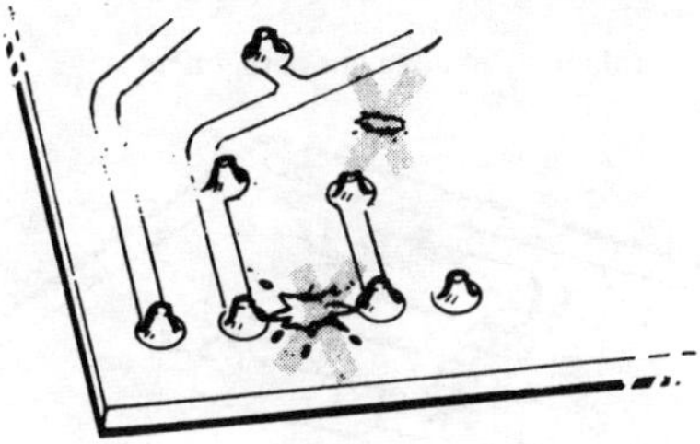

The correct insulation gap. This should be about the diameter of the wire, or as specified. The gap should be small enough to protect against possible short circuits, but the solder must not be in contact with the insulation as this may poison the joint.

Burning and damage. Common forms of damage caused during soldering are:

- cut, nicked or stretched wires

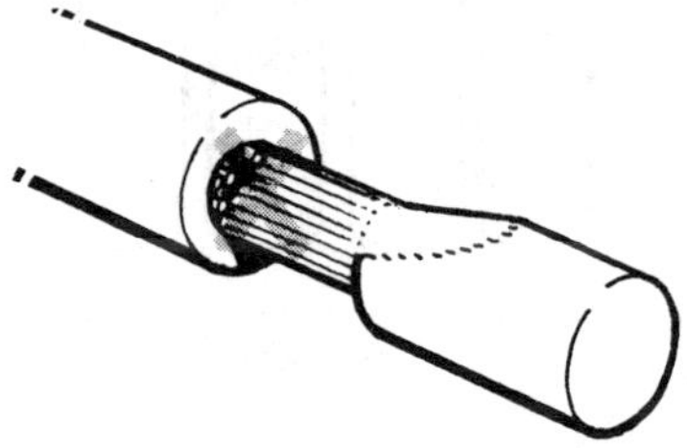

- burning of the conductor insulation

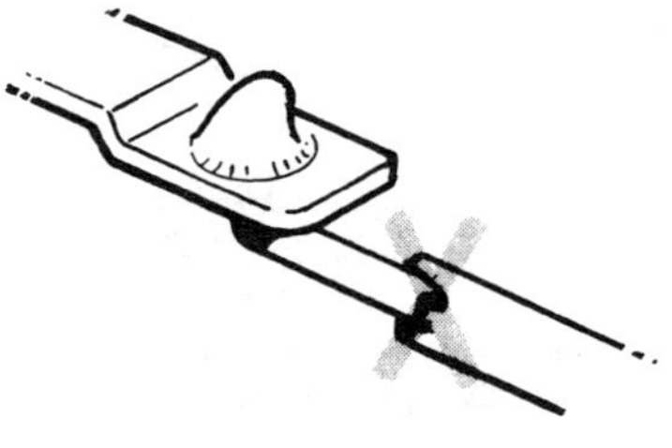

- burning or damage to adjacent components

● track lifting.

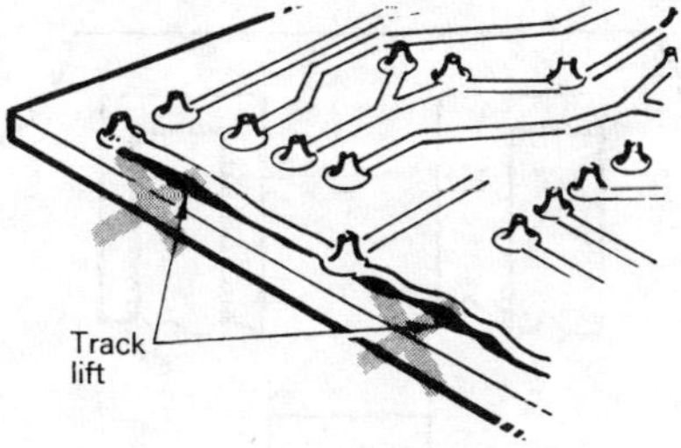

Reflow soldering

A reflow soldered joint is formed by placing a tin coated lead on a tinned connection and applying heat.

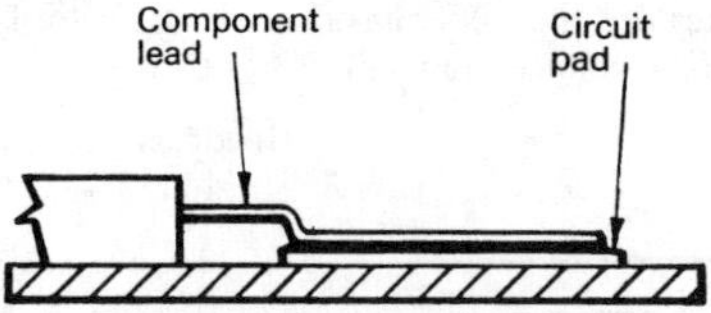

The heat melts the tinning which runs between the surfaces to form a strong joint.

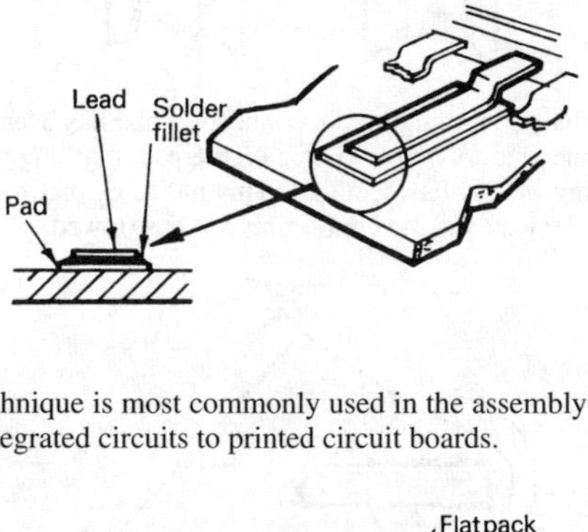

This technique is most commonly used in the assembly of flat-pack integrated circuits to printed circuit boards.

Ensure that the lead is positioned so as to allow a fillet to form down both sides.

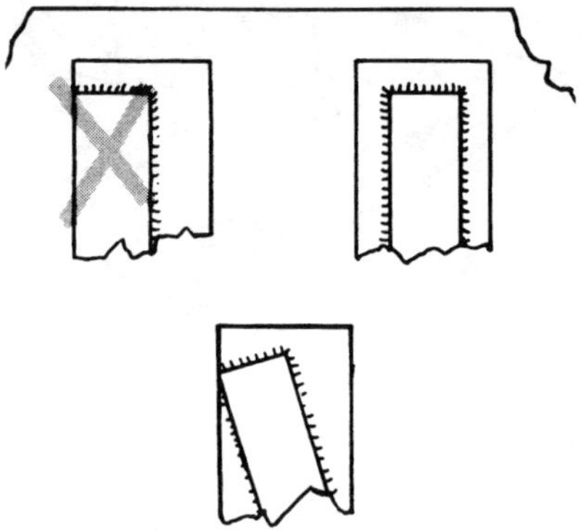

Check that there is a good fillet along each side of the lead. This indicates that the solder has flowed satisfactorily between the lead and the connecting point.

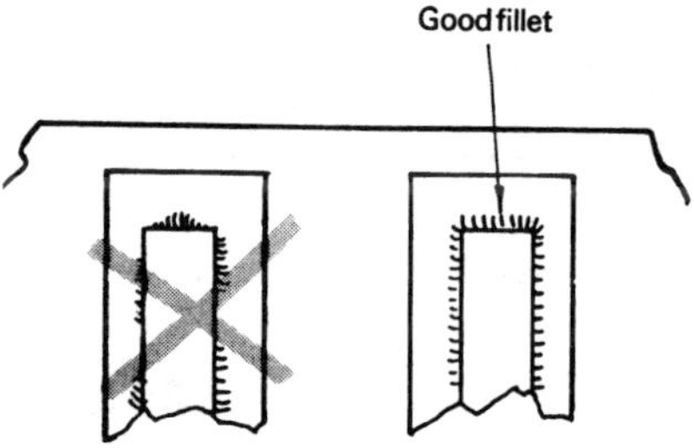

Ensure that the tinning on the connection pad has been melted around the lead. A line is visible on the pad, marking the limit of melting, if this has occurred. This indicates that the solder on both the lead and the connecting pad has flowed.

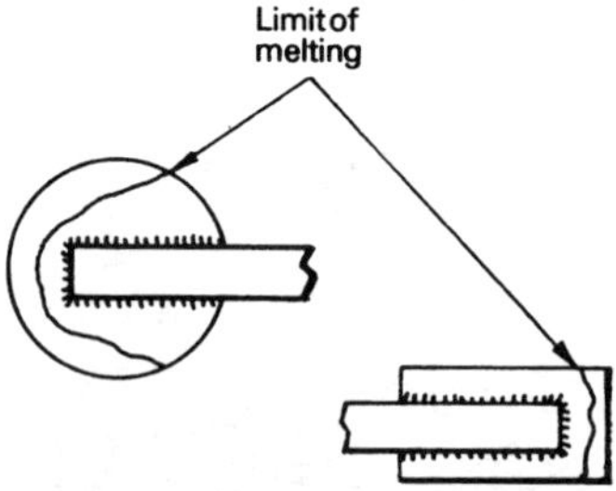

Inspection of wrapped joints

Types of wrapped joint

- wrapped joint with insulation on the last turn

- wrapped joint without insulation on the last turn

- bound joint

- twin bound joint.

Good wrapped joints depend on the use of the correct terminal and wrapping wire.

Ensure that these have been used.

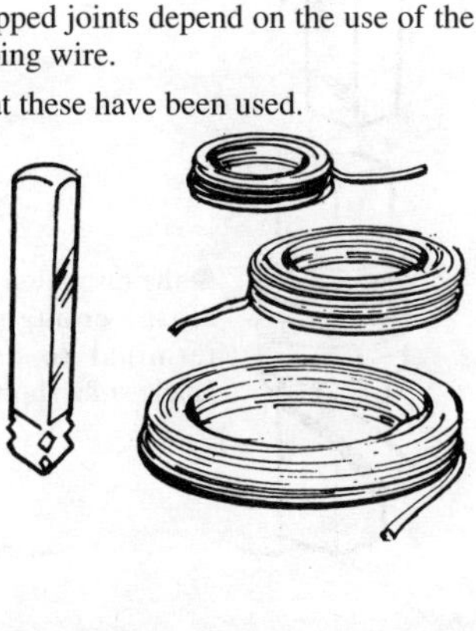

Criteria for a good wrapped joint

● ideally there should be no gaps between the turns. In practice up to three gaps of less than 0.125 mm may be visible from any one side of the terminal

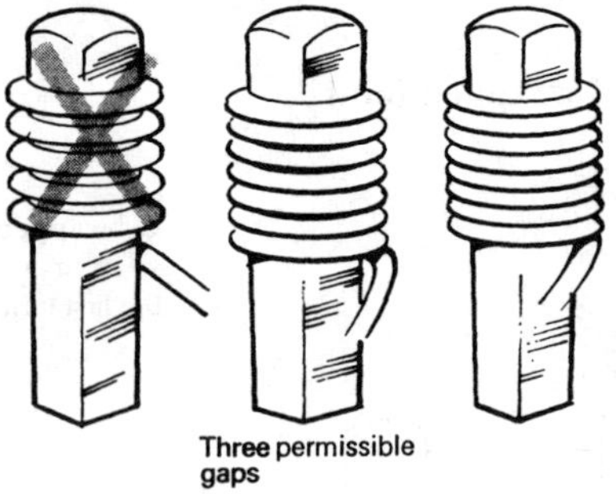

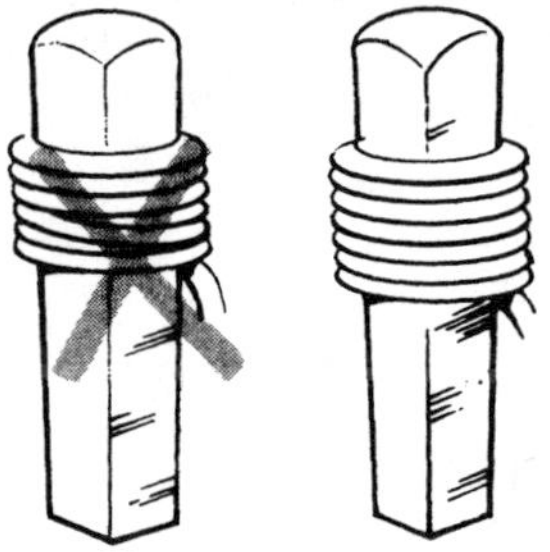

● the turns must not overlap

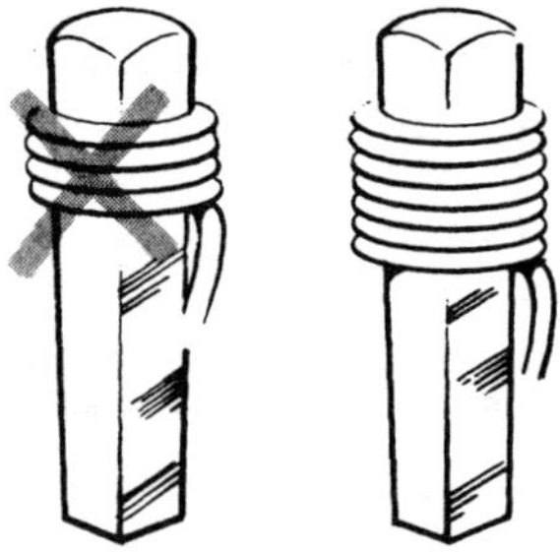

● there must be 6–7 turns of wire in metallic contact with the terminal

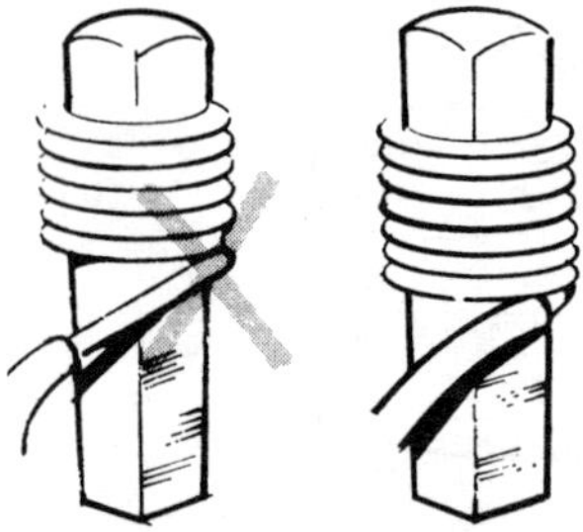

● the insulation must start close enough to the terminal to avoid any chance of a short circuit

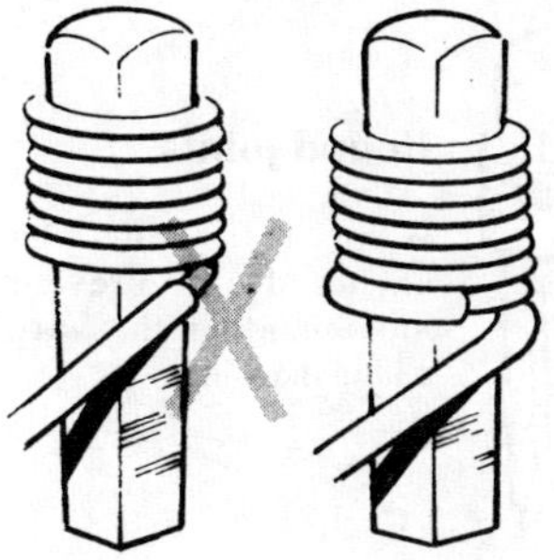

● where the joint may be subjected to vibration an extra 0.5–1 turn of insulated wire is usually specified. Ensure that this is present

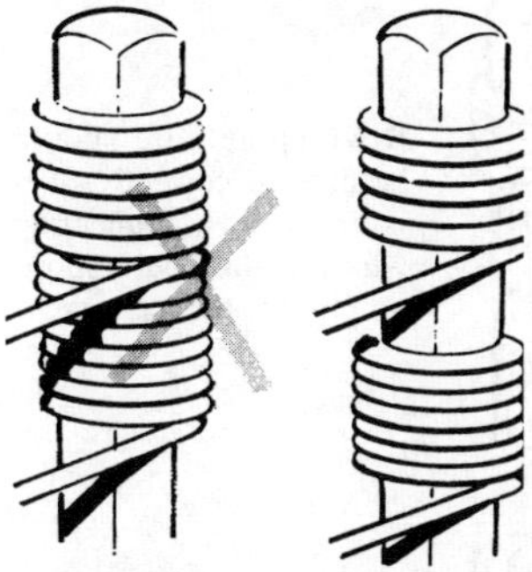

● where two or more joints are made on the same terminal, no two joints must touch or overlap

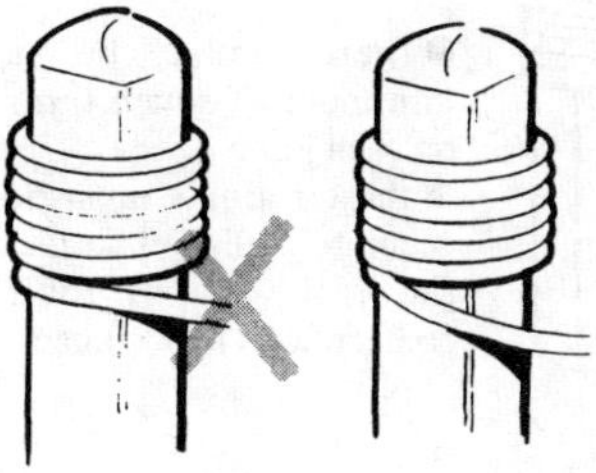

● there must be no strain on the terminals from tight leads. There should be sufficient slack in the wire to allow dressing

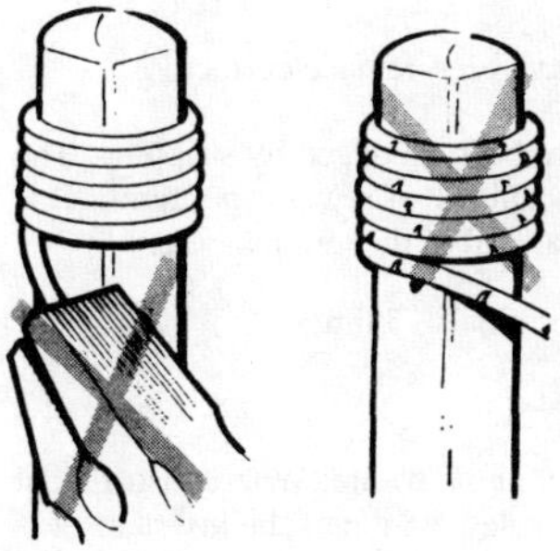

● joints must not have been disturbed once they have been made. They must not have been tightened with pliers. Unwrapped wire must not have been reused.

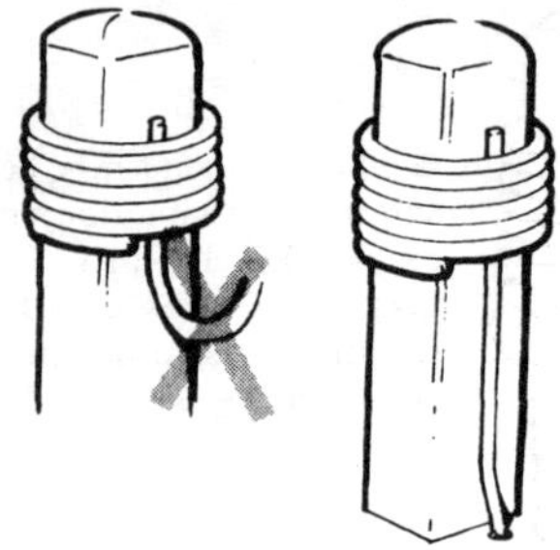

Bound joints

● the lead must be anchored to prevent movement of the wire within the joint

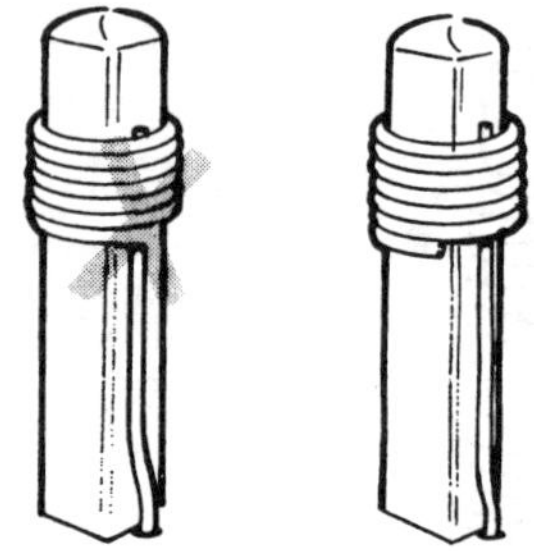

● wrapping wire should be dressed correctly. If it is not itself a connecting wire it should be cropped off close to the terminal.

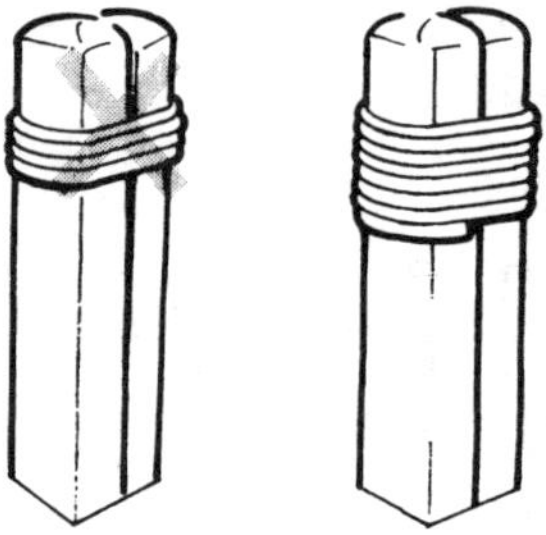

Twin bound joints

● there must be a minimum of seven turns on each joint

● the terminals must be accurately aligned so that the full area of their contact faces is in contact.

Reworked joints

Ensure that rejected joints have been remade correctly.

● the terminal must not have been damaged by stripping. The joint must be unwound. The number of rewraps per terminal is limited, depending on the material of the terminal, e.g.

For brass and nickel silver terminals 15 times.

For monel terminals 100 times.

● the joint must be remade with new wire. Where insufficient wire remains for a new wrap a new wire must be inserted.

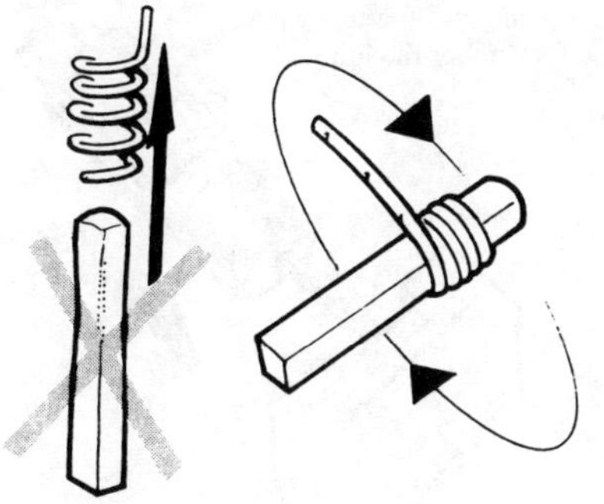

Stripping test

The quality of wrapped joints depends on the efficiency of the wrapping tool. Check the efficiency of each tool at specified intervals using a stripping test.

Prepare a number of wrapped joints on a test piece. Measure on a tension gauge the force required to strip the wire off each terminal. The force required must be above the specified limit in each case.

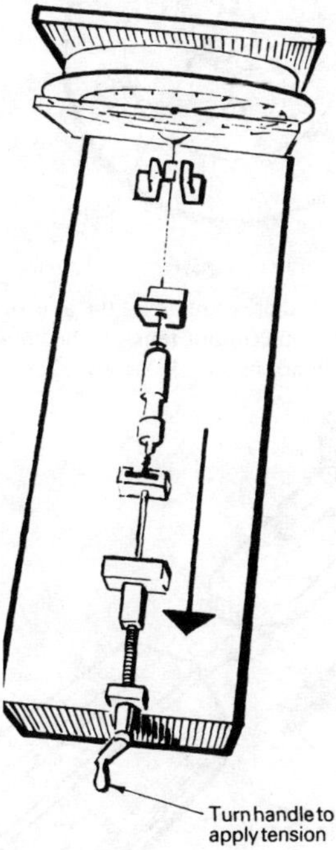

Fit the test piece into the test gauge and fit the fork under the joint.

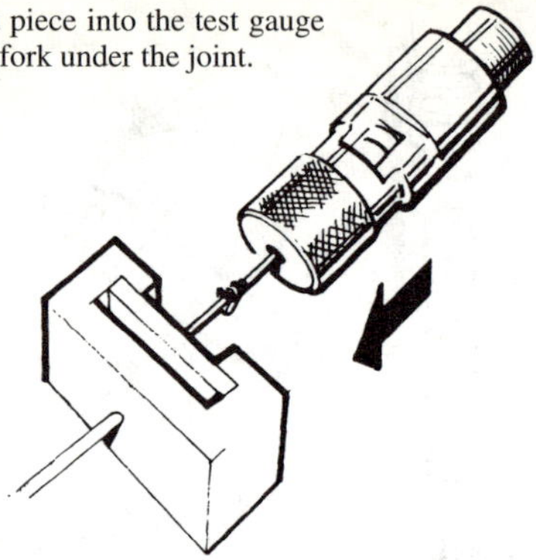

Ensure that the pointer is set at zero. Increase tension on the joint by winding the winch handle until the joint gives. The indicator hand will have moved the pointer to the maximum tension the joint withstood.

Note the pointer reading. Return the pointer to zero.

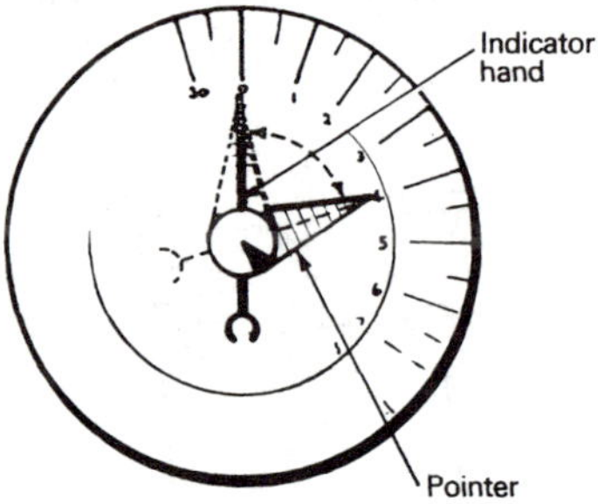

If a hand held test gauge is used:

- take care to exert the pressure along the axis of the terminal
- note carefully the maximum tension the joint withstands. Hand held gauges do not usually have a marker.

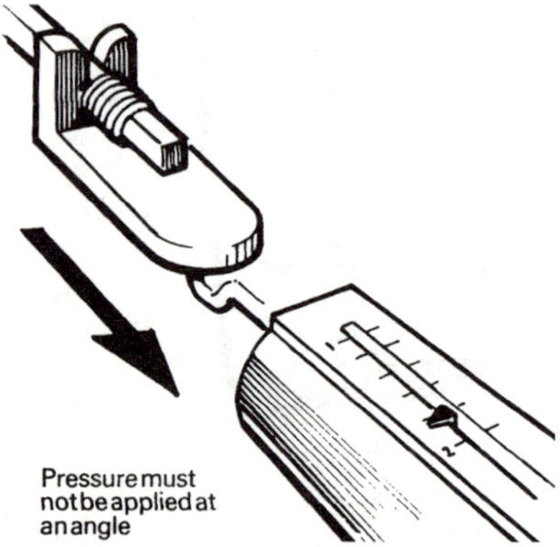

Unwrapping test

Carry out unwrapping tests at specified intervals to ensure that the wrapping tool is not binding the wire so tight that it is weakened. Unwrap the wire from a sample using a pair of pliers. Use a rotary motion. It must be possible to remove the wire without breaking it.

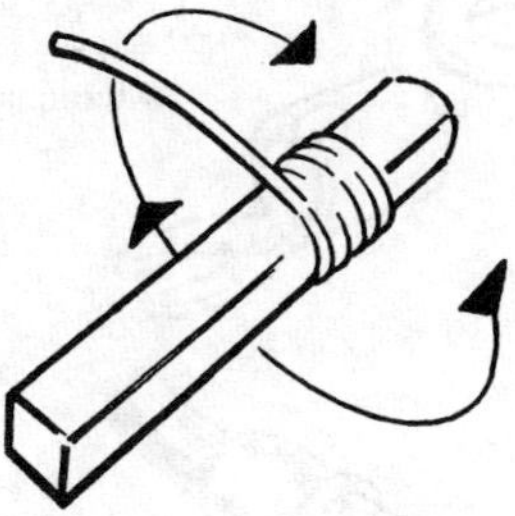

Inspection of crimped joints

Some common types of crimped connector are:

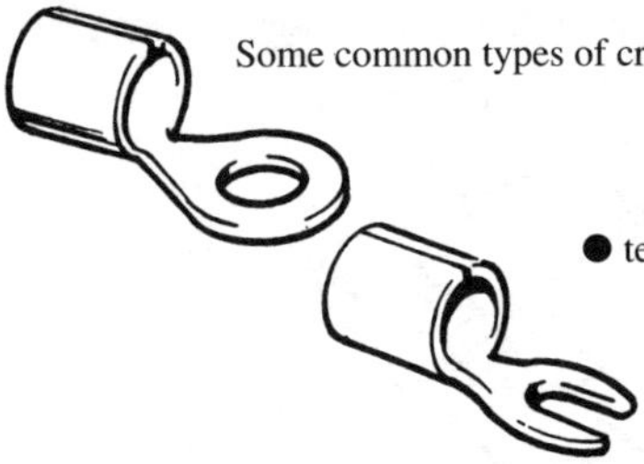

● terminal tags and lugs

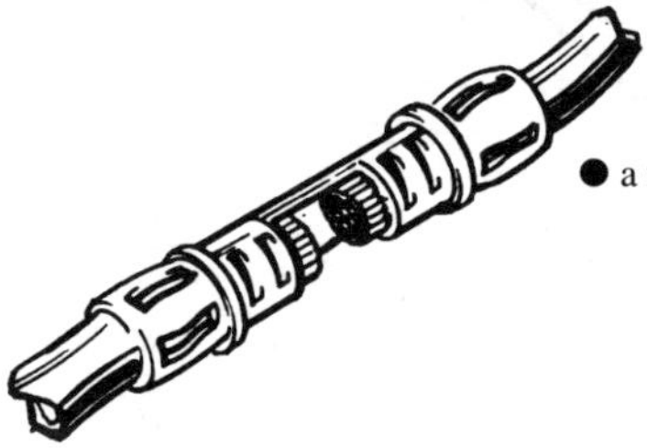

● a ferrule or splice

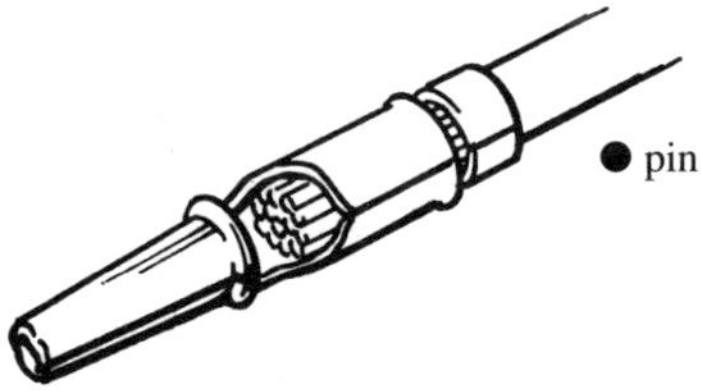

● pin or socket contacts.

Some crimping shapes.

Use the assembly drawing and parts list to ensure that:

● the correct size and type of connector has been used

● the correct crimping shape has been obtained. This depends on the crimping die used.

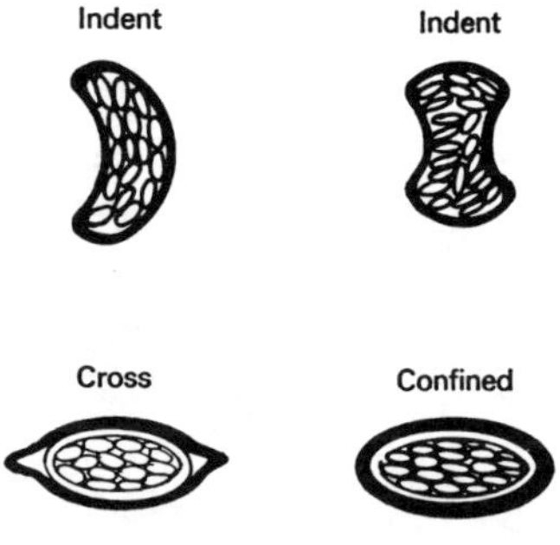

Criteria for a good crimped joint

The conductor must be positioned correctly in the connector. Most connectors have an inspection window in the crimped area. If bare wire is not visible through this window the joint is suspect.

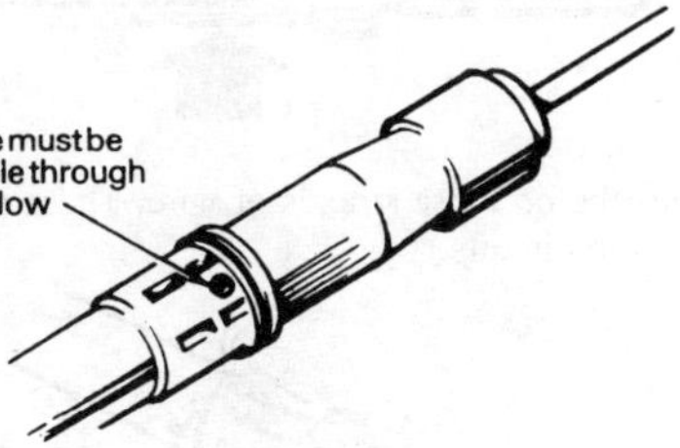

The connector must be free from fractures, rough edges and flashes.

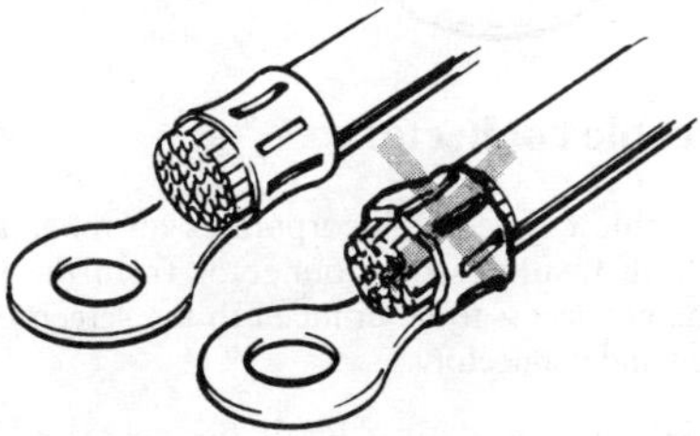

The crimp must be in the correct position on the connector to give maximum strength to the joint, and to avoid damage.

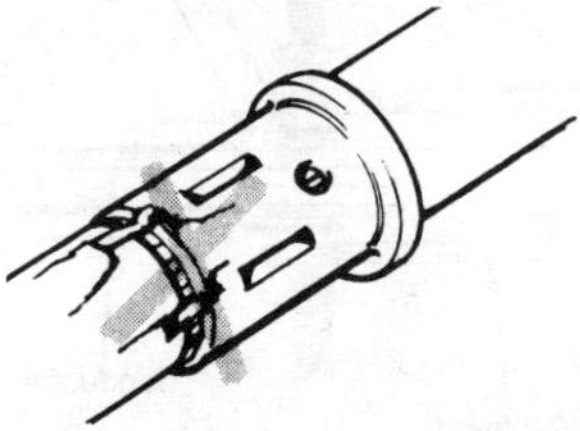

The insulation gap must be as specified.

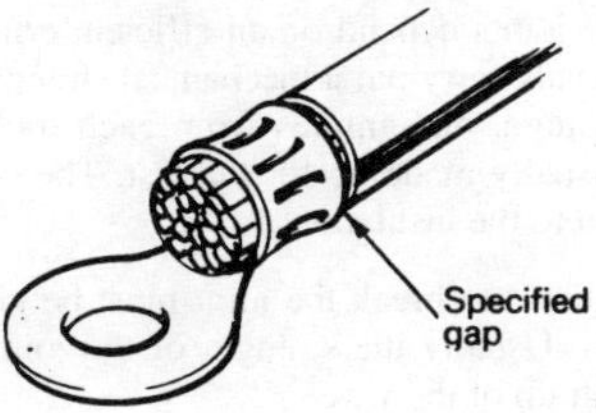

Many crimped connectors are also crimped to the insulation. The insulation must be held firmly.

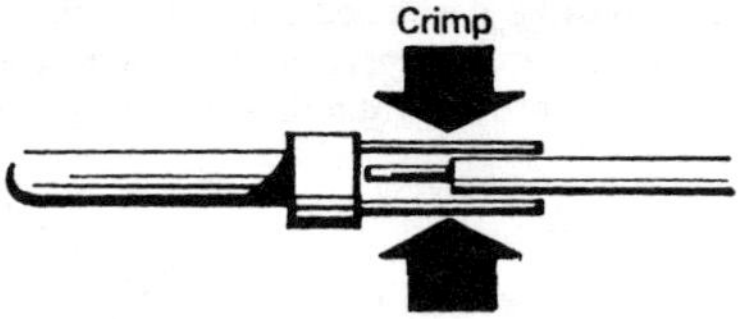

There must be no loose strands of wire. These are a common cause of short circuits.

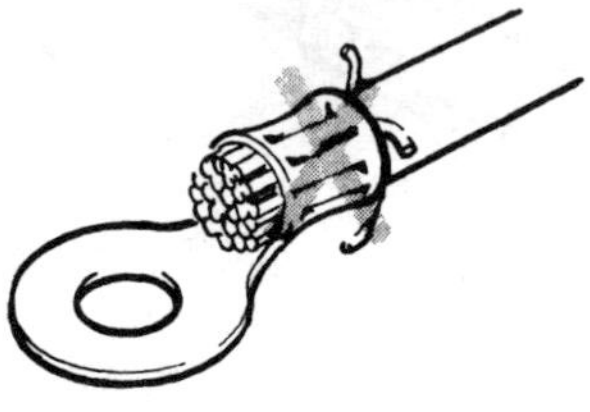

Screen cable connectors

A screen cable connector incorporates an inner insulation sleeve which insulates the connector from the screen. It permits the connector to be crimped to the screen only, or to both screen and connector.

The insulation sleeve must be of the specified type and correctly fitted to prevent short circuits.

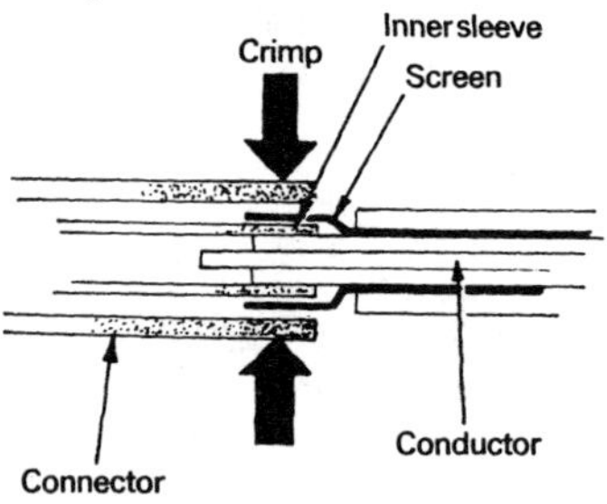

Mechanical strength test

Good crimped joints depend on an efficient crimping tool. At specified intervals carry out a mechanical strength test, using a tension test gauge, on samples from each tool. Special test samples are usually made up for the test. The connector must not be crimped to the insulation.

The force required to break the joint must be greater than the specified limit. Usually the strength of the joint must exceed the breaking strain of the wire.

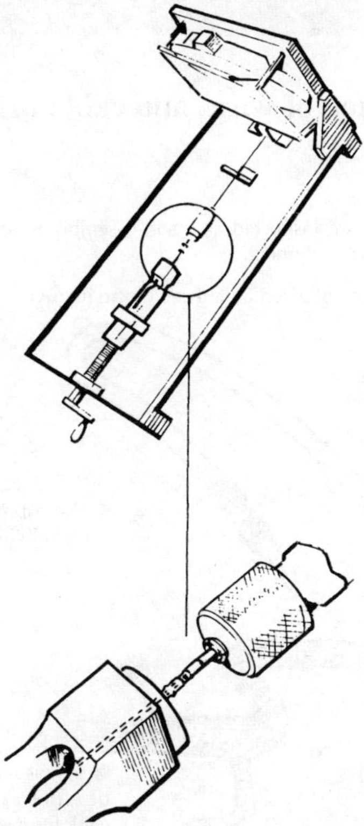

Continuity test

Where specified, carry out a continuity test on crimped joints.

Clip the leads of a buzzer, oscillator, or continuity tester to the connector and the lead. Continuity is indicated by a buzz or other sound. The pitch of the sound may indicate the degree of resistance of the joint.

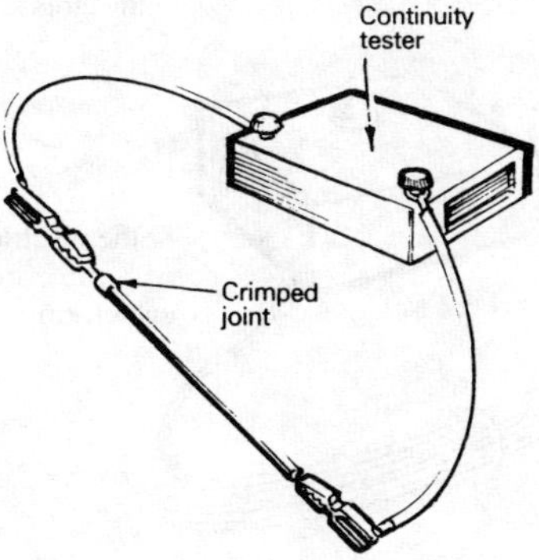

Inspection of wires and cableforms

General

Cableforms are inspected as a sub-assembly before they are fitted into the equipment.

The inspection of a completed cableform involves:

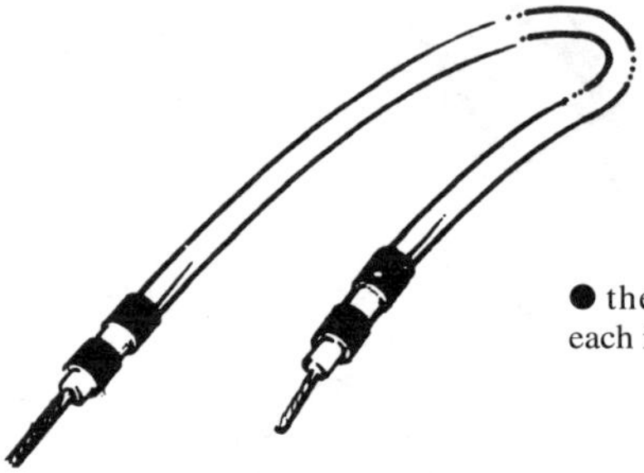

- the inspection of each individual wire

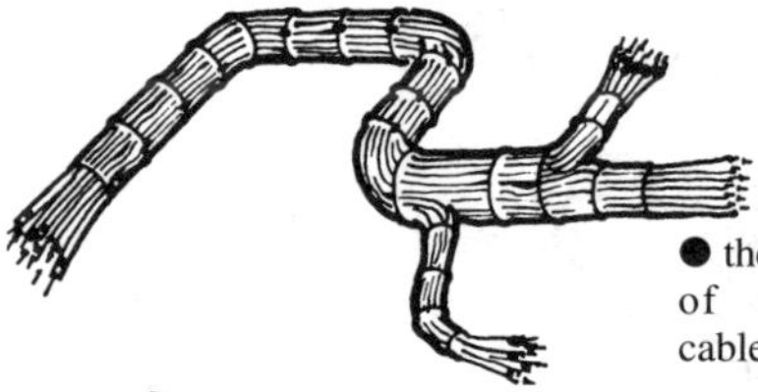

- the visual inspection of the completed cableform

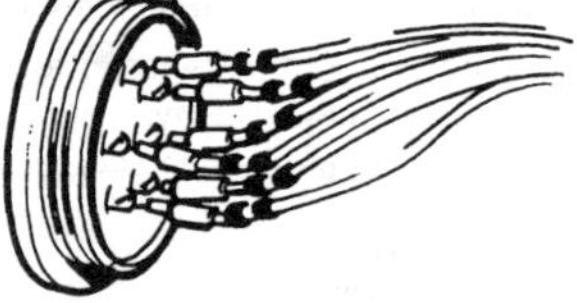

- inspecting the assembly and wiring of plug and socket connectors

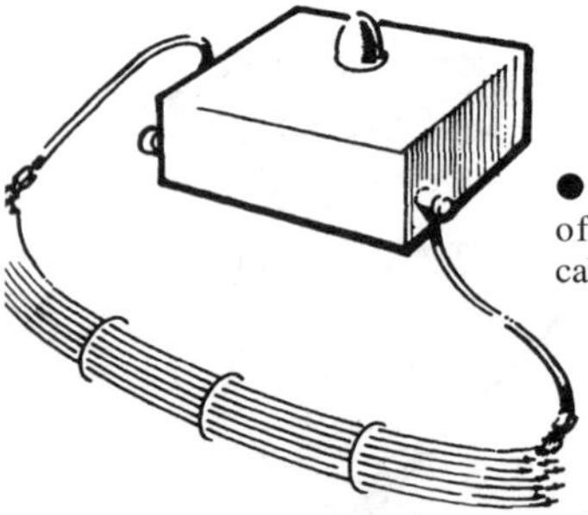

- the electrical testing of the completed cableform.

Individual wires

Inspect each wire against the wiring schedule. Check for correct length, type and gauge.

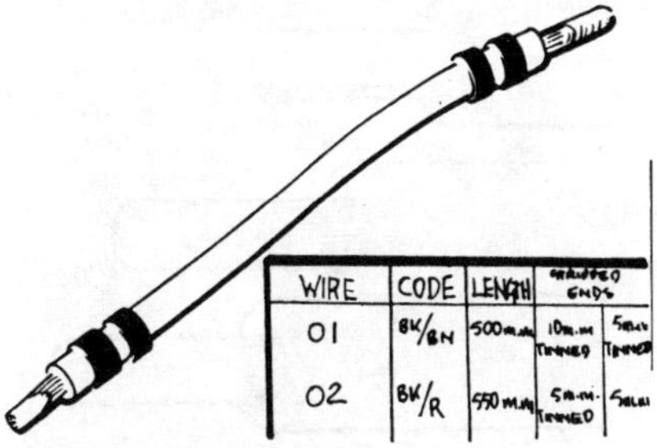

Stripping and tinning

Ensure that the correct length of insulation has been stripped from each end of the wire.

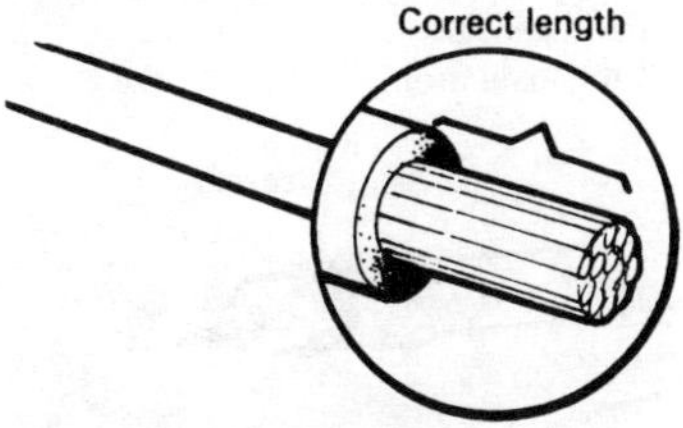

Check for damage to the wire and insulation.

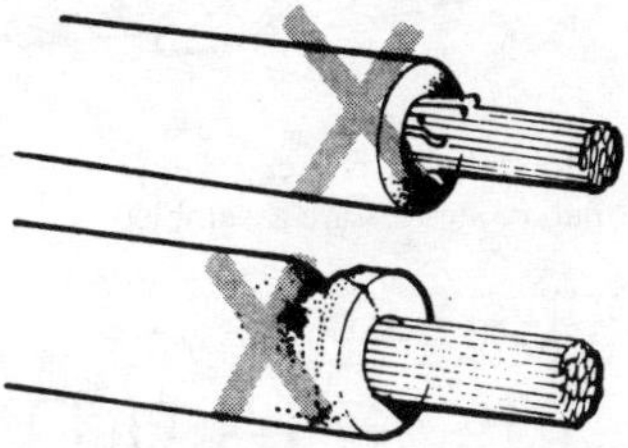

Pigtails and stranded conductors should be adequately twisted. There must be no loose, broken or doubled up strands.

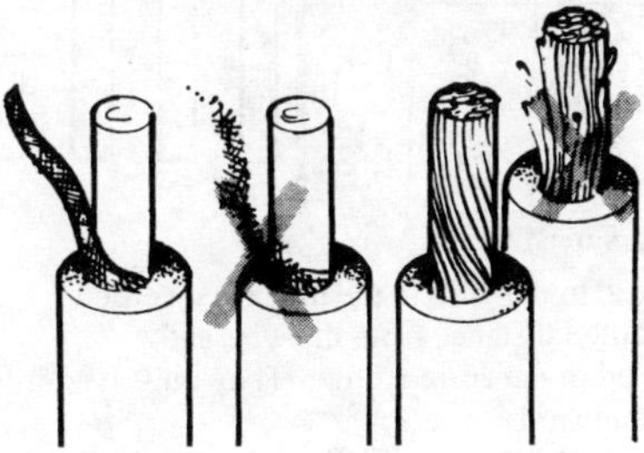

Ensure that where specified the wire ends have been tinned.

There must be sufficient tinning to form a joint, but the tinning must not touch the insulation.

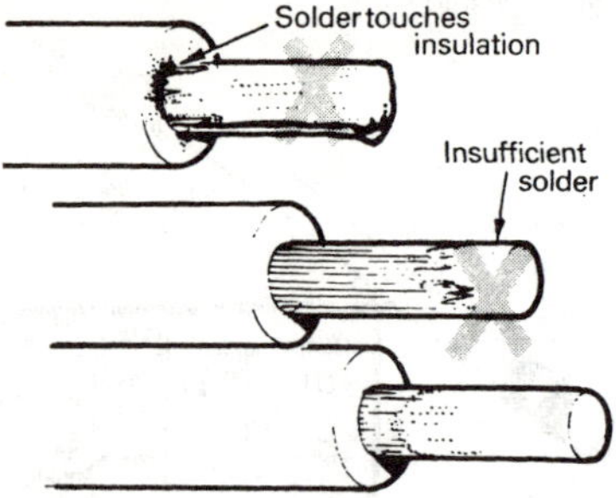

Identification

Wires are identified by:

- coloured marker sleeves
- the colour of the insulation.

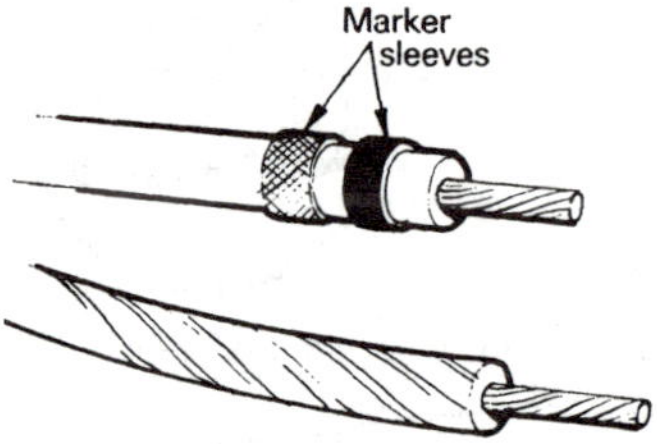

Ensure that:

- the correct coloured wire is used
- the correct marker sleeves are assembled.

ITEM	LENGTH	MARKER
1	40	B/3
2	50	B/G
3	50	B/R
4	60	B/G

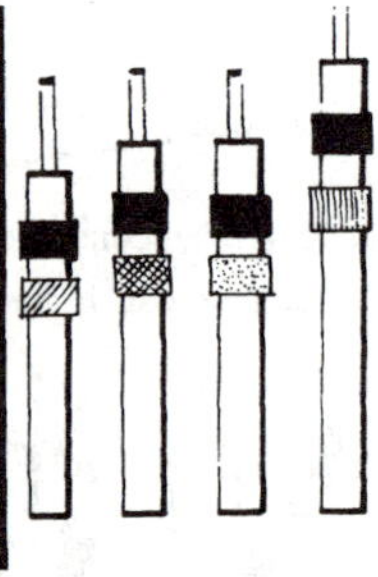

The markers must be:

- assembled to the wire in the manner specified
- the specified distance from the wire end
- assembled in the correct order. They must read from the end of the wire inwards.

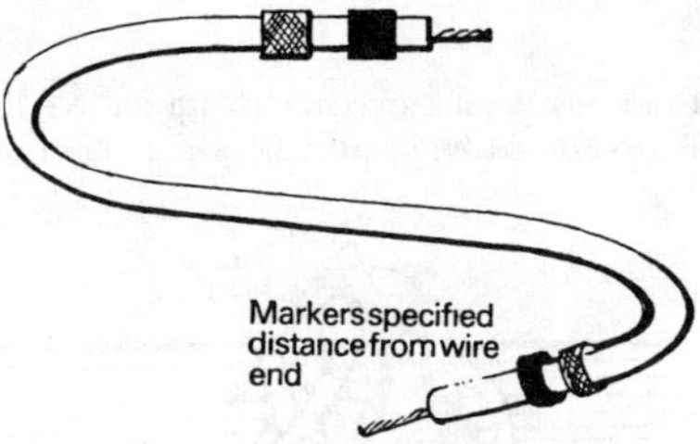

Cableforms

Ensure that the cableform has been laid so that no wire is:

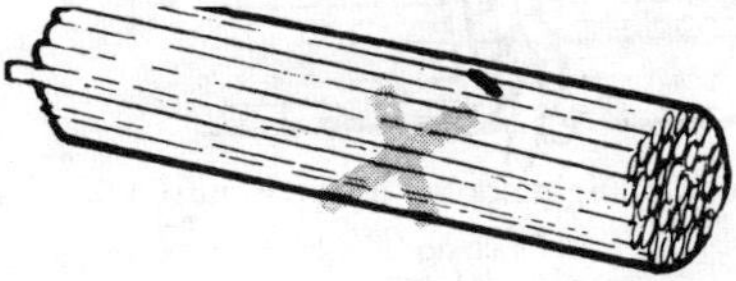

- stretched

- loose

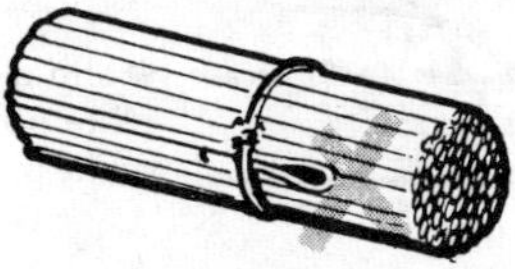

- folded

- twisted

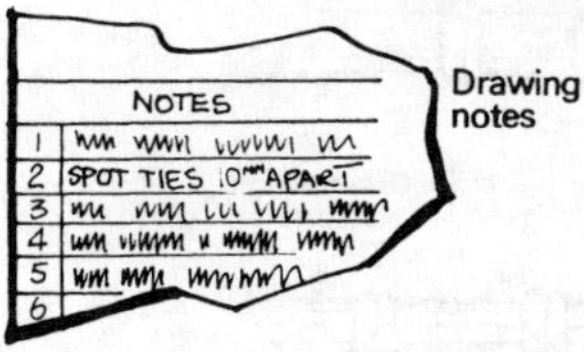

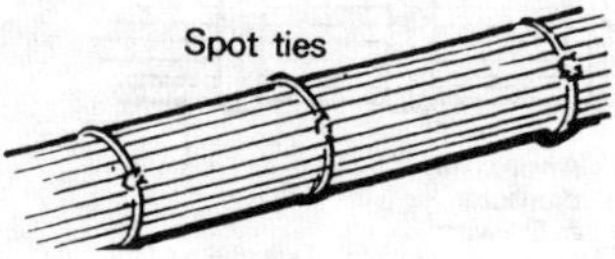

Binding

Ensure that:

- the correct form of binding is used
- the materials used are as specified.

Spot ties

Each spot tie must be a clove hitch with an overhand knot added. The overhead knot must be a reef knot and not a *granny*.

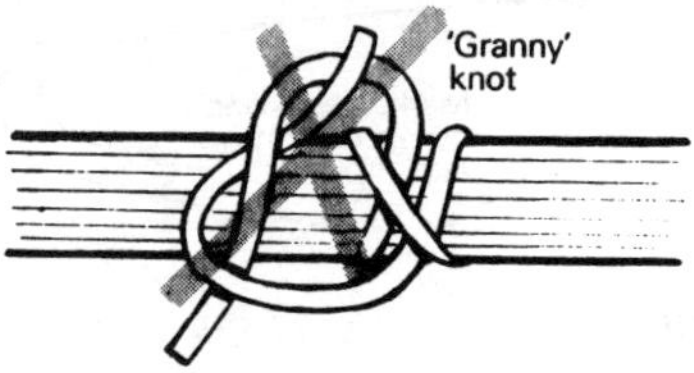

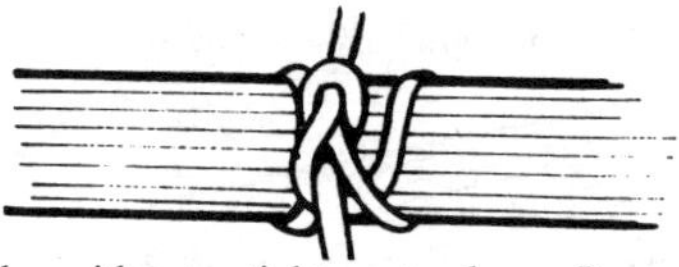

The tie must be neither too tight nor too loose. It must not:

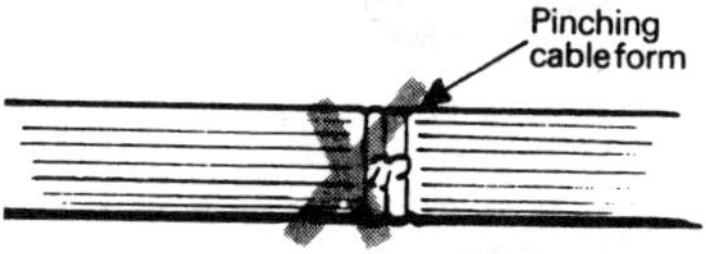

- pinch the cableform

- slip along the cableform under gentle pressure.

Check the spacing of the ties.

On cableforms of less than 25 mm diameter they should usually be 12 mm to 18 mm. On larger cableforms this distance should be roughly the diameter of the wire, unless otherwise specified. The spacing is closer on bends.

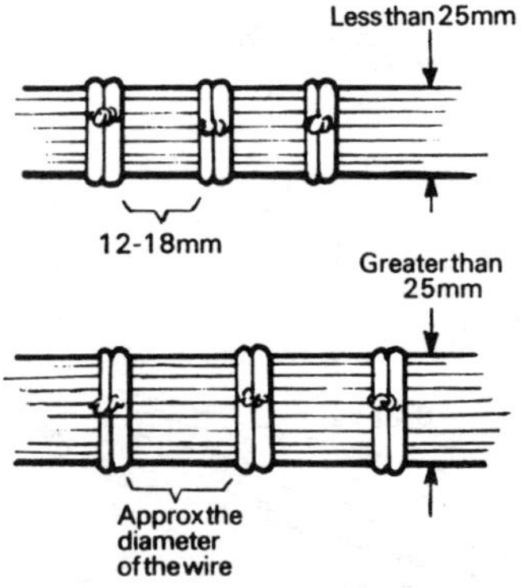

Lacing

Ensure that the correct knot has been used.

Check starting ties.

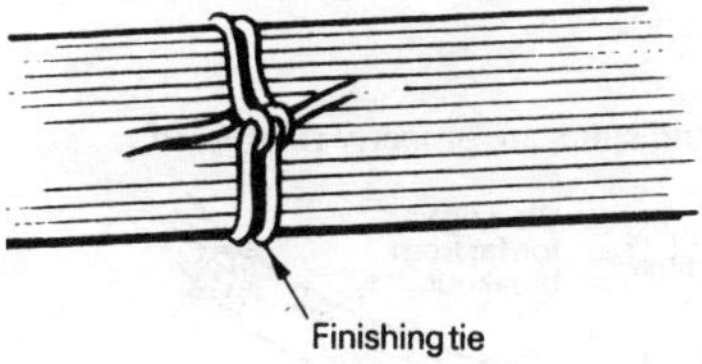

Check finishing ties.

The correct locking stitch must be used for the lacing. For cableforms of greater than 25 mm diameter a double locking stitch should be used.

The tightness and spacing of the stitches should be as for spot ties.

The lacing must not float round the bends.

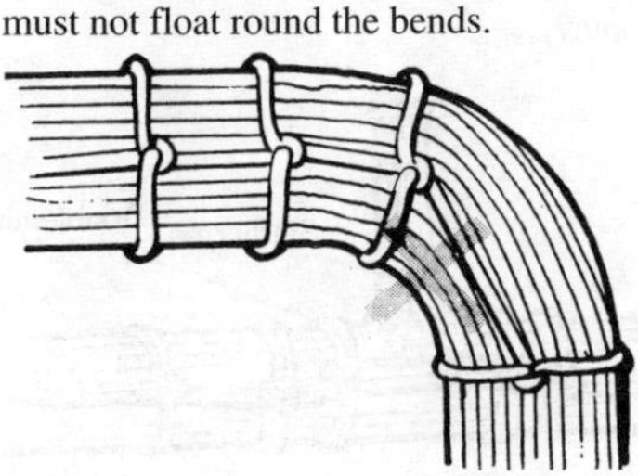

The stitches should be closer together to avoid this.

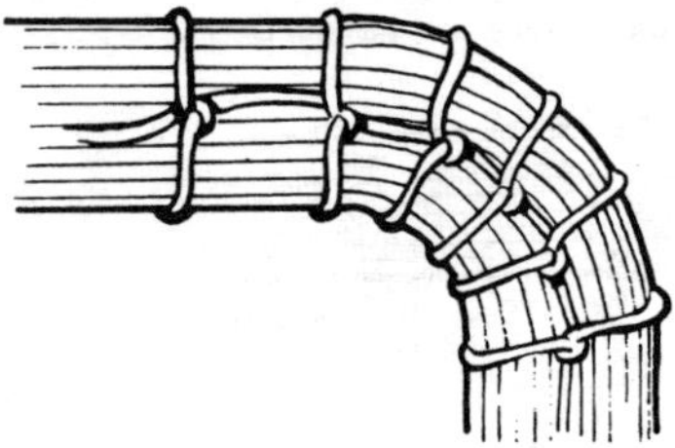

Ensure that branches and spurs are properly reinforced.

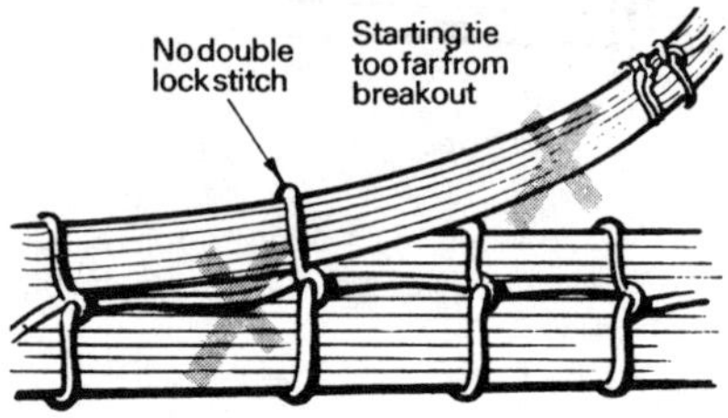

Multistrand breakouts should have a double lock stitch before the branch. The starting tie on the branch should be close to the start of the branch.

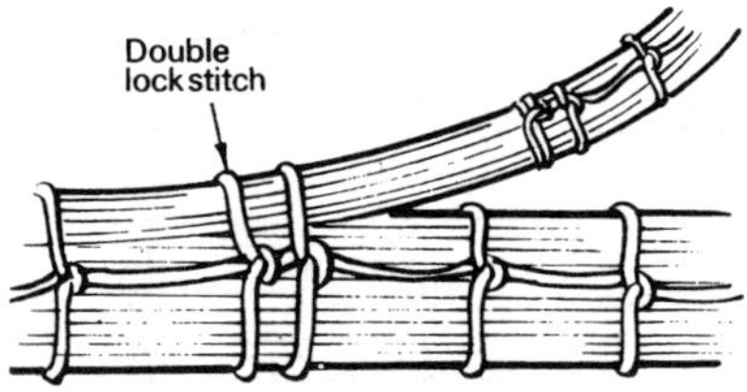

Spurs should be reinforced by double locking stitches on both sides of the spur.

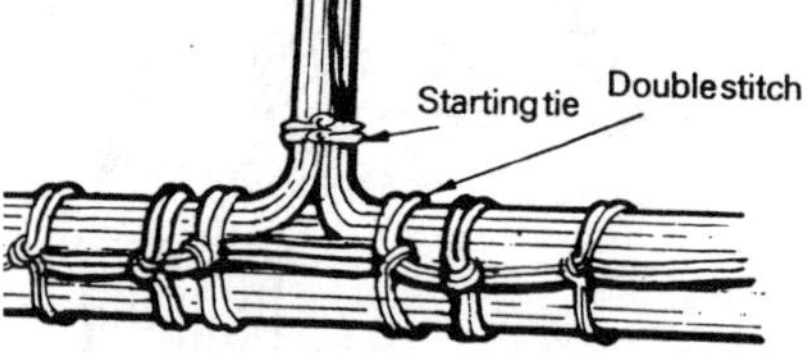

Lacquer

Ensure that all spot ties, and the start and finish ties of lacing, have been lacquered with the specified lacquer.

Straps

Check that each strap has been secured correctly, and lacquered with adhesive where required. They must meet the same conditions of spacing and tension as spot ties and lacing.

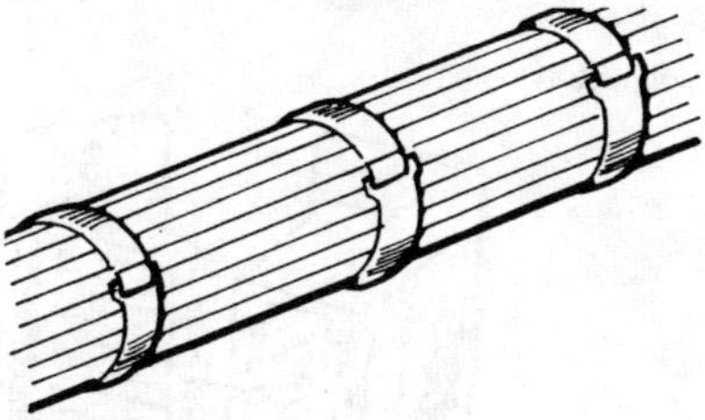

Inspection of connectors

Plus and socket connectors are usually assembled to the cableform during cableforming.

Some typical plugs and sockets are:

- blade connectors

- spring connectors

- shrouded pin connectors

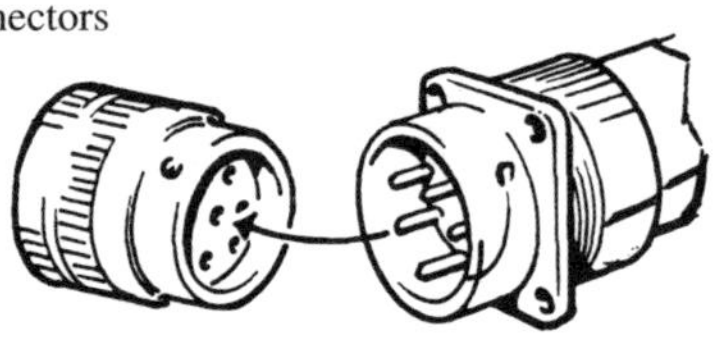

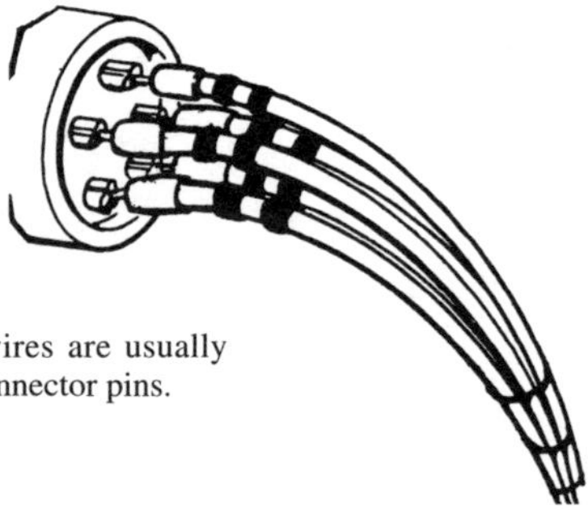

The cableform wires are usually soldered to the connector pins.

It is usual to inspect plug and socket assemblies in stages because earlier work is obscured by later work.

1st stage

The 1st stage inspection is primarily concerned with the soldering of the cableform to the connector.

Ensure that:

- the correct type and size of connector is being assembled
- insulation sleeves and rear sections of the plug have been assembled to the cableform prior to soldering.

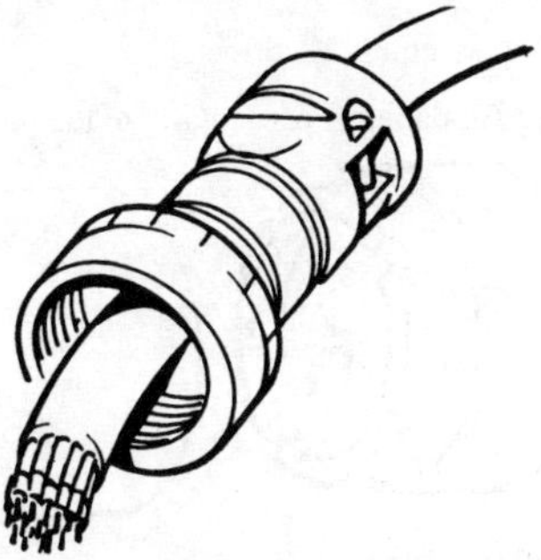

Check against the wiring diagram that each wire is soldered to the correct contact.

WIRE ITEM	MARKER SLEEVES COLOUR	PIN NO
4	BK/BN	PL4/7
4	BK/R	PL4/8
4	BK/O	PL4/9
5	BK/Y	PL4/10
4	BK/G	PL4/11
4	BK/B	PL4/12

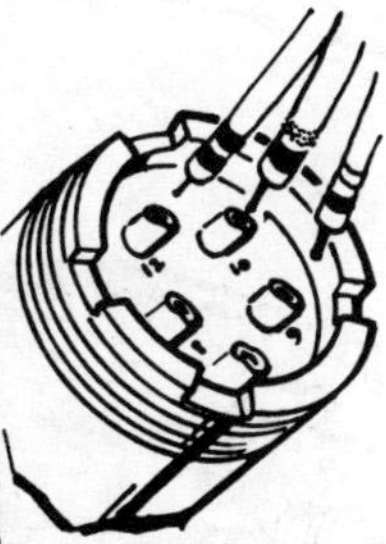

Inspect the soldered joints. The insulation sleeve may be pushed into place over each joint as it is inspected.

Look particularly for:

- wicking
- solder splashes on the wires
- nicked and loose strands of wire.

Ensure that no insulation sleeves are missing.

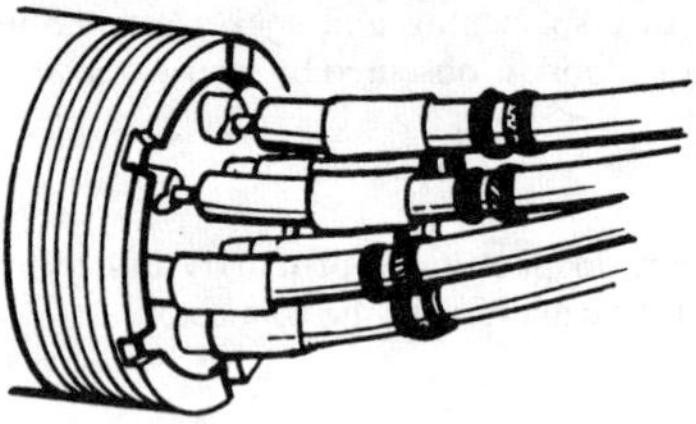

2nd stage

Second stage inspection is of the plug assembly.

Ensure that plug assembly is correct.

Check the plug for damage, particularly to the moulding.

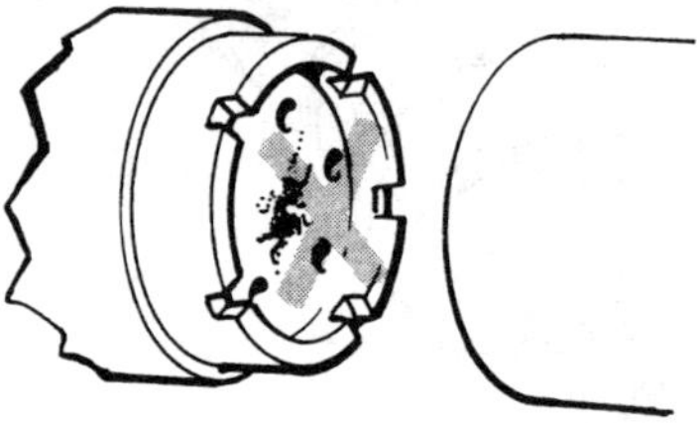

Coaxial connectors

Coaxial connectors normally require inspecting at three stages.

- inspect the stripping of insulation and combing and trimming of the braid. Ensure that there are no loose strands
- inspect the soldering of the conductor to the connector barrel. Check especially for overheating of the insulation on the male contact assembly

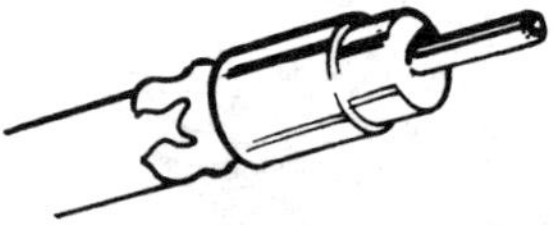

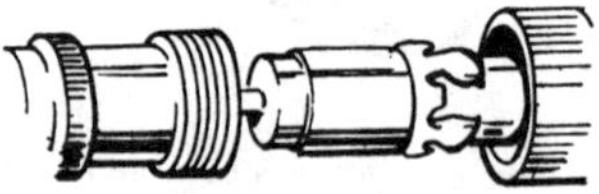

- inspect the assembly of the connector.

Crimped connectors

Crimped connectors are assembled by inserting the pins into the body of the connector after crimping.

This is done with a special insertion tool.

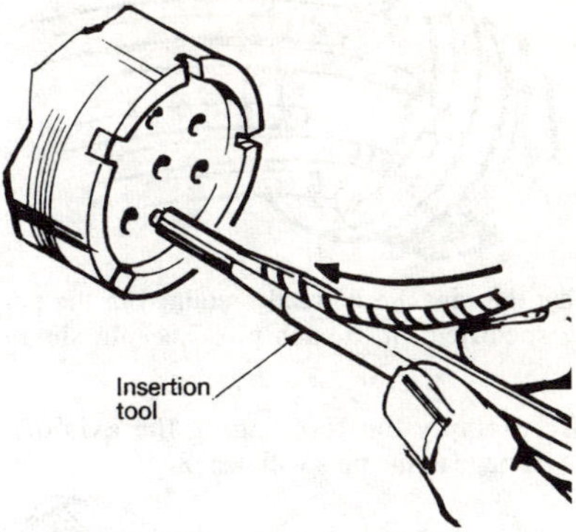

The pin is held in place by a retaining clip.

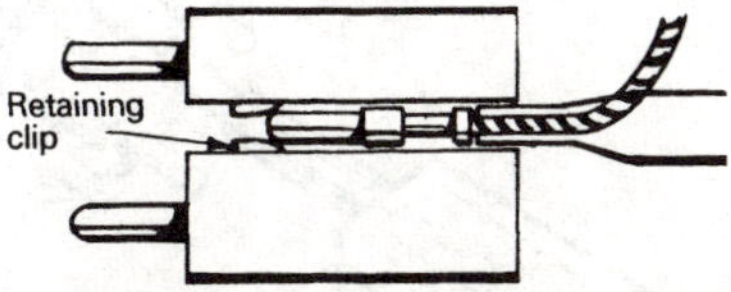

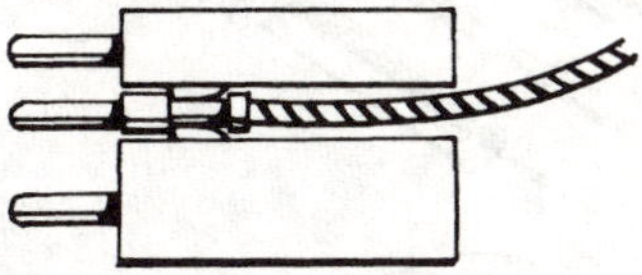

Inspection is in two stages:

● inspect the crimping of the pins to the cableform. Ensure that the correct pin is crimped to each lead

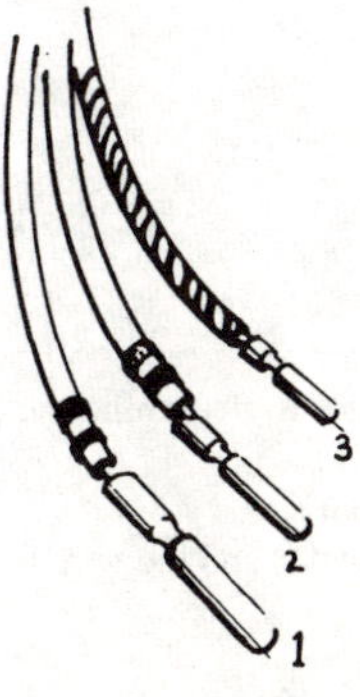

NO	PIN SIZE	LEAD
1	3	Bn/Y
2	4	Bn/G
3	6	Bn/O
4	2	Bn/R
5	2	Bn/BK
6	2	Bn/BK
7	2	Y/R
8	2	Y/G

● inspect the insertion of the pins. Ensure that the pins have been inserted into the correct holes as specified on the drawing.

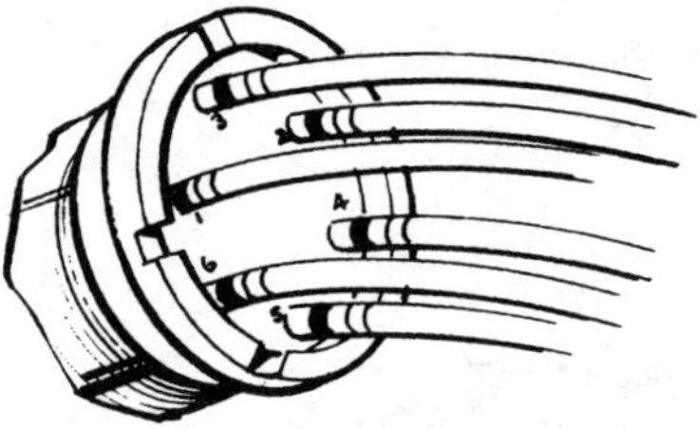

Ensure that the pins are correctly retained in the plug. Apply the force specified on to each pin. The pin should remain secure.

Take care to apply the force along the axis of the pins, otherwise damage to the pins will occur.

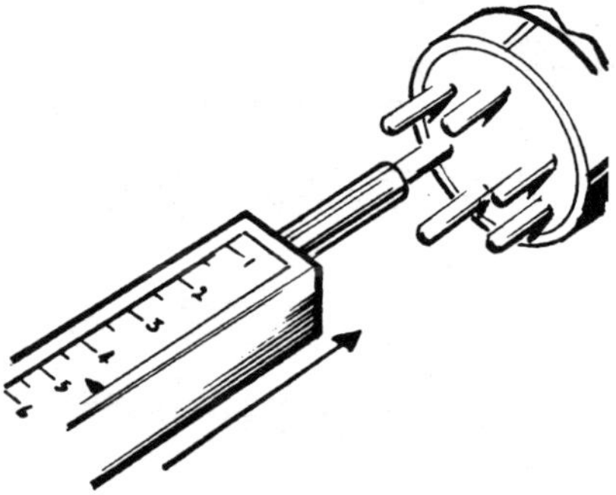

Ensure that there is no damage to the plug or pins.

Electrical tests

Continuity test

Use the prescribed continuity tester to determine the presence of electrical continuity

● between the two ends of each wire
● between each pin of each connector and the wire to which it is attached.

Work according to the wiring schedule.

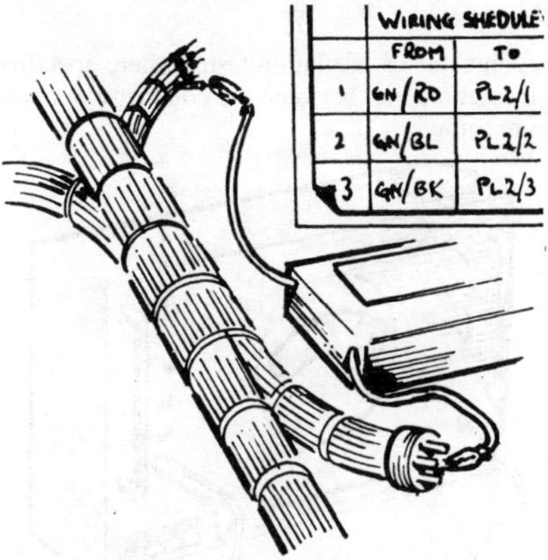

Automatic continuity test

Automatic continuity testing equipment can be devised, e.g.

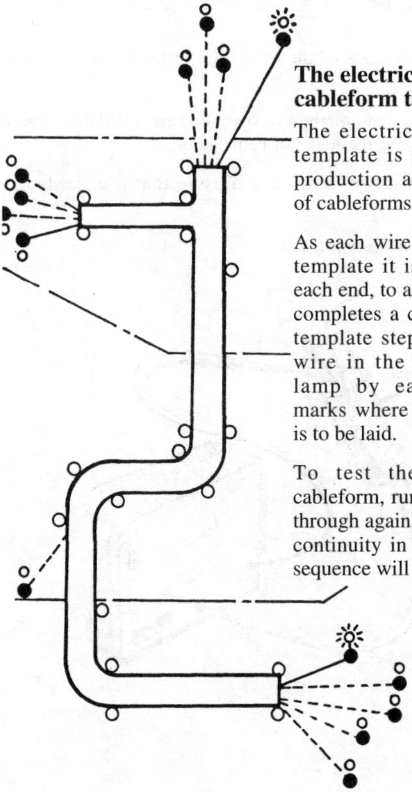

The electrical cableform template

The electrical cableform template is an aid to the production and inspection of cableforms.

As each wire is laid on the template it is attached, at each end, to a terminal. This completes a circuit and the template steps to the next wire in the sequence. A lamp by each terminal marks where the next wire is to be laid.

To test the completed cableform, run the sequence through again. If there is not continuity in each wire the sequence will be stopped.

Insulation test

Use a megohmeter or insulation tester where specified to test for poor insulation, e.g. between the conductor and screen of a coaxial connector.

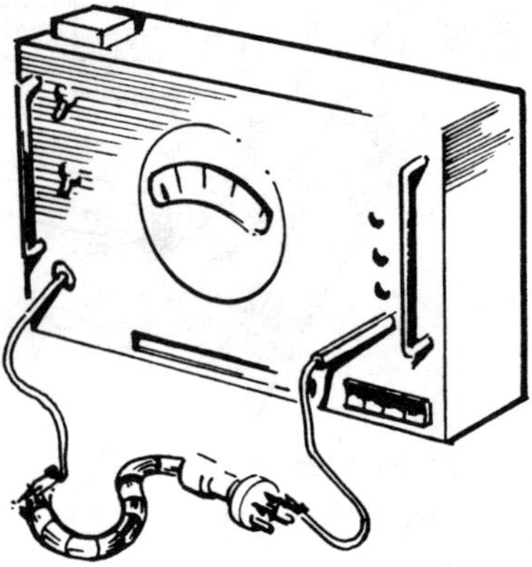

Connect the two leads of the insulation tester to the insulated conductors.

Apply a voltage. The method of operating the insulation tester varies according to the make of instrument.

Note the resistance in megohms. Ensure that it exceeds the specified limit.

Switch off the insulation tester.

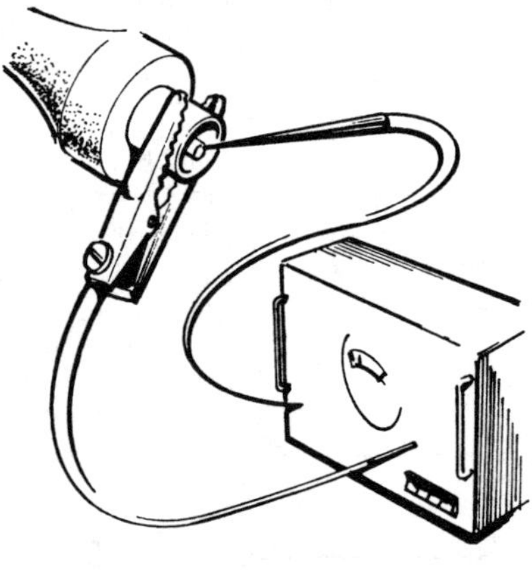

Inspection of equipment assembly

Inspect assemblies for:

- compliance with the drawings
- the required standard of work
- damage.

Inspect systematically. Use the relevant inspection schedule.

The example below shows part of an assembly, and its representation on an assembly drawing.

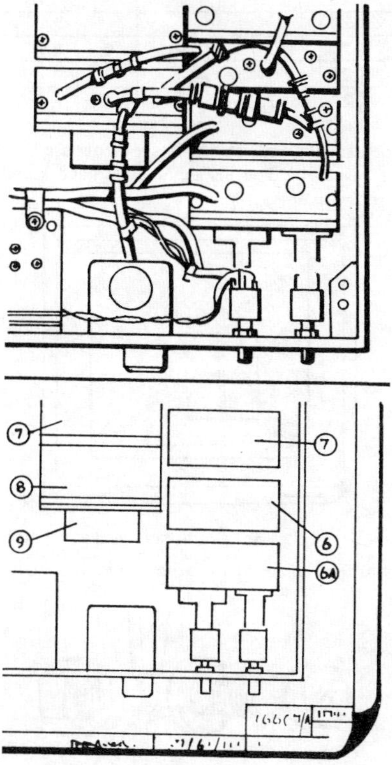

Fixing devices

Check each nut and screw fixing for:

- correct size and type of nut, screw and washer
- correct assembly.

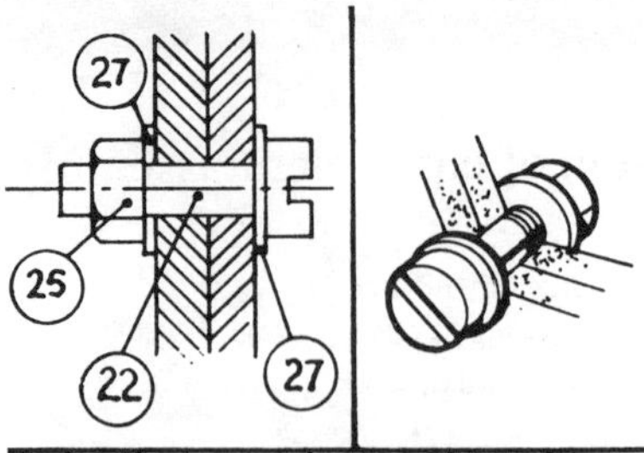

Item	Partno.	Name	No.off	Remarks
21				
22	72/4271	screw	3	15mm
23				
24				
25	47/6742	nut	6	BR/T
26				
27	6B721	washer	1	

Usually the drawing indicates only the items used and not the method of assembly. The inspector must have knowledge of assembly requirements.

Some examples of the use of nuts, screws and washers:

● plain washers are used to protect finished surfaces or as spacers.

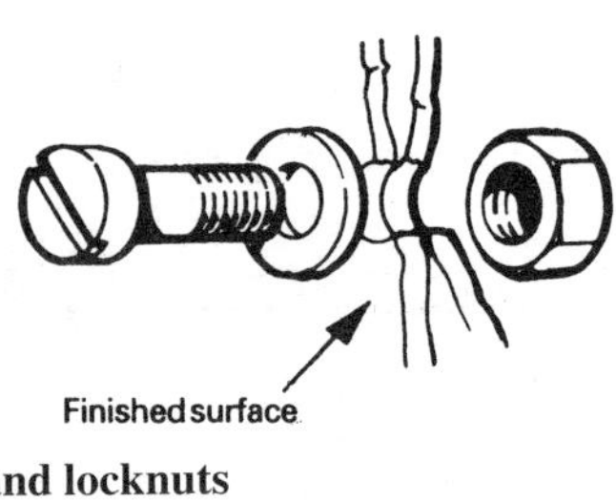

Stiffnuts and locknuts

● stiffnuts and locknuts increase the friction between threads to reduce the likelihood of their working loose. They are used in place of ordinary nuts.

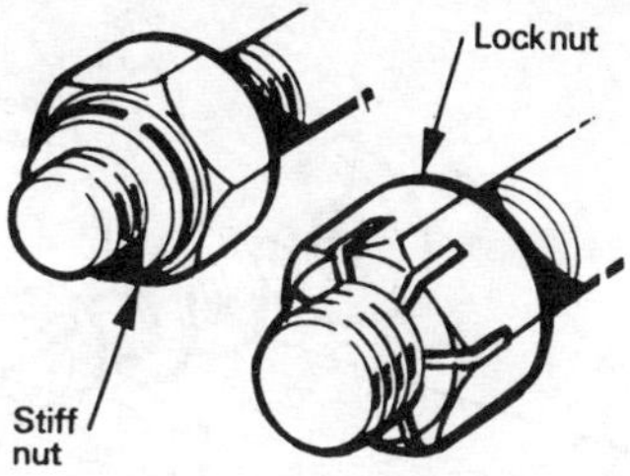

Locking washers

- there are three main types of locking washers:

  spring washers
  crinkle washers
  shakeproof washers.

Shakeproof washers are not recommended for high quality products.

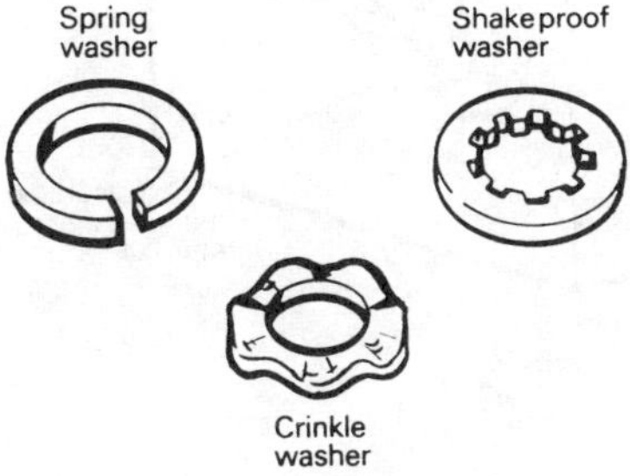

Locking washers may be used against the nut, or the screwhead.

Plain washers are used to protect panels from damage caused by spring or shakeproof washers.

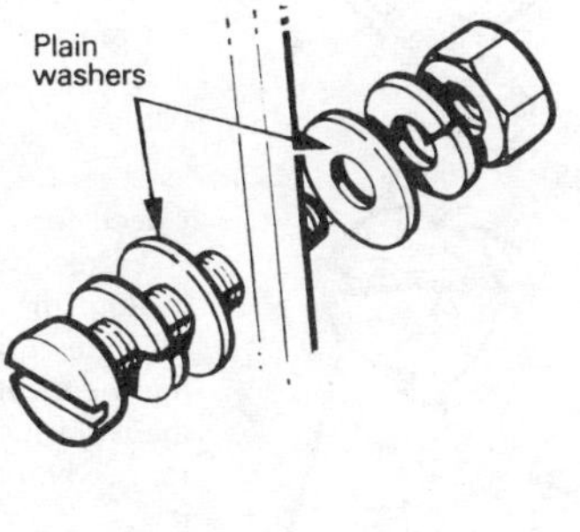

The use of locking washers is usually mandatory where the screw is screwed into a tapped hole.

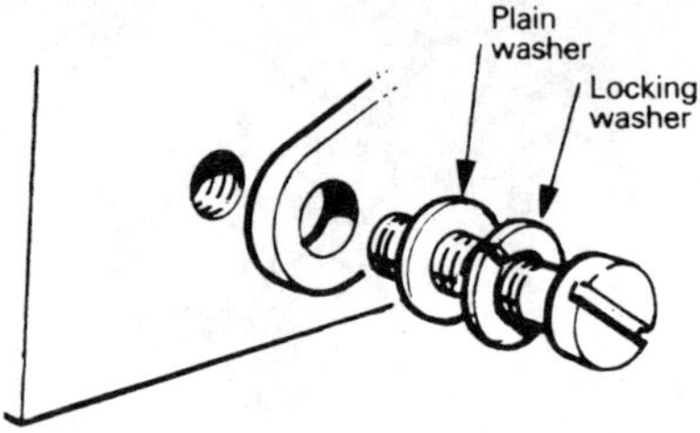

Ensure that all nuts and screws are tight.

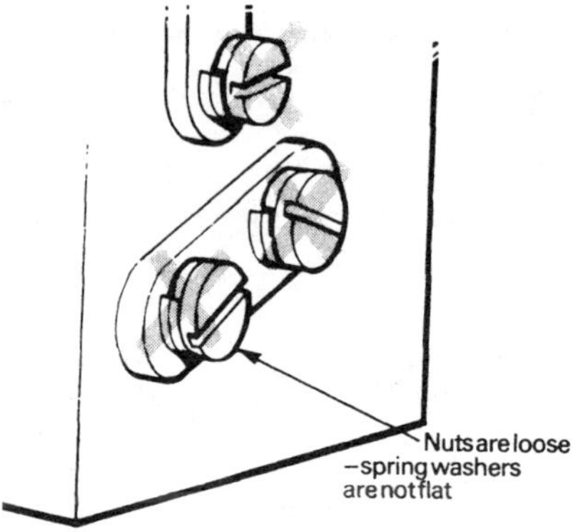

Locking varnish

Ensure that the specified locking varnish is used as required.

The varnish should usually be applied to the screw thread before assembly.

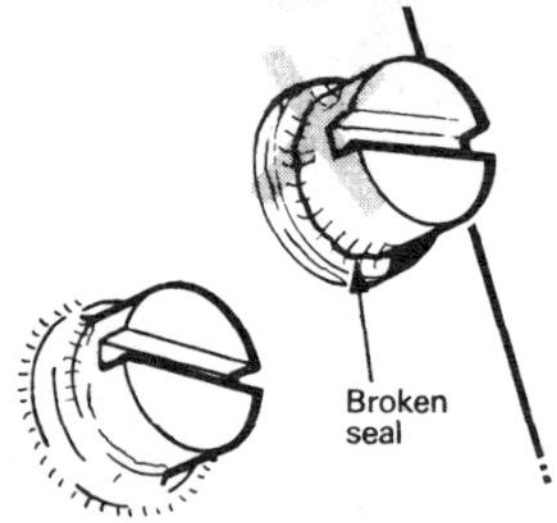

Check for broken seals. A break in the varnish around the screw head indicates that the screw has been moved since the varnish set, and is no longer being locked by it.

Ensure that:

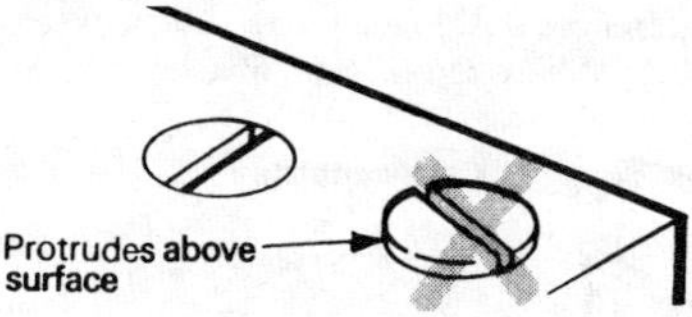

● the rims of countersunk screws do not protrude above the surface of the panel

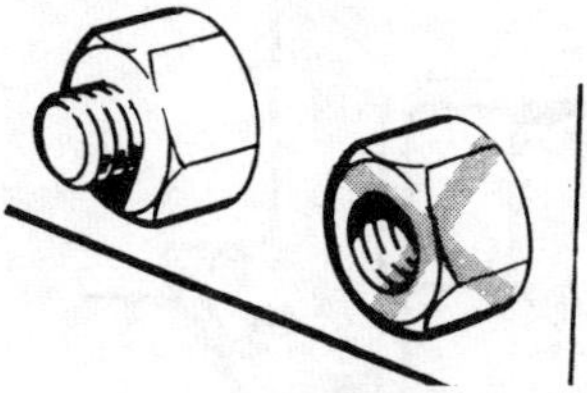

● screws protrude through their nuts by at least one thread, or as specified

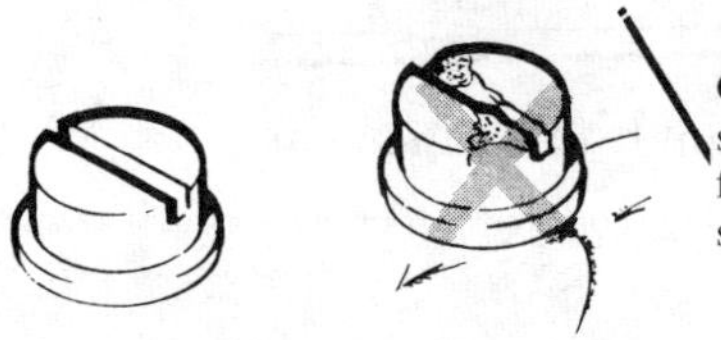

● screw slots and surrounding areas are free from burrs and scratches.

Burred screwslots are a common defect, frequently caused by the use of an unsuitable screwdriver.

The intimate contact of dissimilar metals

When two dissimilar metals come into contact a potential difference is developed between them. If this is greater than 0.5 V (0.25 V in the tropics), corrosion may occur.

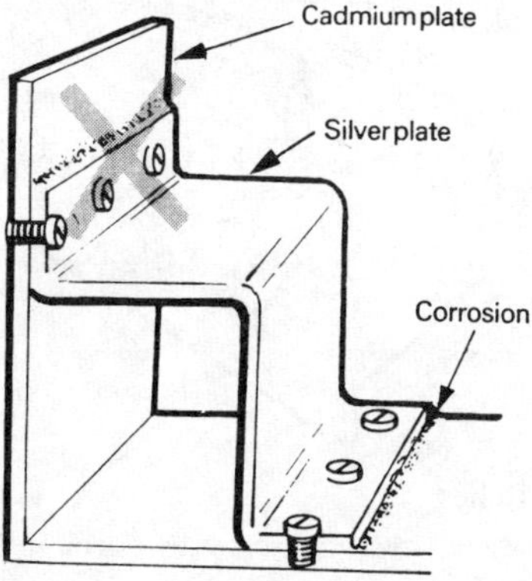

Example:

A silver plated bracket must not be screwed directly on to a cadmium chassis as a potential difference of 0.79 V is created.

If tinned washers are introduced between the bracket and the chassis the potential differences developed are 0.31 V between tin and cadmium, and 0.48 V between tin and silver. Both are within the limit.

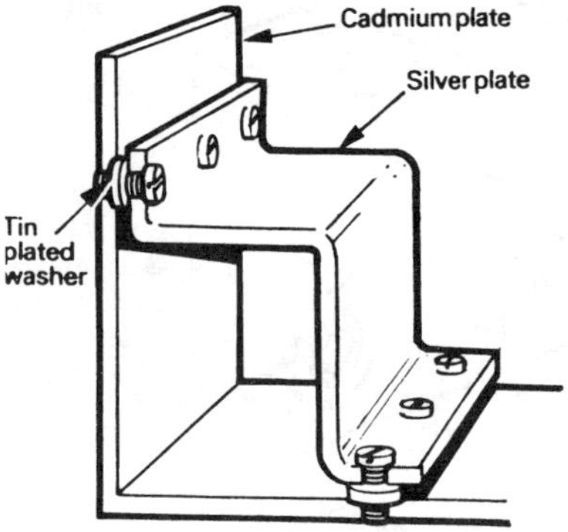

Mechanical assembly of components

Ensure that the assembled component is of the correct size and type.

Gears

Check the play or *backlash* between the gears. This must be within specified limits.

Make sure that gears are free of swarf, dirt and solder splashes. Fine gears can be ruined by foreign bodies in their teeth.

Ensure that the correct lubricant is applied.

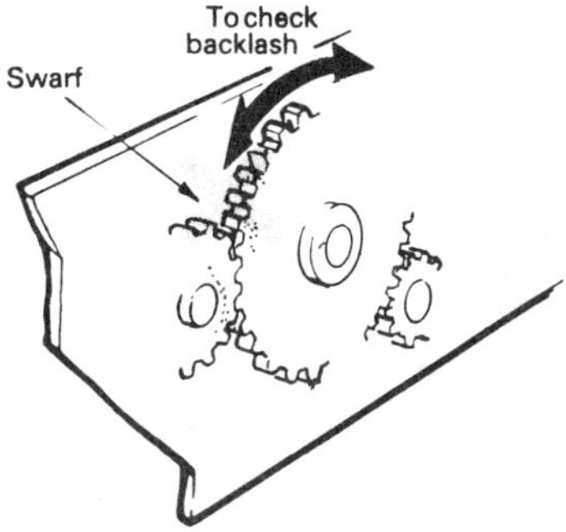

Wafer switches

Ensure that wafer switches are correctly assembled.

Common errors are:

- wafers assembled in the wrong order
- wafers incorrectly orientated on the drive spindle, e.g. turned 180° from the required orientation

Table of contact potential differences (volts)

	Gold	Rhodium	Silver	Nickel	Copper	Brass	Stainless Steel	Tin	Chromium	Steel	Alumin-ium	Cadmium	Zinc
Zinc	1·40	1·40	1·11	0·96	0·92	0·80	0·65	0·63	0·60	0·40	0·35	0·32	0·00
Cadmium	1·08	1·08	0·79	0·64	0·60	0·48	0·33	0·31	0·28	0·08	0·03	0·00	
Aluminium	1·05	1·05	0·76	0·61	0·57	0·45	0·30	0·28	0·25	0·05	0·00		
Steel	1·00	1·00	0·71	0·56	0·52	0·40	0·25	0·23	0·20	0·00			
Chromium	0·80	0·80	0·51	0·36	0·32	0·20	0·05	0·03	0·00				
Tin	0·77	0·77	0·48	0·33	0·29	0·17	0·02	0·00					
Stainless steel	0·75	0·75	0·46	0·31	0·27	0·15	0·00						
Brass	0·60	0·60	0·31	0·16	0·12	0·00							
Copper	0·48	0·48	0·19	0·04	0·00								
Nickel	0·44	0·44	0·15	0·00									
Silver	0·29	0·29	0·00										
Rhodium	0·00	0·00											
Gold	0·00												

- spacers omitted from between the wafers
- damage to wafers.

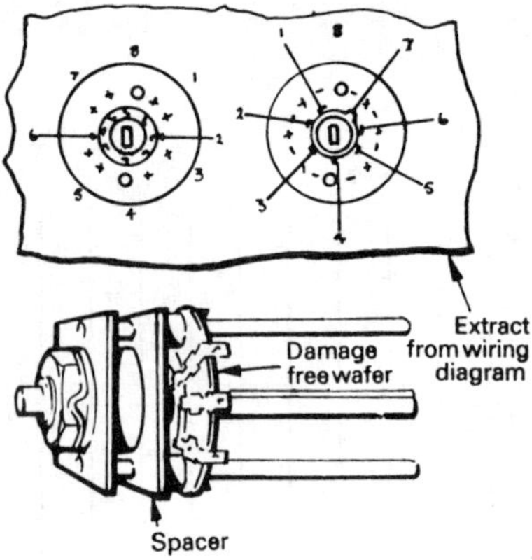

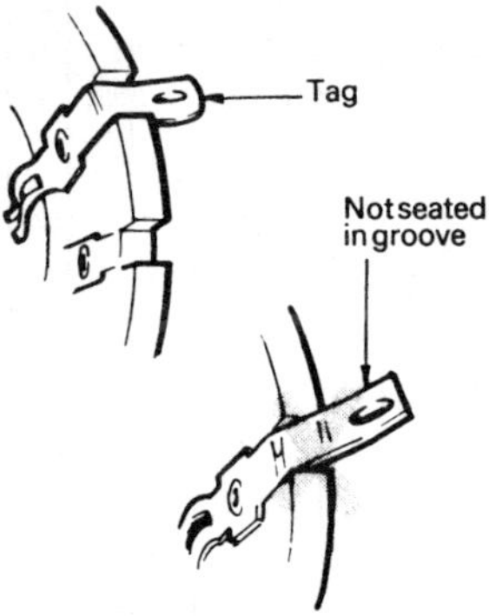

Ensure that the switch operates smoothly. It should move crisply and positively into each position. There must be no play on the contacts, and no loose contacts.

Inspect the assembly for damage.

Other switches

All switches must be correctly orientated and assembled securely. Switch mountings are sometimes keyed to ensure correct orientation.

Check the mechanical operation of the switch.

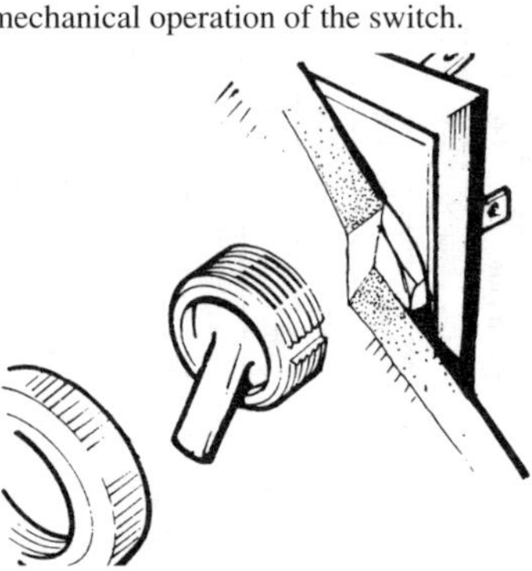

Potentiometers

Inspect for correct orientation and secure assembly. Ensure that locking devices are used where required.

Check that the spindle rotates freely. Do not operate the wiper hard against the end stops in either direction. Damage may result.

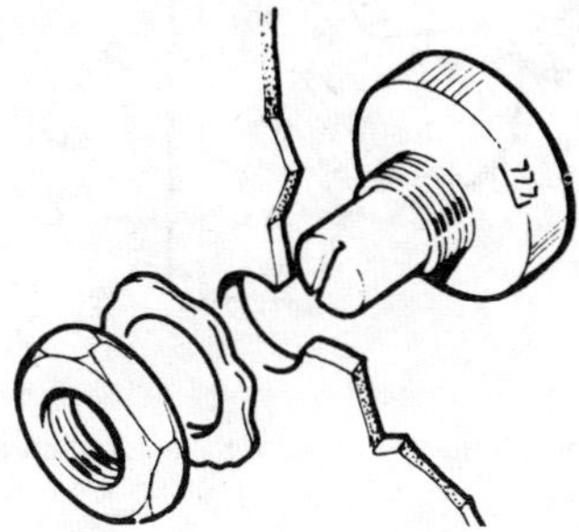

Plugs and sockets

Ensure that plugs and sockets are securely and squarely assembled. Check particularly for damaged pins.

Control knobs

Where required, knobs must be assembled to align with coding on the front panel.

Ensure that the control turns smoothly and that there is no play. Limits of play may be specified.

The knob must be securely fitted to the spindle.

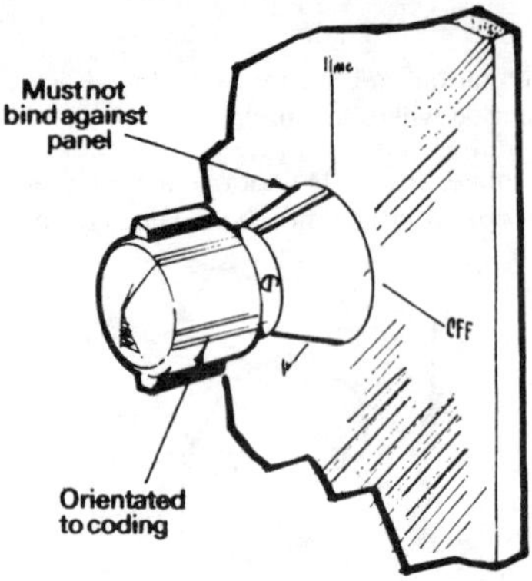

Fuse holders

Ensure that fuse holders are securely assembled and correctly orientated.

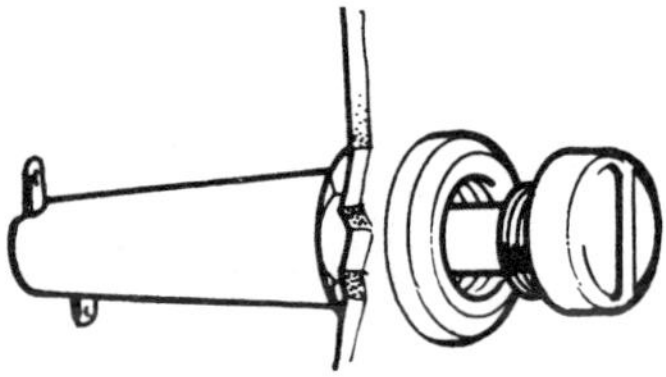

Electrolytic capacitors

Check polarity. Ensure that the shelf-life of the capacitors has not been exceeded.

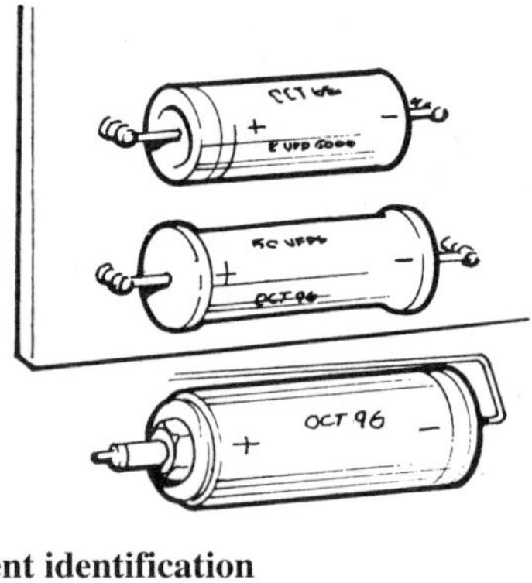

Component identification

Ensure that component identification has not been unnecessarily obscured.

Test instructions and drawings

Units are tested in accordance with instructions given in a *test schedule* which is prepared from a *test specification*. The latter is a document which details the tests necessary to ensure that a unit is working correctly, and the required results of those tests.

The test schedule gives step-by-step instructions on how to test a unit.

The schedule incorporates:

- a list of test equipment required
- a list of associated drawings and specifications to be obtained
- test instructions
- graphs, tables, waveform data
- illustrations to assist connecting and testing the equipment.

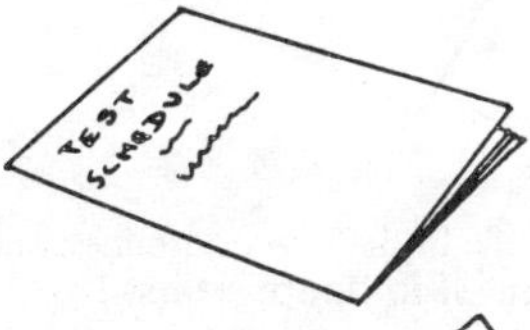

The schedule may incorporate a section to record test results or this may require a separate *test report form.*

Before starting a test using a test schedule, check the following details:

- the test schedule *is* for the unit concerned
- the test schedule is for the corresponding model number and/or drawing issue number of the unit
- all additional drawings and specifications are obtained and are to the correct issue number.

Failure to check these points may result in the unit being incorrectly connected to the test gear. *Damage to the unit and personal injury may result.*

In some cases it may be difficult to determine issue numbers, modifications states etc; **if in doubt → ask!**

During a test

Record all relevant details concerning the test in the spaces provided on the test report form, i.e. unit serial number, date of test, name of tester, etc.

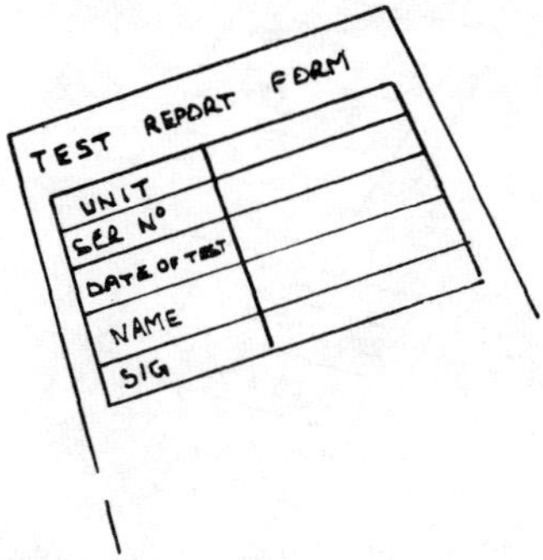

Read a test clause or paragraph completely and make sure you understand it before making any connections, adjusting controls or taking readings. Failure to do this may lead to a dangerous or costly mistake.

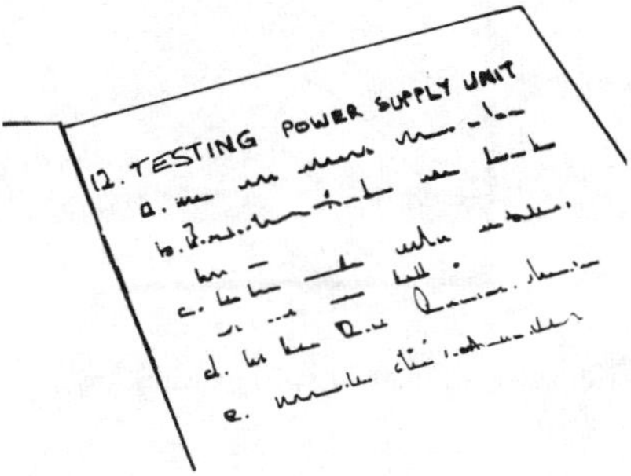

Always write legibly, preferably in CAPITAL LETTERS.

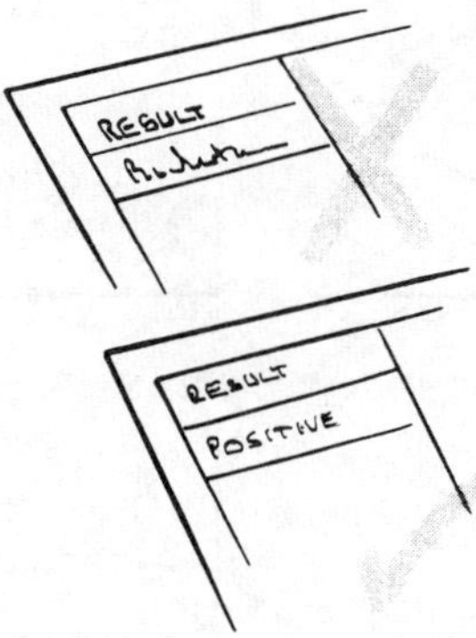

Always record results in decimal figures.

Remember — other people will have to read and interpret your completed test report forms and associated graphs.

Certain types of test result, i.e. frequency response, servo amplifier gain, drift rates, etc., may need to be presented in graphical form.

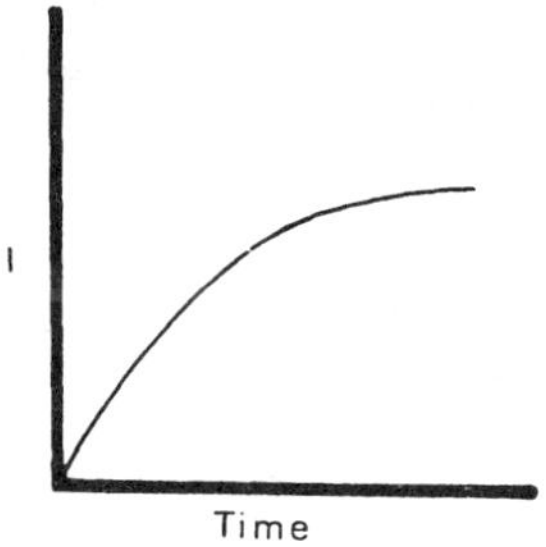

There are three main methods of recording the points on a graph:

- dot
- circled dot
- cross.

Do not mix the three methods of marking.

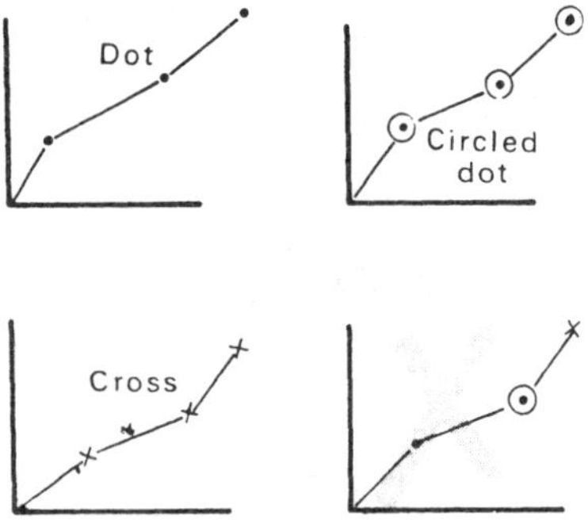

It may be necessary to place more than one curve on a graph. Use either colour coding or coded broken lines to differentiate between curves.

Coded broken lines

Do not use the same broken line coding for two curves on the same graph. This can be misleading.

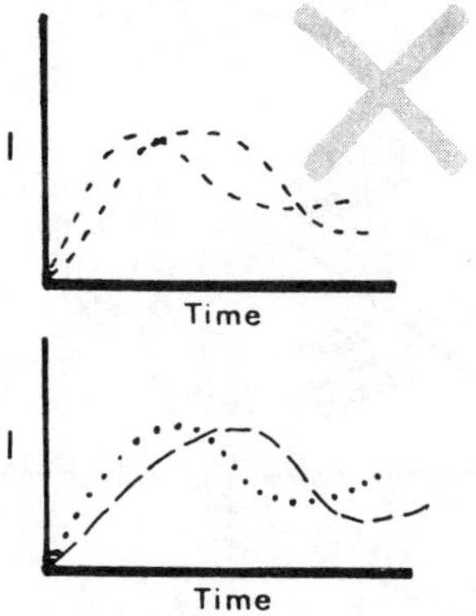

Make sure that you clearly indicate the coding method used in a prominent place on the graph.

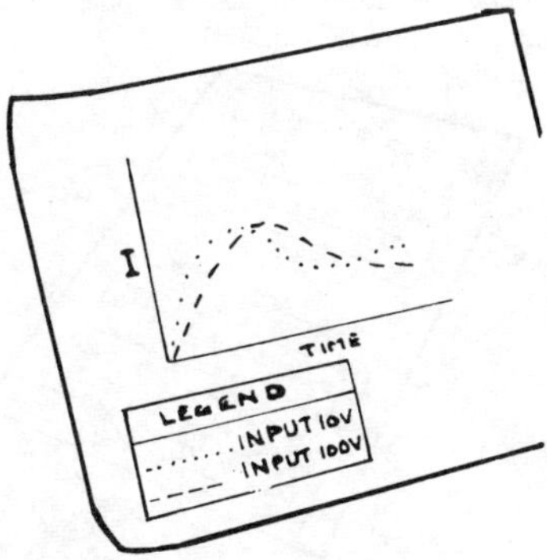

After testing

If the test results are acceptable, sign to this effect in the space provided.

The test report form will then be filed and the unit will be either stored or delivered to another department for further inspection or another stage of assembly.

If the test results are not acceptable sign to this effect in the space provided.

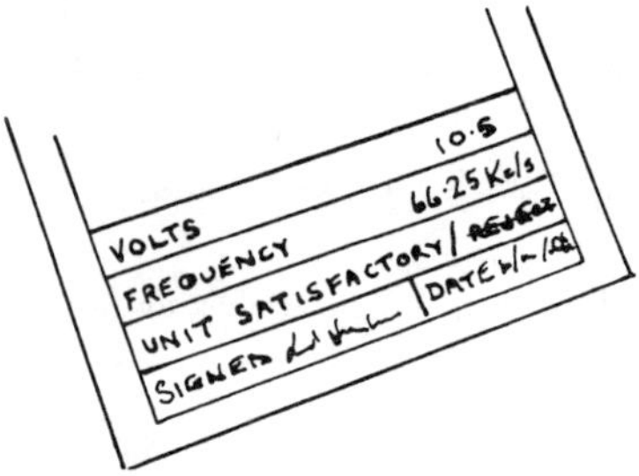

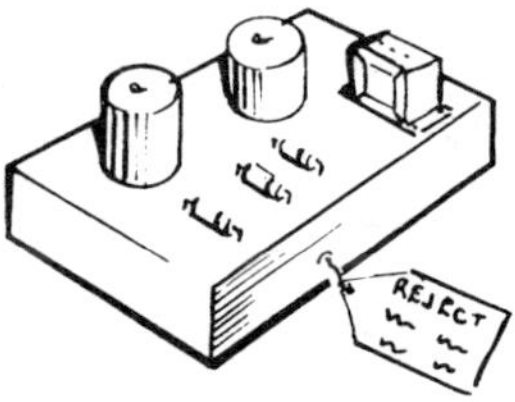

Attach a reject ticket to the unit, making sure that you fill in all required details.

Make out a *defect report form.* This will require the following information:

- name of unit
- serial number
- drawing number and issue
- modification state
- reason for rejection, with details
- additional details regarding the test equipment used, e.g. last calibration date, may also be required.

The completed defect report form, the test report form and the unit will then be sent to an engineering department for scrutiny.

Provide *all* the information required clearly and correctly. This will save time and money.

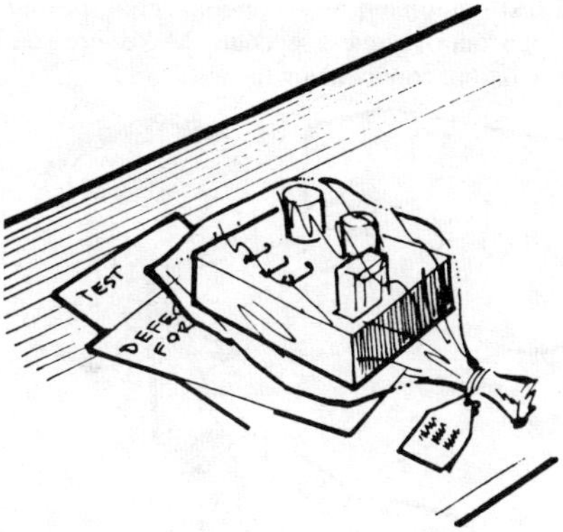

Connecting equipment

A list of test equipment and interconnecting wiring will be found at the front of the test schedule. Make sure you have all you require before commencing the tests.

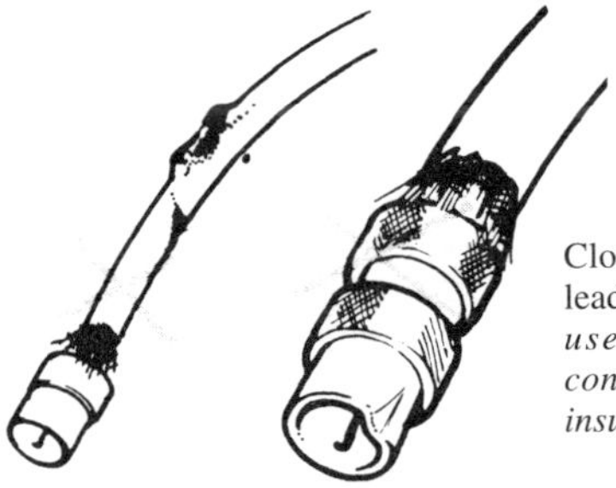

Closely examine all test leads for damage. *Do not use leads with faulty connectors or damaged insulation.*

The test schedule will provide instructions for connecting the unit under test to the test equipment. Read these instructions carefully and completely before you make any connections. *Make sure you understand before you act* — failure to do so may result in personal injury and/or damage to the equipment.

Quality

If a particular item of test equipment is not available, do not use alternative equipment without permission. Authority from an engineering department is usually required.

If permission is granted to use alternative equipment, details of the equipment must be entered on the test report form, i.e. name, type number, serial number, last calibration date.

Note. If you find the unit under test to be defective the above information will be vitally important.

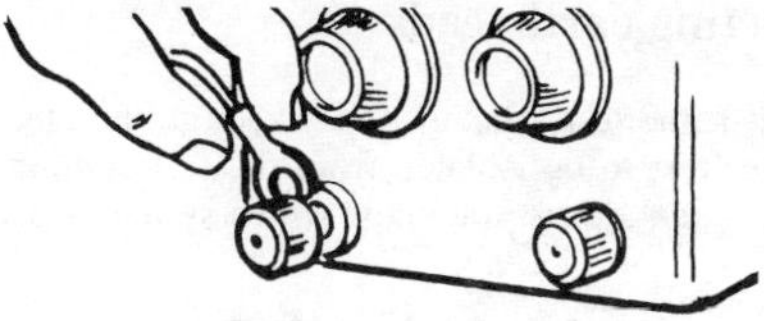

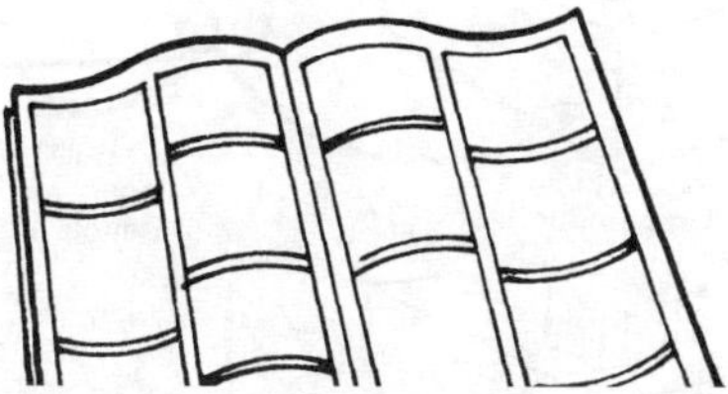

Safety

Before connection, make sure that all switches on the unit under test and the test equipment are off.

Switch off the mains supply to the test equipment and withdraw the plug, unless otherwise specified.

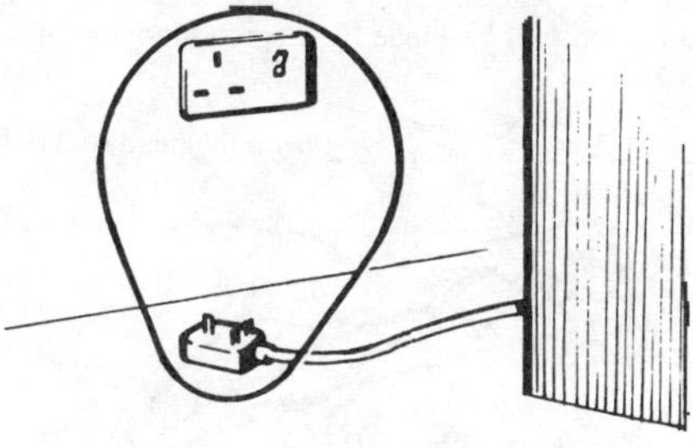

Connecting earth leads

Note that some test schedules may require the chassis of the unit under test to be isolated from the test equipment, some may not — make *sure* you know which system is used in your test.

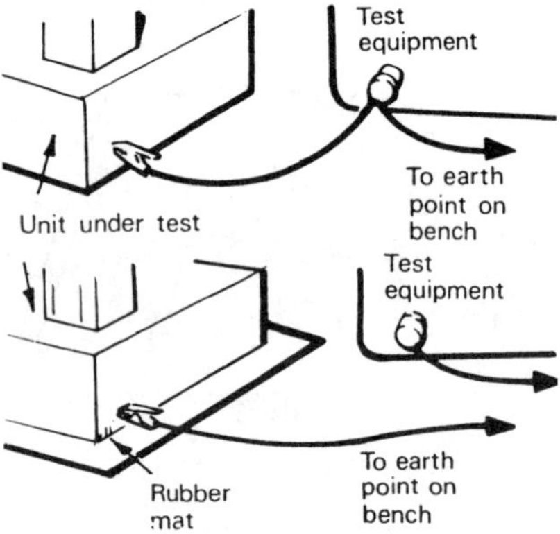

Chassis isolated

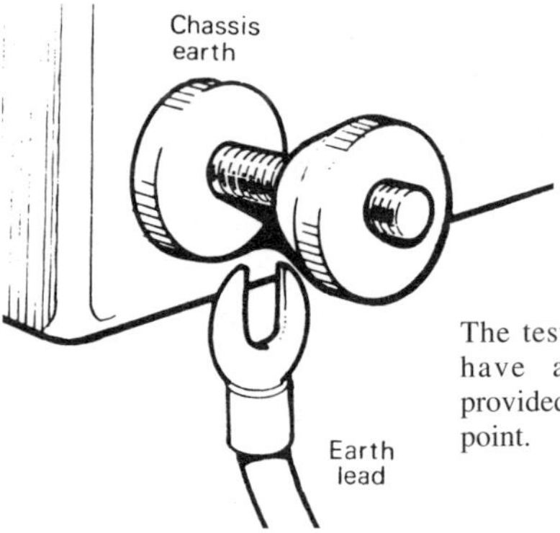

The test equipment will have an earth point provided, always use this point.

If the unit under test does not have a chassis earth point the earth connection will be made by a crocodile-type clip placed either:

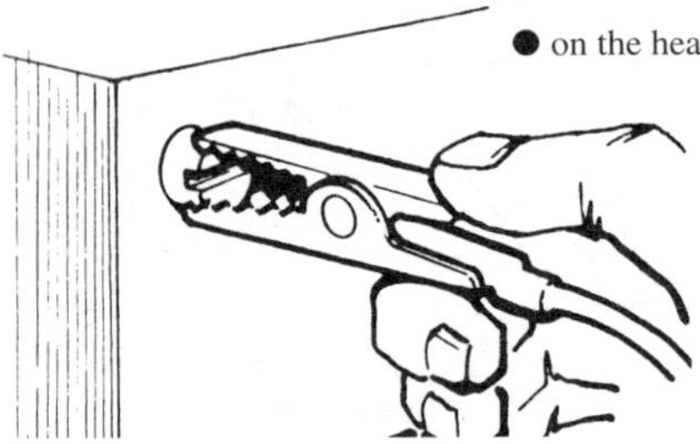

- on the head of a bolt, or

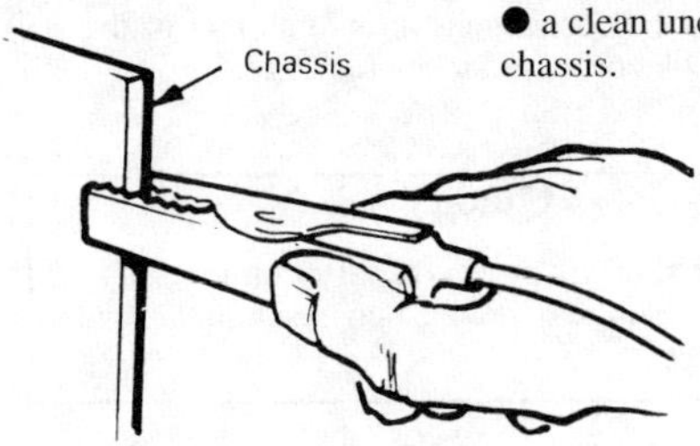

● a clean uncoated part of the chassis.

> **Quality**
>
> Do no scrape off paint or plating to make an earth connection.

Connecting power supply leads

Ensure that power plugs and sockets are correctly mated and that the retaining collar is screwed tight.

In certain cases, particularly with low-voltage d.c. supplies, you may be required to make individual connections. You should:

● use insulated wire

● observe the colour-code convention, *brown* for *live*, *blue* for *neutral* and, if necessary *green* and *yellow* for *earth*

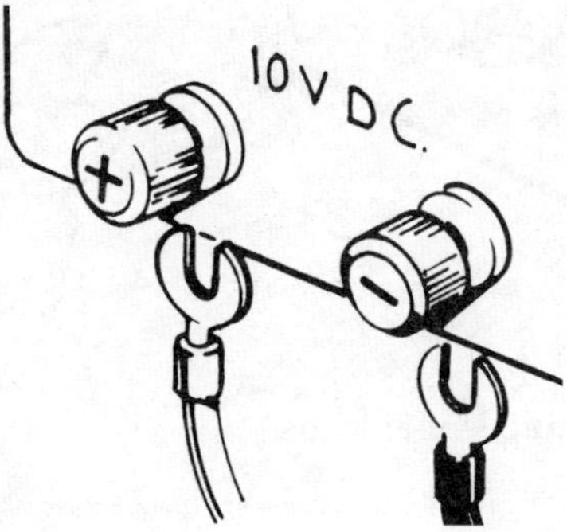

● make sure that the connections at the unit end of the lead are insulated and that the clips are not too large.

> **Quality**
>
> Double-check that all power supply connections are correctly made. Failure to do this may result in damage to the unit under test.

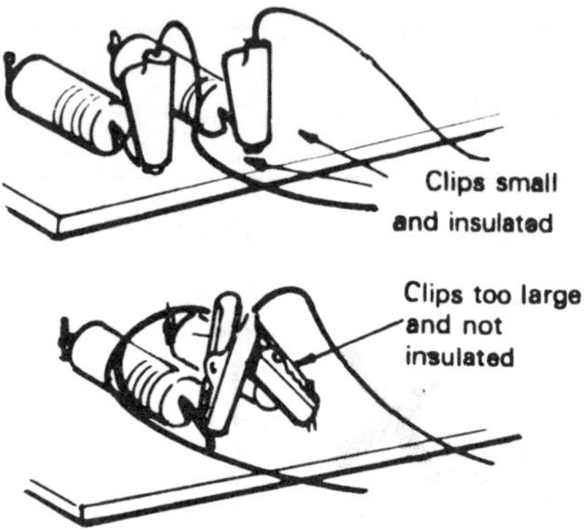

Connecting monitor leads

If individual permanent connections are required, to a multimeter or power meter for instance, observe the rules detailed above.

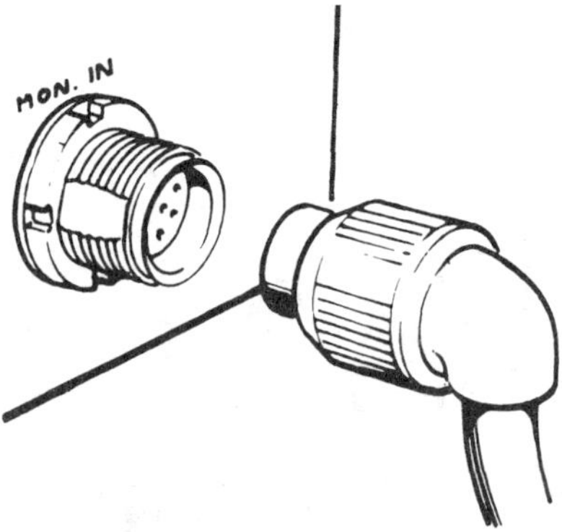

Connecting signal leads

Make sure that plugs and sockets are mated correctly.

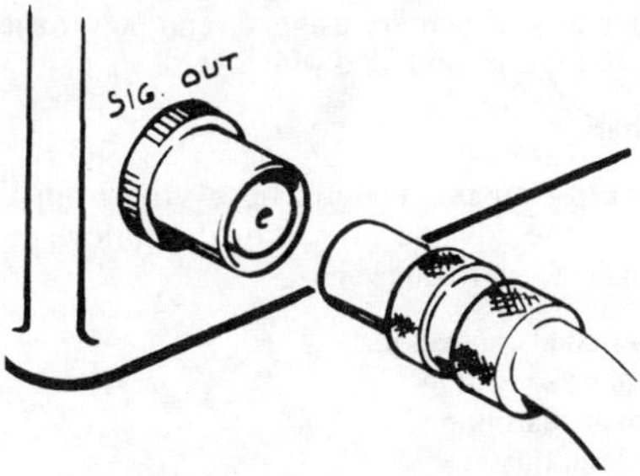

If the unit end of the lead is terminated in a crocodile-type clip, ensure that the screen is stripped well back and that the clip is insulated.

If you are required to make temporary connections during the schedule using multimeters etc., it may be advisable to use the leads supplied with the instrument. Make sure the leads are in good condition; change or repair where necessary.

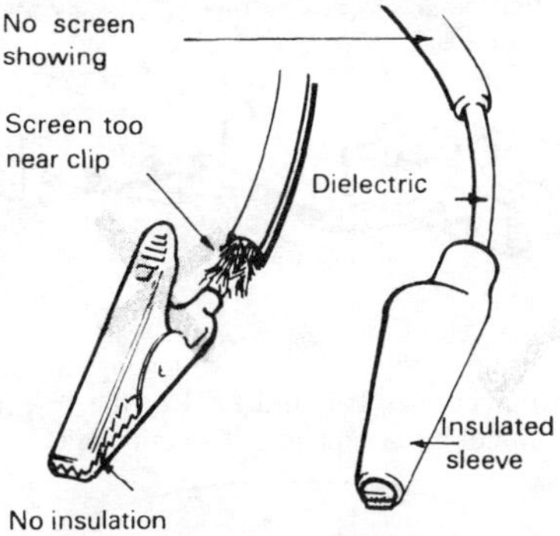

Take care, particularly when dealing with high frequencies, to keep leads, especially signal leads as short as possible.

It may be necessary to stand the unit under test on a suitable stand.

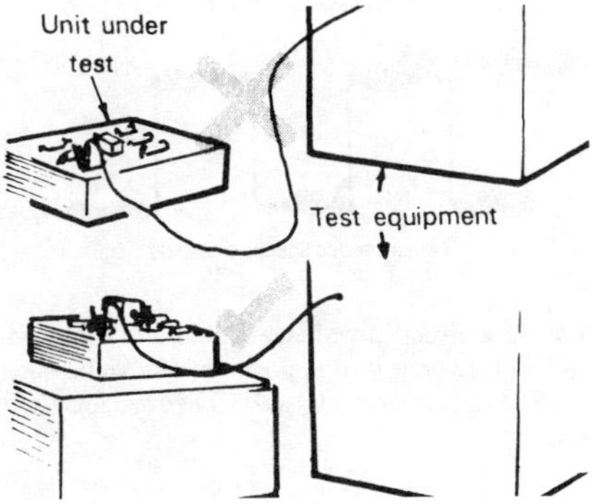

Keep leads tidy; if you are using several power supply leads, group them together and bind with tape.

Remember

When taking measurements involving high frequency equipment, the adverse affects of the following are more serious than at low frequencies:

- poorly mated connectors
- inadequate screening
- inadequate earthing
- general untidiness.

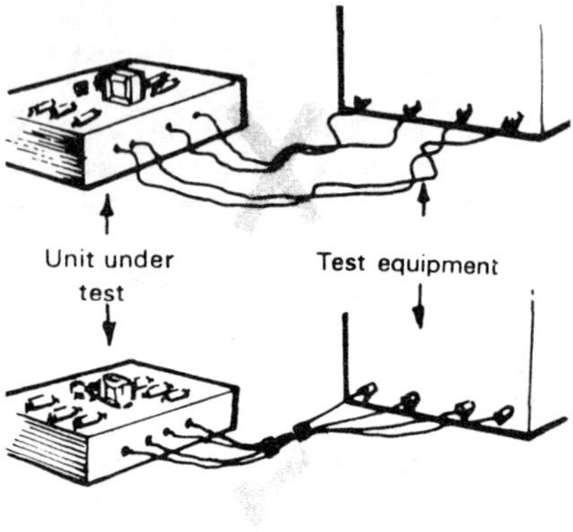

Inattention to those points may make testing difficult and/or the results inaccurate. It could lead to the rejection of a satisfactory unit or the acceptance of a faulty unit.

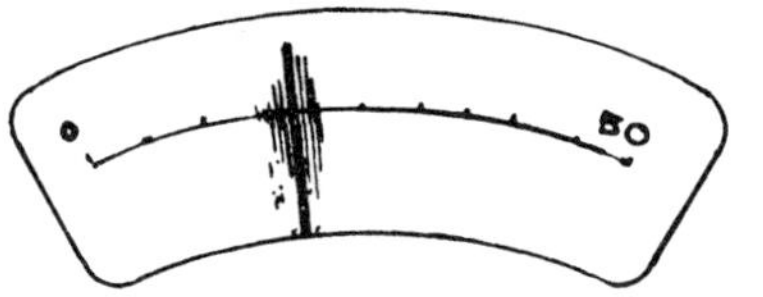

Erratic needle wander on meter

Noise on oscilloscope display

The use of the correct impedance signal and monitoring leads is dealt with elsewhere in this pocketbook. Make sure that the connections you have made are in accordance with those directions.

Conventions used on circuit diagrams

Apart from the British Standards circuit symbols, there are certain general rules to which circuit diagrams are laid out which assist the user in *reading* the diagram.

● the diagram is surrounded by a grid system which may be used to locate elements or components

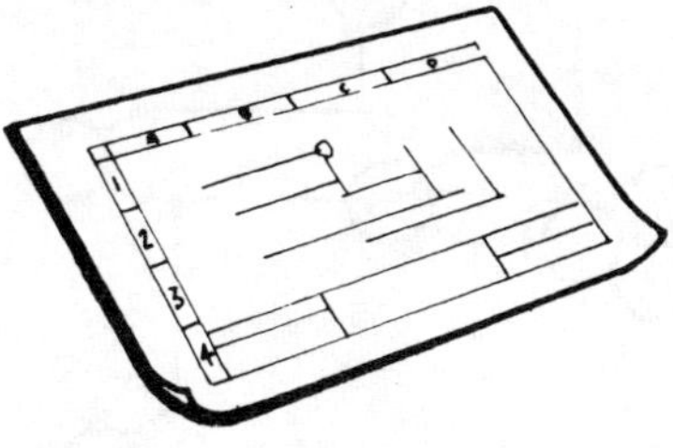

● the bottom of the diagram has an area to give precise details of the circuit, i.e.

drawing number
unit title
mod. state.

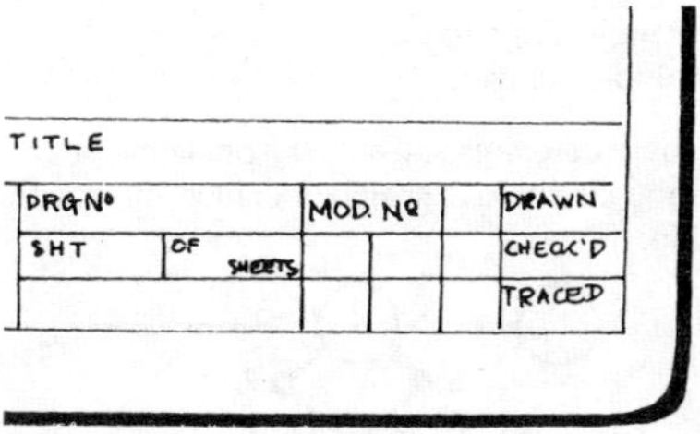

● by convention circuit diagrams signal flow is from left to right, i.e. input to the unit on the left, output from the unit above

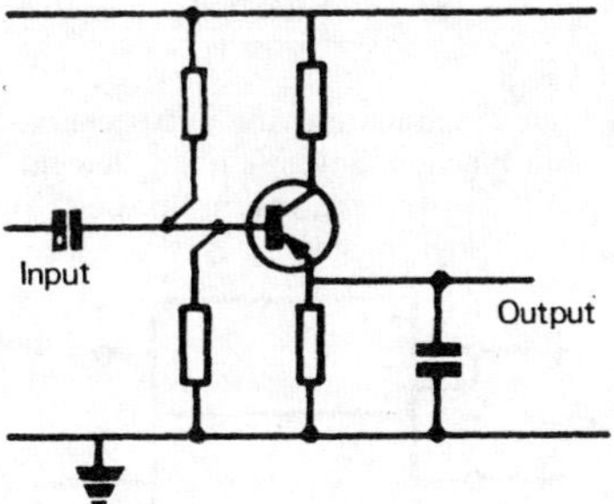

- on logic diagrams, data, or signal flow, is from left to right

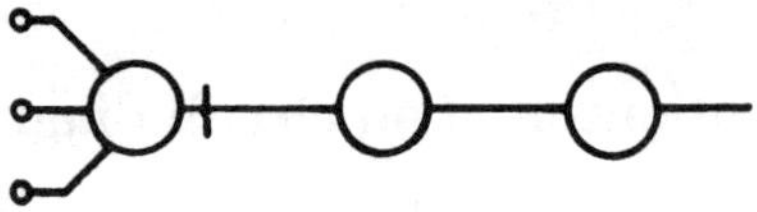

- control flow is from top to bottom.

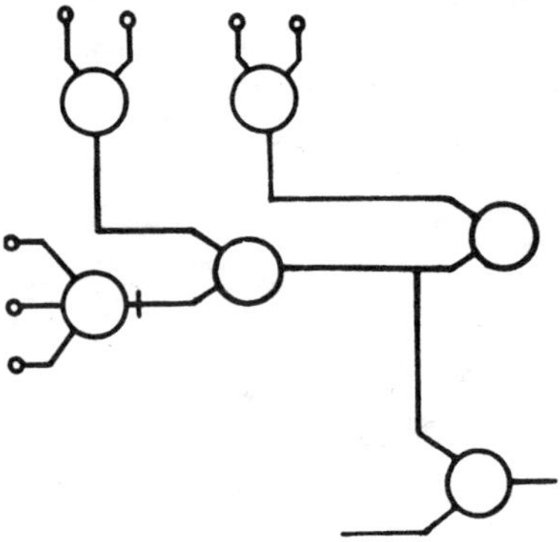

If you are testing a circuit for which there is a logic diagram remember that the diagram can be one of two types:

- hardware logic diagram
- functional logic diagram.

Hardware logic diagrams show every pin connection and logic element in the circuit and provide detailed information on the circuit action.

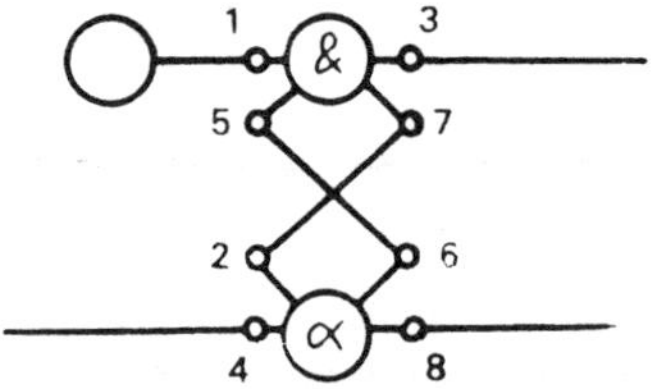

Functional logic diagrams provide only general information about the circuit function. Several circuit elements, or in some cases complete units, are often lumped together and represented by one logic symbol.

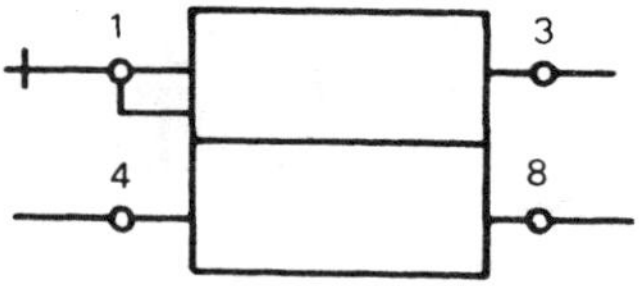

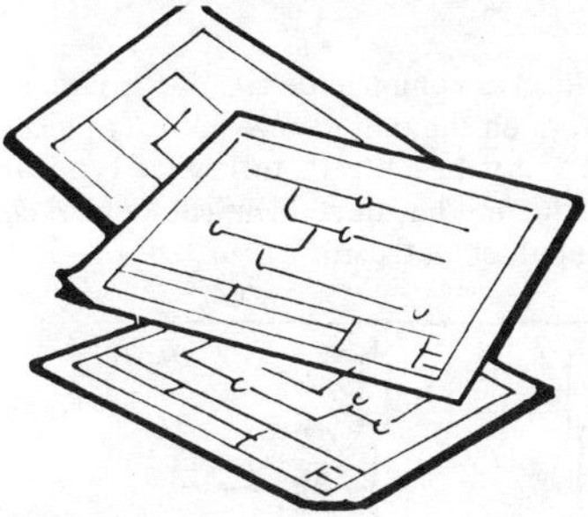

If you use a logic diagram, make sure which type it is, usually it will be marked. **If it is not marked and you are in doubt → Ask!**

Connecting from information given in diagrams

The information may be presented in two ways.

Block diagrams

This type of diagram will give details of connections to be made but will not give circuit details. This may be adequate for simple units with no adjustment points.

It is usually possible to determine the location of a particular plug or socket on the unit from a block diagram.

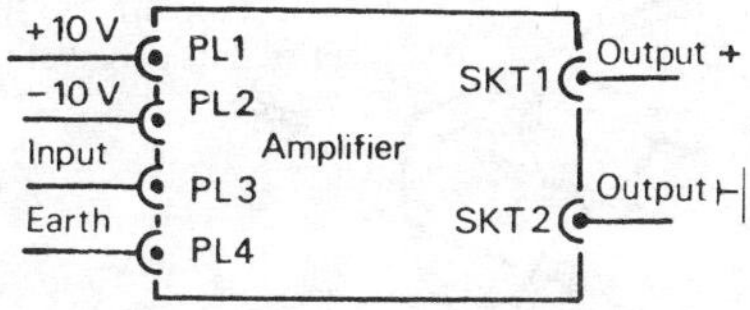

The symbols used on block diagrams conform to the recommendations in BS 3939.

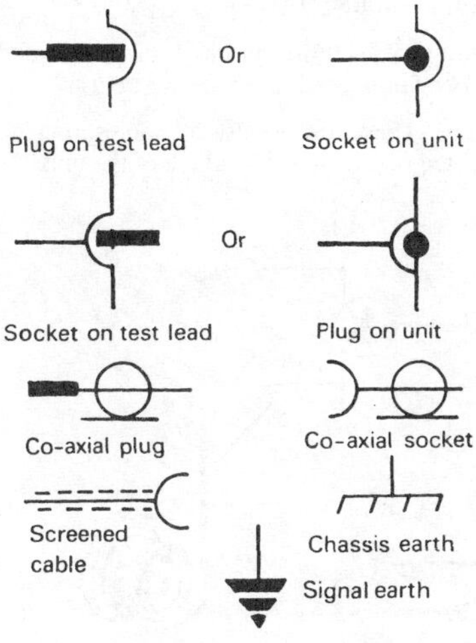

Circuit diagrams

This type of diagram will give complete details of the circuit. The position of connectors on the unit cannot be determined. In some cases you may not be able to tell what type of connector is used, i.e. if the unit has three identical, unmarked plugs on it — which is input, output, earth?

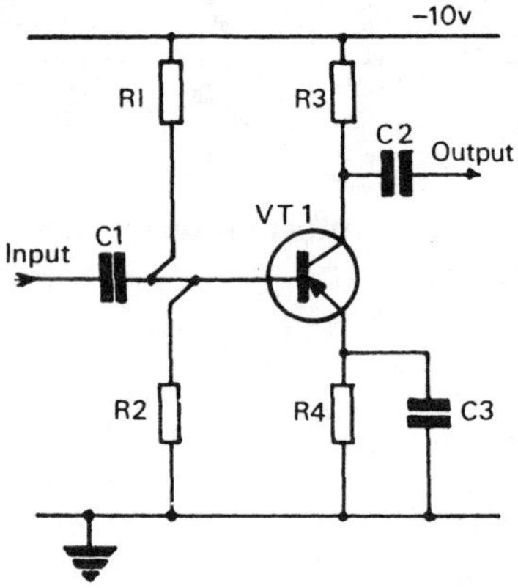

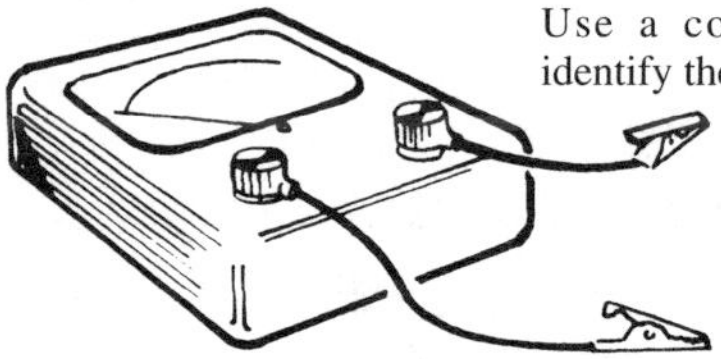

Use a continuity tester to identify the connectors.

The symbols used in circuit diagrams conform to the recommendations in BS 3939.

Symbols additional to those on block diagrams, and which a tester should be familiar with, are shown below.

Monitor points. These are usually for monitoring waveforms. They are located on the outside surface of the unit.

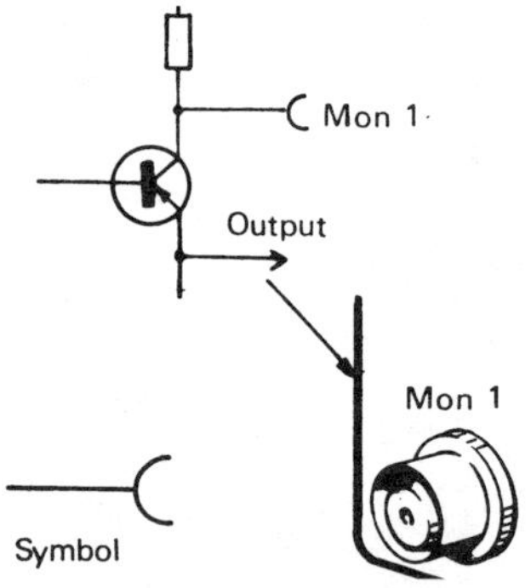

Test points. These are usually for monitoring voltages or waveforms but can also be used for a point to inject a voltage or signal for testing purposes.

Test points are usually located inside a unit in the form of a terminal post on a sub unit.

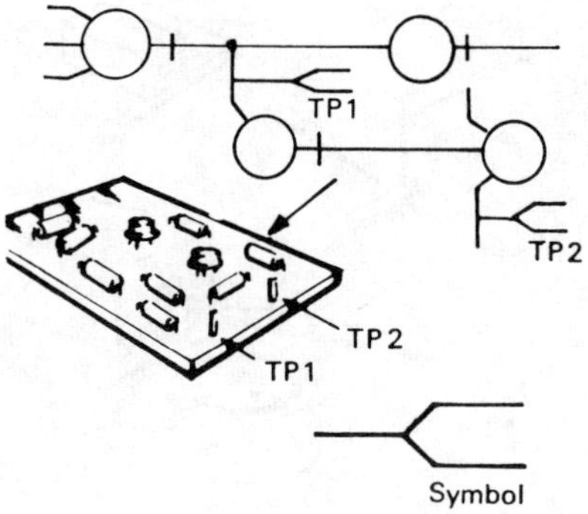

Variable resistors and capacitors. These components can be of two types:

● continually-in-use variable, i.e. amplitude and frequency controls on transmitters. These are usually located on the unit front panel and have a control knob

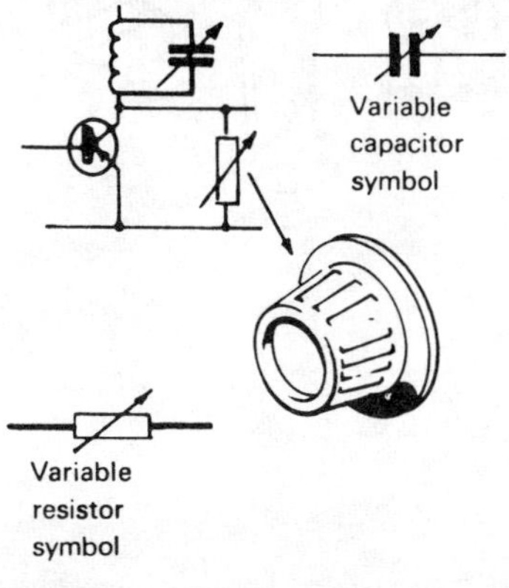

● preset variable, i.e. i.f. trimmers and input level settings.

These are usually located inside the unit, adjacent to their associated components.

Adjustment is by means of a screwdriver. Once these controls have been set to their correct position, they are not touched again, in most cases they are locked with varnish.

In some cases you may need to use a non-metallic screwdriver, e.g. on i.f. coil cores. **Before you act → make sure you have the correct tool.**

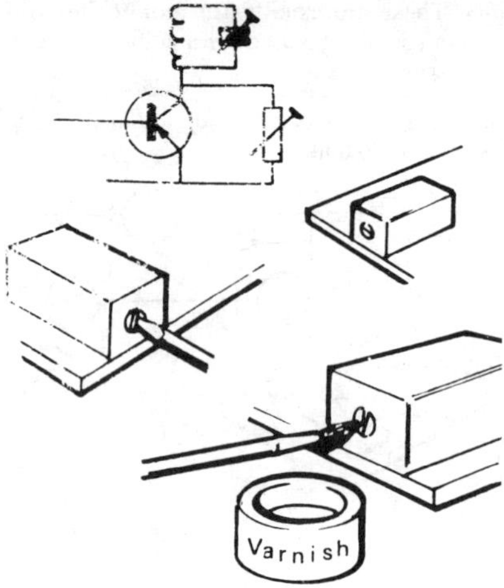

When adjusting pre-set controls, use the correct size screwdriver otherwise you may cause damage.

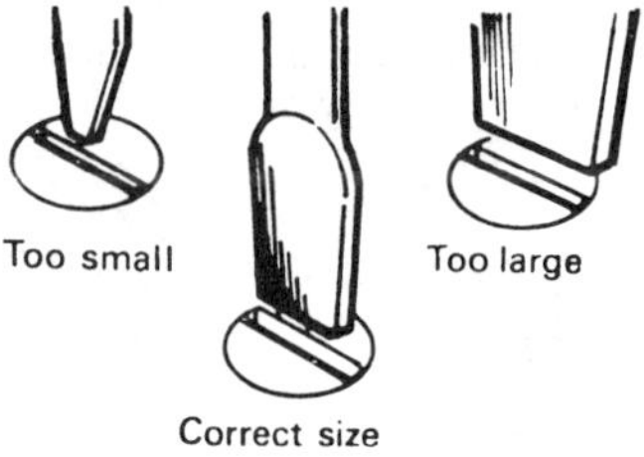

Special test jigs

Test jigs are used for the production testing of quantities of board assemblies or small modules.

There are three basic types of special test jig, with many variations:

- edge connector type
- probe type
- connector on flying lead type.

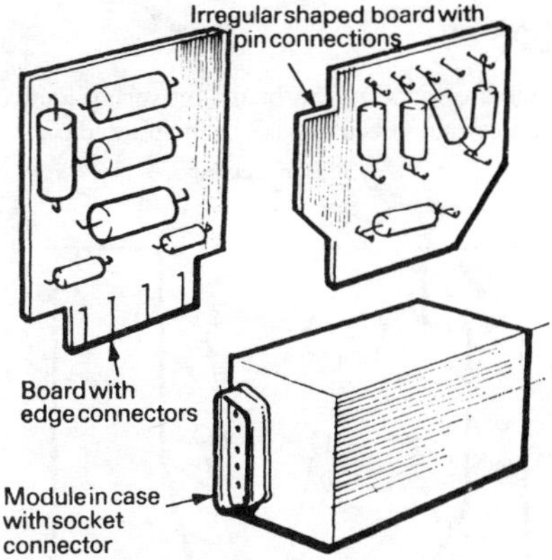

Edge connector jig

This type of jig has an edge connector socket mounted on the front panel, guide rails to hold the board in position and a control panel.

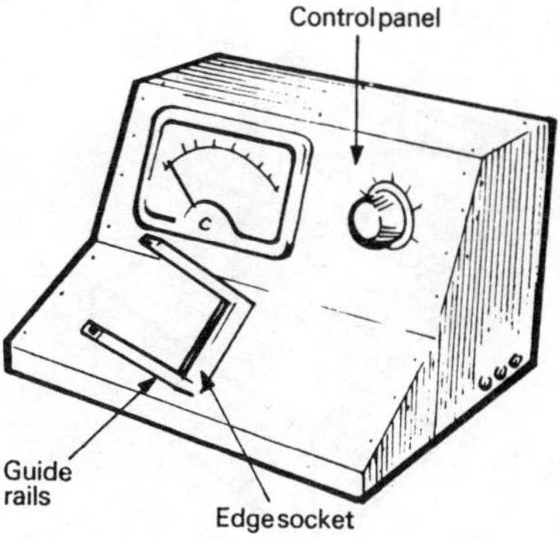

Probe type

This type of jig is for testing boards of irregular shape and/or without edge connectors.

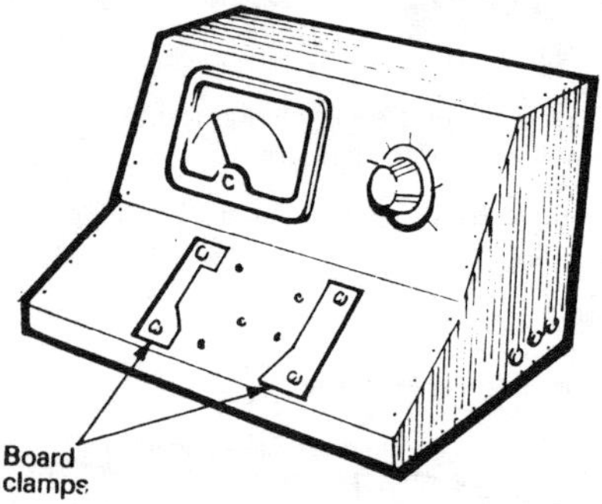

Connections are made to the board by spring-loaded probes which touch certain pins and parts of the printed track.

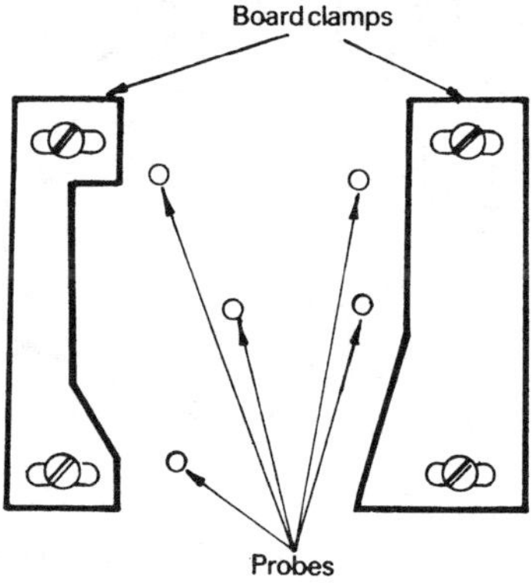

A probe type can be used to carry out insulation and continuity tests on a frame assembly which has only the cableform installed.

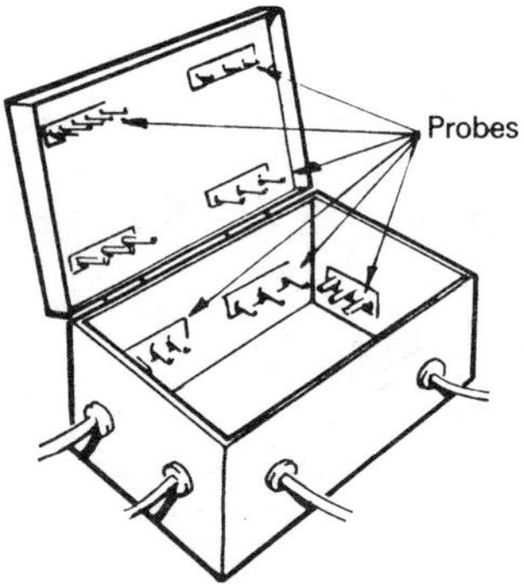

The continuity and insulation tests may be carried out by a programmed control box.

Connector on a flying lead

These are usually co-axial or screened leads and are used in addition to either the edge connector or probes.

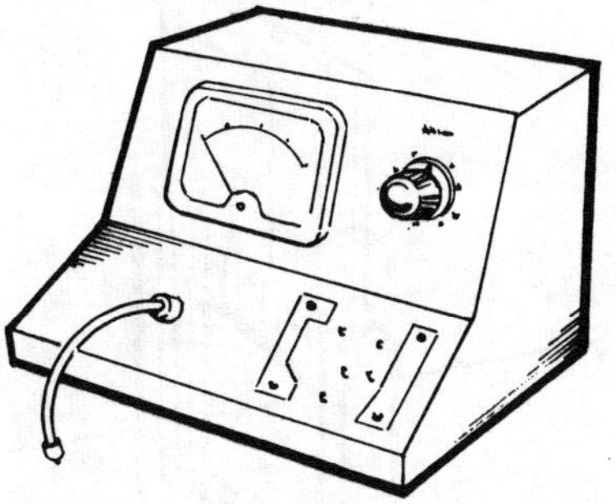

Most test jigs have plugs and sockets mounted on the sides and rear to connect additional test equipment.

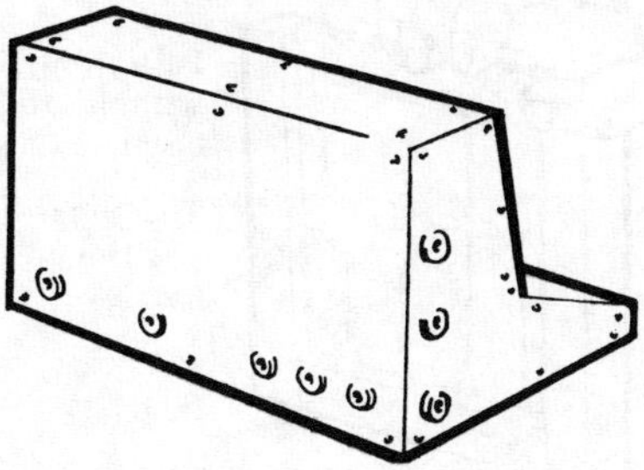

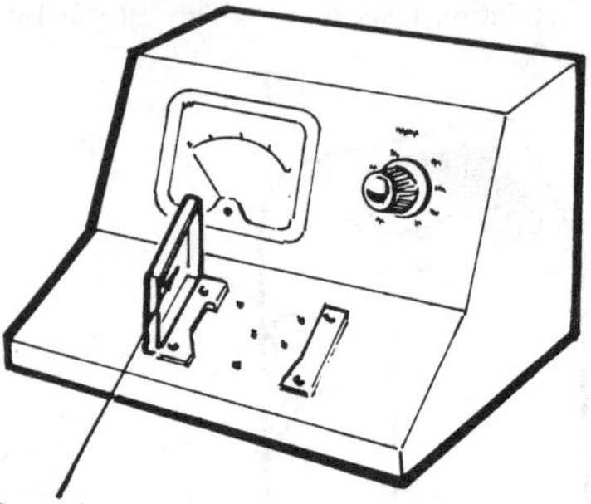

Any jig type may be fitted with a screening can.

Using the edge connector jig

Make sure you have read and understood the test instructions and that you have the test jig and all associated equipment at hand.

Ensure that all switches are in the off position.

Connect the equipment in accordance with the test instructions.

Locate the sides of the board into guide rails.

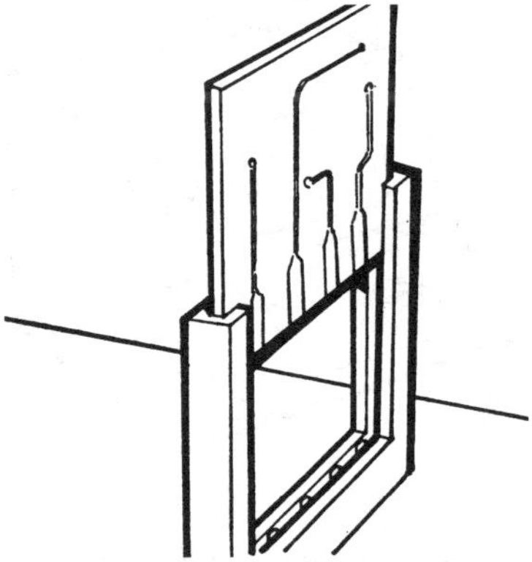

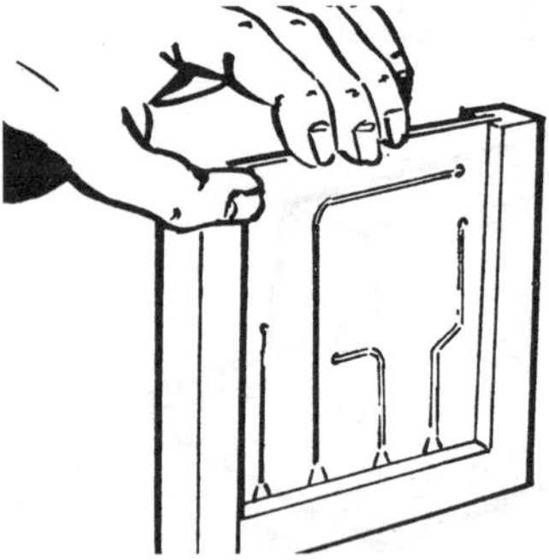

Push firmly into place without *rocking* the board.

The board is now positioned ready for test.

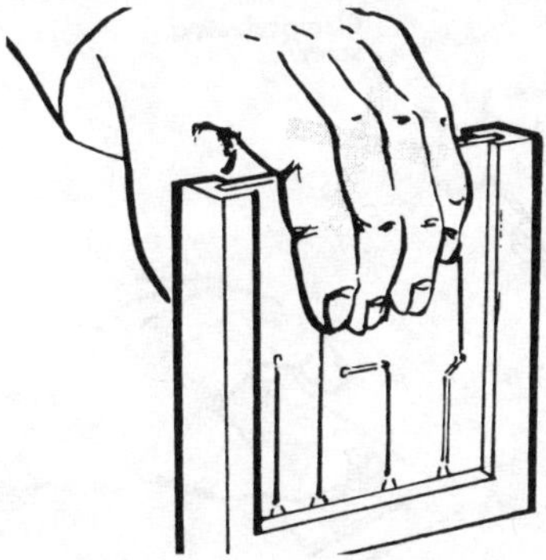

When testing make sure that you do not touch the printed track or the board components.

Using the probe type jig

Make sure you have read and understood the test instructions and that you have the test jig and associated test equipment in hand.

Ensure that all switches are in the off position.

Connect the equipment in accordance with the test instructions.

Make sure that the springs in the probes are working correctly.

Probe should move in and spring out freely

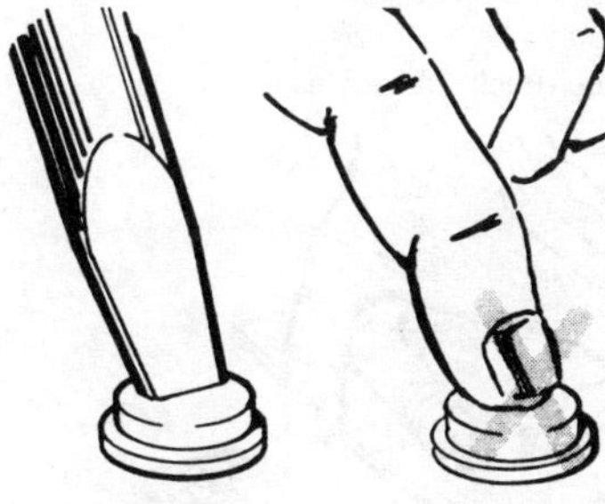

Use a screwdriver or prod, *not* your finger.

Release the clamp retaining screws and pull the board clamps outwards.

Clean the board contact points by the approved method.

Care must be taken.

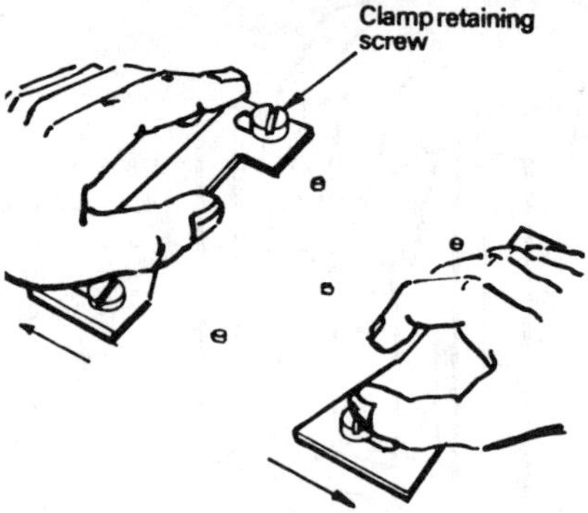

Ensure that none of the board wiring will prevent the probe from making proper contact. Care must be taken.

If necessary *dress* the wiring to provide clearance.

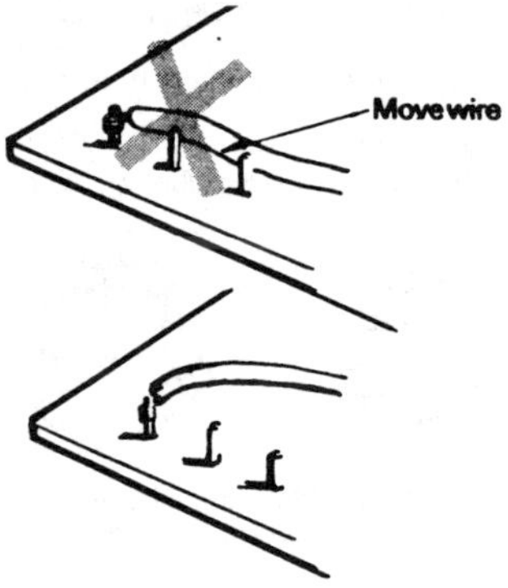

Place the board in position (track side down).

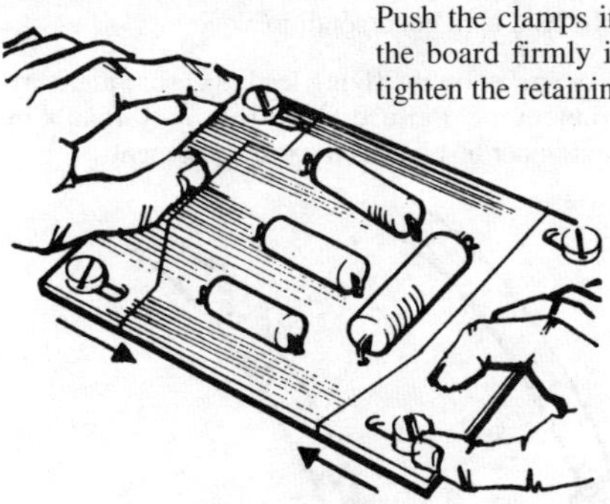

Push the clamps inwards to hold the board firmly in position and tighten the retaining screws.

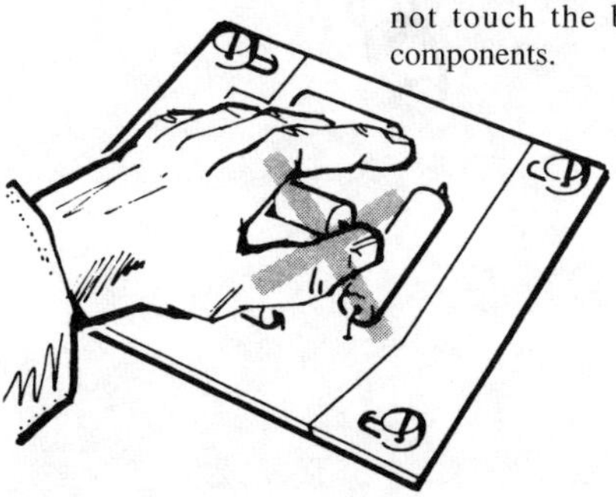

When testing, make sure you do not touch the board track or components.

Using the flying lead type jig

Make sure you have read and understood the test instructions and that you have a test jig and all associated equipment at hand.

Ensure that all switches are in the off position.

Connect the equipment in accordance with the test instructions.

Make sure that the lead is in good condition.

Inspect the plug or socket on the flying lead to ensure there are no damaged connectors. Failure to do this may result in damage to the unit under test and/or the test equipment.

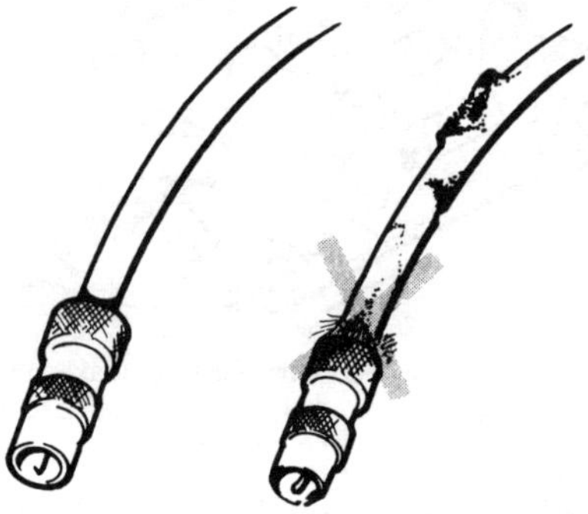

Measurands in test and inspection

Common multiples and sub-multiples of each unit are given.

Current

The unit of current is the Ampere (A).

1 Milliampere (mA)

$$= \text{one thousandth of an ampere}$$

$$= \frac{1}{1{,}000}\ \text{A} = 10^{-3}\ \text{A}$$

1 Microampere (µA)

$$= \text{one millionth of an ampere}$$

$$= \frac{1}{1{,}000{,}000}\ \text{A} = 10^{-6}\ \text{A} = 10^{-3}\ \text{mA}$$

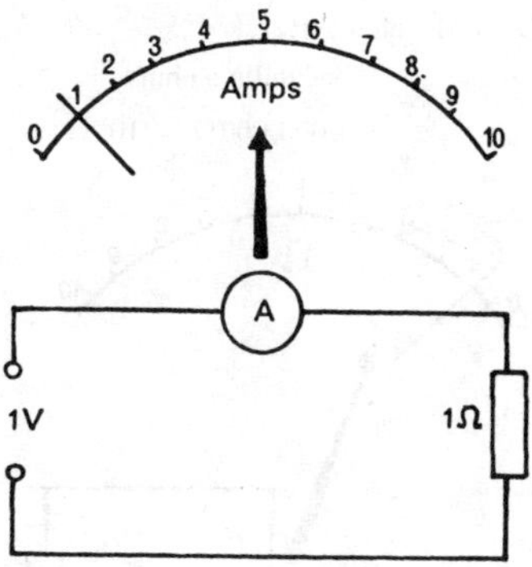

Potential difference

The unit of potential difference is the Volt (V).

1 Millivolt (mV)

$$= \text{one thousandth of a volt}$$

$$= \frac{1}{1{,}000}\ \text{V} = 10^{-3}\ \text{V}$$

1 Microvolt (µV)

$$= \text{one millionth of a volt}$$

$$= \frac{1}{1{,}000{,}000}\ \text{V} = 10^{-6}\ \text{V}$$

$$= 10^{-3}\ \text{mV}$$

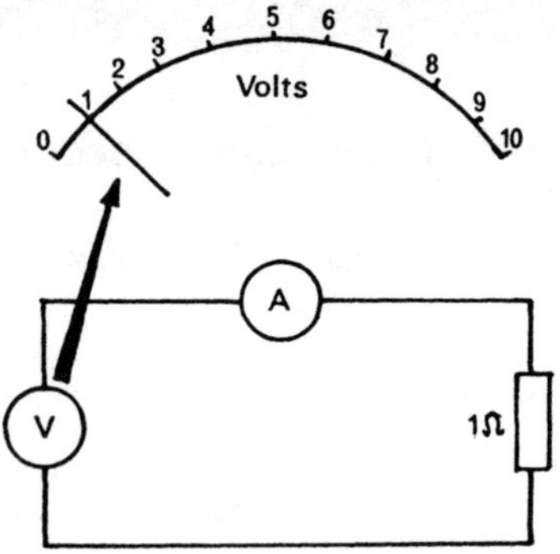

Resistance

The unit of electrical resistance is the Ohm (Ω).

1 Kilohm (kΩ)

$$= \text{one thousand ohms}$$

$$= 1{,}000\,\Omega = 10^3\,\Omega$$

1 Megohm (MΩ)

$$= \text{one million ohms}$$

$$= 1{,}000{,}000\,\Omega = 10^6\,\Omega$$

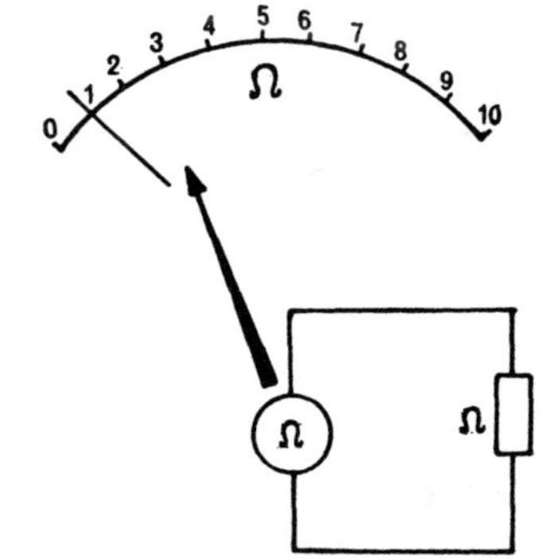

In some circuit diagrams the resistor symbol is accompanied simply by a number, this is its value in ohms. Resistor values of the order of kilohms and megohms bear the suffix k and M respectively.

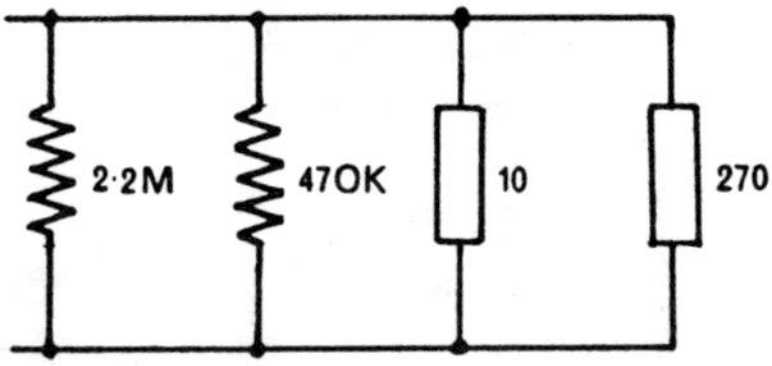

Capacitance

The unit of electrical capacitance is the Farad (F). This unit is impractically large.

1 Microfarad (μF)

$$= \text{one millionth of a farad}$$

$$= \frac{1}{1{,}000{,}000}\ \text{F} = 10^{-6}\ \text{F}$$

1 Nanofarad (nF)

$$= \text{one thousand millionth of a farad}$$

$$= \frac{1}{1{,}000{,}000{,}000} = 10^{-9}\ \text{F} = 10^{-3}\ \mu\text{F}$$

1 Picofarad (pF)

$$= \text{one million millionth of a farad}$$

$$= \frac{1}{1{,}000{,}000{,}000{,}000} = 10^{-12}\ \text{F} = 10^{-6}\ \mu\text{F}$$

Values of the order of picofarads bear the suffix p, usually all other values are in microfarads.

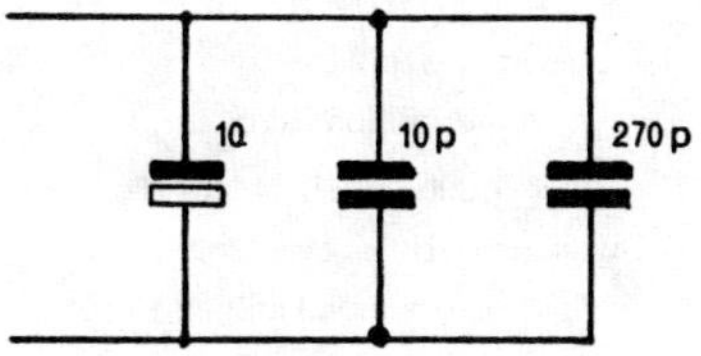

Power

The unit of electrical power is the Watt.

1 Kilowatt (kW)

$$= \text{one thousand watts}$$

$$= 1{,}000\ \text{W} = 10^{3}\ \text{W}$$

1 Milliwatt (mW)

$$= \text{one thousandth of a watt}$$

$$= \frac{1}{1{,}000}\ \text{W} = 10^{-3}\ \text{W}$$

1 Microwatt (μW)

$$= \text{one millionth of a watt}$$

$$= \frac{1}{1{,}000{,}000}\ \text{W} = 10^{-6}\ \text{W}$$

$$= 10^{-3}\ \text{mW}$$

Inductance

The unit of inductance is the Henry (H).

1 Millihenry (mH)

$$= \text{one thousandth of a Henry}$$

$$= \frac{1}{1{,}000}\ \text{H} = 10^{-3}\ \text{H}$$

1 Microhenry (μH)

$$= \text{one millionth of a Henry}$$

$$= \frac{1}{1{,}000{,}000}\ \text{H} = 10^{-6}\ \text{H}$$

Frequency

The unit of frequency is the Hertz (Hz).

This name has replaced cycles per second (c/s).

$$1\ \text{Hz} = 1\ \text{c/s}.$$

Many old instruments are calibrated in c/s and not Hz.

1 Kilohertz (kHz)

$$= \text{one thousand Hertz}$$

$$= 1{,}000\ \text{Hz} = 10^{3}\ \text{Hz}$$

1 Megahertz (MHz)

$$= \text{one million Hertz}$$

$$= 1{,}000{,}000\ \text{Hz} = 10^{6}\ \text{Hz}$$

1 Gigahertz (GHz)

$$= \text{one thousand million Hertz}$$

$$= 1{,}000{,}000{,}000\ \text{Hz} = 10^{9}\ \text{Hz}$$

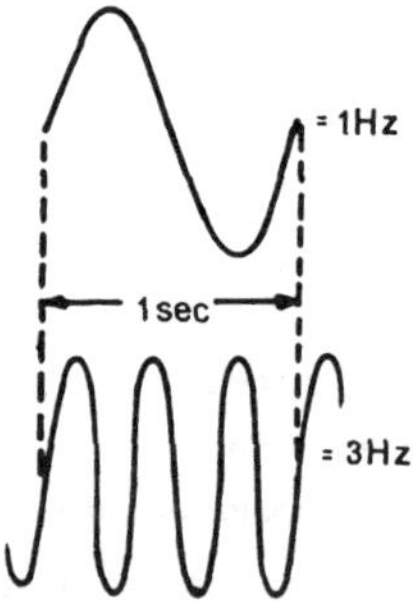

Time and period

The unit of time is the second (s).

1 Millisecond (ms)

$$= \text{one thousandth of a second}$$

$$= \frac{1}{1{,}000}\ \text{s} = 10^{-3}\text{s}$$

1 Microsecond (μs)

$$= \text{one millionth of a second}$$

$$= \frac{1}{1{,}000{,}000}\ \text{s} = 10^{-6}\text{s}$$

1 Nanosecond (ns)

$$= \text{one thousand millionth of a second}$$

$$= \frac{1}{1{,}000{,}000{,}000}\ \text{s} = 10^{-9}\text{s}$$

The period of a waveform is the time between successive cycles and is related to frequency by:

$$\text{Period (s)} = \frac{1}{\text{frequency}}\ \text{(Hz)}$$

e.g. the period of a 1 kHz frequency signal is

$$= \frac{1}{10^3}\ \text{s} = 10^{-3}\ \text{s} = 1\,\text{ms}$$

e.g. the frequency of a signal of period 2.5 μs

$$= \frac{1}{2.5 \times 10^{-6}}\ \text{Hz} = \frac{10^6}{2.5}\ \text{Hz}$$

$$= 4 \times 10^5\ \text{Hz} = 0.4\,\text{MHz}$$

$$= 400\,\text{kHz}$$

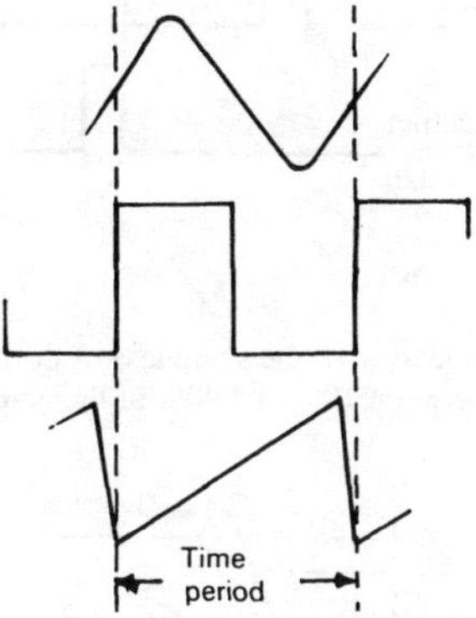

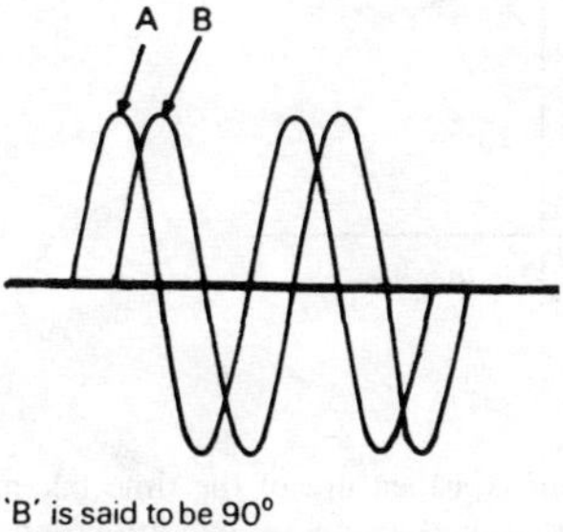

'B' is said to be 90° lagging in phase with respect to 'A'

Phase

Phase comparisons are made between waveforms with the same frequency.

The phase difference between two waveforms is expressed as the angle by which one wave is ahead of, or behind, the other.

Pulse repetition frequency (p.r.f.)

This term defines the number of pulses occurring in one second.

Pulse width

This term defines the width of a pulse in seconds.

Pulse delay

This term defines the delay (in seconds) of a pulse through a circuit. The leading edge of the pulse leaves the circuit in this example 5 ms after the leading edge has entered the circuit.

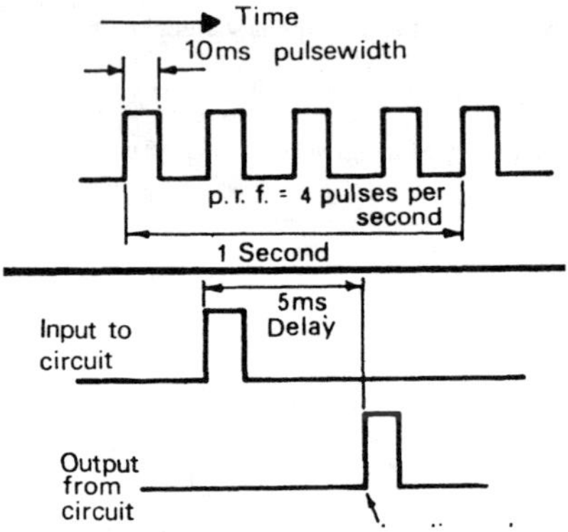

Rise time

Rise time is a measure of the sharpness of the edge of a pulse. It is the time between 10% and 90% of the amplitude ignoring overshoot.

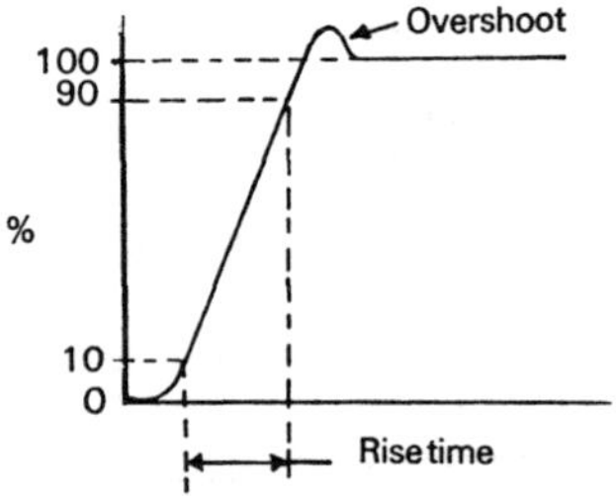

Time constant

The time constant of a circuit is a measure of the time taken for a circuit to respond to an applied change in condition, e.g. for a capacitor to change its voltage.

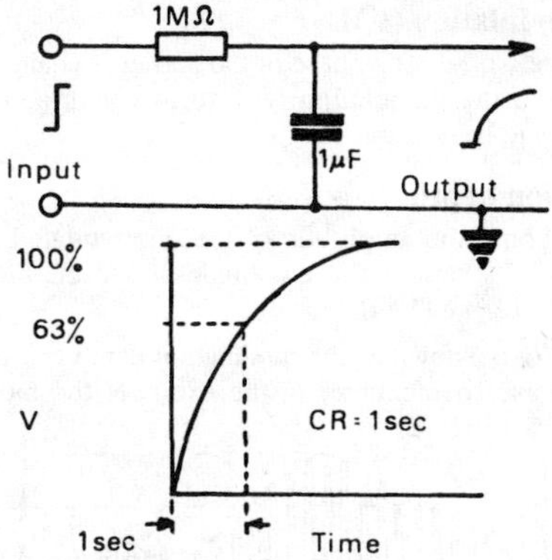

Modulation

If two waveforms of dissimilar frequencies and/or amplitudes are combined in a suitable electronic circuit, one frequency becomes modulated by the other.

Three types of modulation are commonly used:

Amplitude modulation (AM)

The instantaneous amplitude of the carrier wave is changed by the amplitude of the modulating waveform which contains the information to be transmitted.

Frequency modulation (FM)

The instantaneous frequency of the carrier wave is changed by the amplitude of the modulating waveform which contains the information to be transmitted.

Modulating waveform

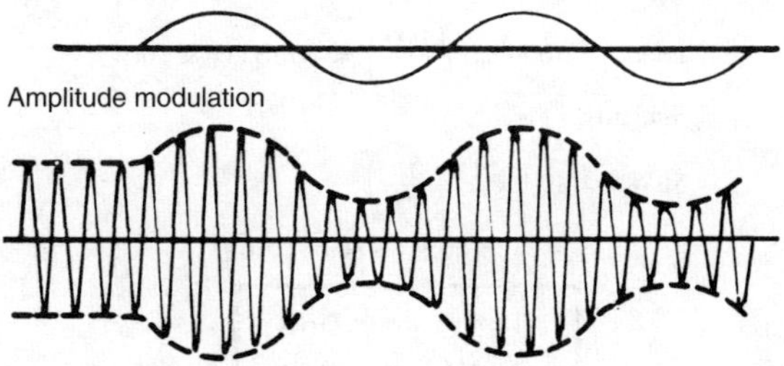

Frequency modulation

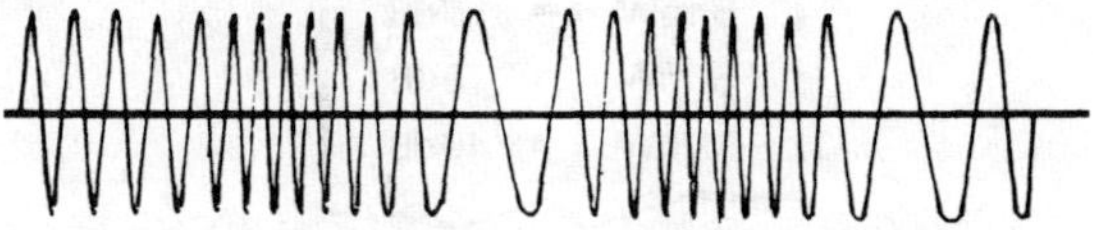

Phase modulation (PM)

Not so widely used. The phase of the carrier is changed by the amplitude of the modulating waveform which contains the information to be transmitted.

Modulation depth

In the diagram, the amplitude of the unmodulated signal is given by A. Its variation in amplitude due to the modulating signal is represented by B.

The ratio $^{B}/_{A}$ is known as the modulation depth and is usually expressed as a percentage. In the example the modulation depth is 50%.

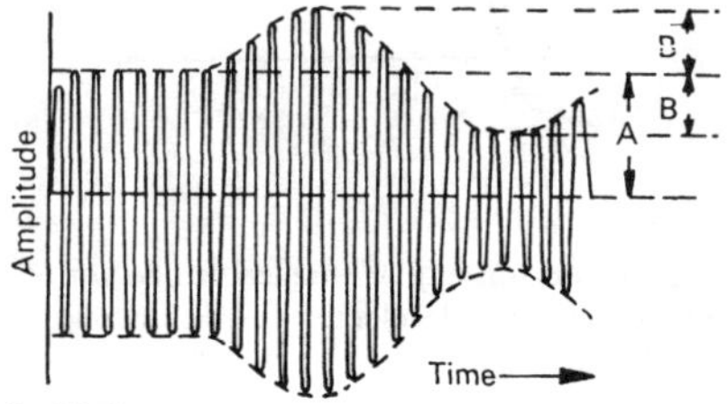

Decibels (dB)

The decibel is fundamentally a unit of power *ratio*, not of absolute power.

However, absolute power can be expressed in dB by assuming some standard power level reference and quoting dB above or below this reference.

The reference standard adopted in Britain is 1 mW.

Using this standard, any power can be found by the formula

$$10 \log_{10}\left(\frac{P}{1\,\text{mW}}\right) \text{dB}$$

e.g.

$$1 \text{ watt} = 10 \log_{10}\left(\frac{1{,}000}{1}\right) = 30\,\text{dB above } 1\,\text{mW}$$

Similarly,

$$5\,\mu\text{W} = 10 \log_{10}\left(\frac{5}{1{,}000}\right)$$

$$= -23\,\text{dB or } 23\,\text{dB below } 1\,\text{mW}$$

1 mW	=	0dB
2 mW	=	3dB
4 mW	=	6dB
5 mW	=	7dB
8 mW	=	9dB
10 mW	=	10dB

Q factor

'Q' is a measure of the selectivity of a tuned circuit.

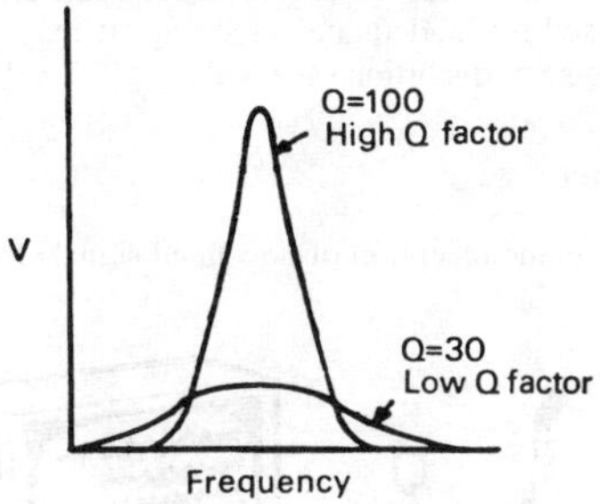

Bandwidth

The bandwidth of a circuit is the frequency range over which the response of the circuit does not fall below the response at a specified reference frequency by more than 3 dB.

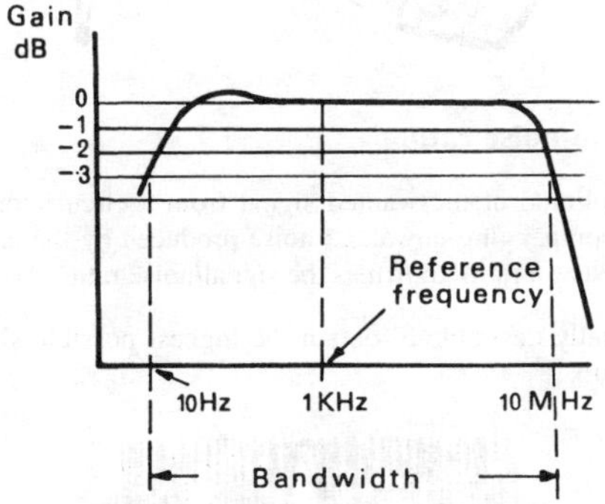

Power factor

The power developed in an a.c. circuit depends on the values of current and voltage and their phase difference. The cosine of the phase difference (Cos∅) is called the power factor. It can be said that the power factor represents the proportion of the current flowing which produces power in the circuit. In heavy electrical work the power factor is particularly important and has to be kept as near unity as possible.

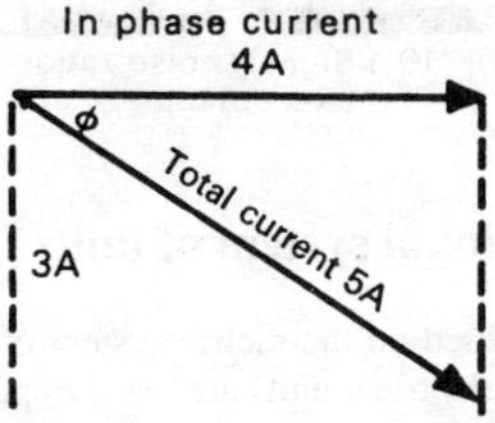

Crosstalk

Crosstalk is unwanted interaction between circuits. It is usually caused by inadequate screening. It may give rise to excessive noise or distortion of signals.

Interference

Interference is the reception of unwanted signals from external sources.

Signal-to-noise ratio

The amplitude of the wanted signal from a circuit compared with accompanying unwanted noise produced by the circuit is expressed as a ratio known as the signal/noise ratio.

It is usually desirable to obtain the highest possible signal to noise ratio.

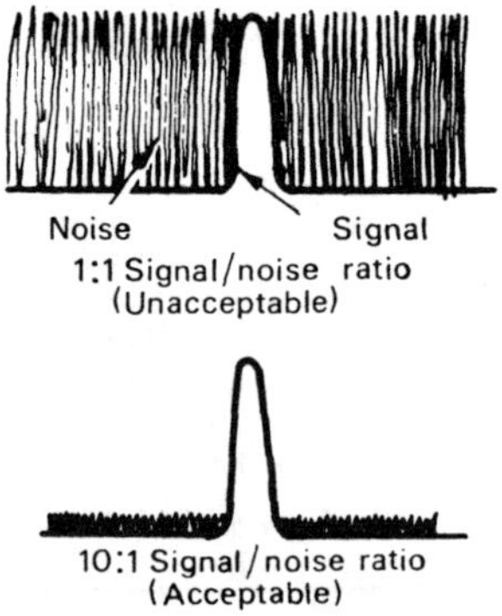

The international system of units

This system is based on the metric system of units. The most commonly used electrical units are listed opposite.

Quantity	Unit	Symbol
Frequency	Hertz	Hz
Work Energy Quantity of heat	Joule	J
Power	Watt	W
Quantity of electricity	Coulomb	C
Potential difference Electromotive force	Volt	V
Electric field strength	Volt per metre	V/m
Electric resistance	Ohm	Ω
Electrical capacitance	Farad	F
Magnetic flux	Weber	Wb
Inductance	Henry	H
Magnetic flux density	Tesla	T
Magnetic field strength	Ampere per metre	A/m
Magnetomotive force	Ampere	A

The names of multiples, and sub-multiples, of basic units are related to the unit and are formed by prefixes. The names of prefixes and the most commonly used examples are:

Factor by which unit is multiplied		Prefix	Symbol	Example
One million million (billion)	10^{12}	Tera	T	
One thousand million	10^{9}	Giga	G	GHz
One million	10^{6}	Mega	M	Megawatt
One thousand	10^{3}	Kilo	k	Kilometre
One hundredth (0.01)	10^{-2}	Centi	c	Centimetre
One thousandth (0.001)	10^{-3}	Milli	m	Milliampere
One millionth	10^{-6}	Micro	μ	Microampere
One thousand millionth	10^{-9}	Nano	n	Nanosecond
One million millionth	10^{-12}	Pico	P	Picofarad

Glossary

Acceptable Quality Level (A.Q.L.) The maximum number of defects per hundred units (%) of a batch sample which for purposes of sample inspection can be considered satisfactory as a true process average.

Acceptable Conforming to standards required.

Acceptance Sampling Plan The rules to be followed to determine the size of a sample or samples to be taken and the maximum number of rejects permitted for acceptance in any size of batch.

Average Outgoing Quality The average quality of the outgoing product, including all accepted lots or batches, all rejected lots or batches, and all replacements of defective items.

Batch (Lot) A quantity of a commodity, manufactured by one supplier under conditions that are presumed to be uniform.

Batch Size The number of units produced in a batch.

Classification Grouping into classes or divisions.

Classification of Defectives or of Defects Enumeration of possible defects of the unit of product, classified according to their importance, e.g. critical, major, minor.

Control Chart A chart, showing upper and/or lower control limits, on which values of some statistical measure are plotted, for a series of samples or sub groups. Frequently, a central line is shown to assist detection of a trend towards either control limit.

Control Limits Limits on a control chart used as criteria for action, or as indications of significant variations between samples.

Critical Defect A defect that judgement and experience indicate is likely to result in hazardous or unsafe conditions for individuals depending on, using or maintaining the product.

Critical Defective A unit of product that contains one or more critical defects.

Defect Any deviation of the unit of product from the standards or requirements laid down.

Defective, Defective Item A unit of product that contains one or more defects.

Defective Sample A sample of product that has failed to meet the standards or requirements laid down.

Defects per Unit The number of defects found in a unit of production.

Degree of Inspection The intensity of inspection. It is determined by the desired Outgoing Quality Level.

Double Sampling A type of sampling inspection in which a second sample may or may not be taken, depending on the result of the first.

Final Inspection The last of several inspections at successive stages of manufacture, repair or modification.

100% Inspection Inspection of each unit of product as opposed to any form of sampling inspection.

Inspection Determination of the quality of materials, components, assemblies or finished products for purposes of process control, product control, quality audit or fault diagnosis.

Inspection Level A specified level that governs the amount of inspection for a particular batch or lot size.

Lot Tolerance Per Cent Defective (L.T.P.D.) The number of defects per hundred units in an incoming batch, that the customer is willing to accept.

Major Defect A defect, other than a critical defect, that is likely to result in the failure of the unit of product, or to reduce materially its useability.

Minor Defect A defect unlikely to reduce materially the useability of the unit of product.

Minor Defective A unit of product containing one or more minor defects but no major or critical defects.

Normal Inspection Inspection that is used when there is no statistically significant evidence that the quality of the product being submitted is better or poorer than the specified quality level.

Preventive Inspection Control of Quality on the shop floor at each operation. It involves examination, feed back of information, and corrective action when necessary.

Process Average The average percentage defectives or the number of defectives per hundred units of product submitted by a supplier for original inspection.

Process Capability The inherent quality of manufacture achievable in a particular operation, as limited by current methods of production and which cannot be improved by techniques of process control.

Quality (of a unit of product) The degree to which a unit of product meets the requirements of the specification. It is a combination of the quality of design and the quality of manufacture.

Quality Assurance Overall supervision by the manufacturer of design and the quality control tasks to ensure that the quality required is obtained.

Quality Audit The monitoring of quality levels at any stage to provide information for management.

Quality Control Chart A chart showing upper and/or lower control limits, on which values of some statistical measure are plotted for a series of samples or sub groups. Frequently a centre line is shown to assist detection of a trend towards either control limit. (See Control Chart).

Quality Engineering The application of specialised managerial, scientific, and technological skills to achieve the desired quality at minimum cost.

Quality of Manufacture A measure of the fidelity with which the product, taken at the point of acceptance, conforms with design.

Random Without choice or control. A process of selection is termed random if each object has an equal chance of selection.

Random Sample A sample of n items taken from a population of N items in such a way that all possible combinations of n have an equal chance of being chosen.

Receiving Inspection Inspection by the customer of materials and manufactured goods as delivered by suppliers.

Reduced Inspection Provision in a sampling procedure for switching to a less severe sampling plan when the average quality offered is high.

Rejection Refusal to accept submitted materials or product.

Sample A group of items, taken from a larger collection of items, which serves to provide information needed for assessing the characteristics of the collection, or as a basis of action on the collection or the process that produced it. The term is also used to describe a portion of material taken for the sample purposes from a larger bulk of material.

To Sample The act of taking a sample.

Sample Size The number of units in the sample.

Sample Inspection Evaluation of the quality of material or units of a product by the inspection of samples, as distinct from 100% inspection.

Shipping Sample A sample of material or units of product taken from the bulk at the time of consignment.

Single Sample One group of items or individuals taken from a larger batch or lot to provide information needed for assessing the characteristics of the batch or lot.

Single Sampling A type of sampling inspection in which only one sample is taken per batch or lot.

Specification A detailed description of the properties and performances of the product including requirements and methods of test.

Specification Tolerance The tolerance quoted on the specification.

Statistical Quality Control The use of statistical methods, particularly frequency distribution charts and sampling plans, for the control of quality.

Tightened Inspection Provision in a sampling procedure for switching to a more severe sampling plan when the average quality offered is poor.

Tolerance The variability allowed in a process, or a characteristic of a unit of production.

Index

Cableforms, 218
  lacing, 223
  spot ties, 222
  straps, 225
Circuit diagrams see drawings
Company standards, 4
Components
  identification, 175
    diodes, 181
    silicon integrated circuits, 178
    thick film circuits, 180
    thin film circuits, 180
    thyristors, 182
    transistors, 175
  mounting,
    inspection, 169, 190
Connecting equipment, 250
  earth leads, 252
  monitor leads, 254
  power supply leads, 253
  signal leads, 254
Connectors, 226
  coaxial, 228
  crimped, 229
  electrical test, 230
Crimped joints, 214
  criteria of good joint, 215
  inspection,214
Diagrams see drawings
Drawings, 243
  conventions in circuit diagrams, 257
Equipment assembly, 233
  fixing devices, 233
  intimate contact of dissimilar metals, 237
  mechanical assembly, 238
Inspection, 1
  cableforms, 218
  component mounting, 190
  connectors, 226
  crimped joints, 214
  equipment assembly, 233
  final, 172
  printed circuit boards, 187
  procedures, 157
    acceptance, 159
    contracts and work orders, 162
    rejecting work, 157

schedules, 1
soldered joints, 198
techiques, 163
100%, 163
final, 164, 172
first-off, 166
patrol, 167
process control, 168
sample, 165
selective, 166
stages, 163
wires, 218
wrapped joints, 207
International standards, 3
Jigs
special tests, 263
Logic analysers, 154
Materials, recognition and application of, 182
insulating, 182
fibrous, 185
metals, 185
plastics, 183
taping, 184
Measurands, 271
amplitude modulation, 277
bandwidth, 279
capacitance, 273
crosstalk, 280
current, 271
decibels, 278
frequency, 274
frequency modulation, 277
inductance, 274
interference, 280
international system of units, 280
modulation, 277
modulation depth, 278
phase modulation, 278
potential difference, 271
power, 273
power factor, 279
pulse delay, 276
pulse repetition frequency, 276
pulse width, 276
Q factor, 279
resistance, 272
rise time, 276
signal-to-noise ratio, 280
time and period, 274
time constant, 276
Meters
A.C. ammeters and voltmeters, 32
accuracy, 47

anti-parallax mirror, 49
care and handling, 11
D.C. ammeters, 21
D.C. voltmeters, 27
E.H.T. multipliers, 46
electronic, 54
using, 60
extending measuring range, 29
moving-coil, 16
multipliers, calculating, 28
multirange, 38
controls, 40
terminals, 45
ohmmeters, 34
series type, 35
shunt type, 37
practical use, 72
resistance, 18, 19
effect of, 31
measurement, 50
sensitivity, 17, 19, 30
voltage and current measurement, 52
National standards, 3
Oscilloscopes
cathode-ray, 106
measuring
frequency, 138
rise time, 144
waveforms, 135
operating, 123
specification, 121
synchronising, 114
automatic, 132
test leads, clips and prods, 117
Printed circuit boards, 187
inspection, 187
Quality control, see inspection
Regional standards, 3
Signal generators, 91
audio frequency, 93
care and handling, 94
radio frequency, 92
test leads and terminations, 103
Soldering,
inspection, 169, 198
defects, 200
reflow, 205
Spectrum analysers, 147
Standards and specifications, 3
aims, 4
British Standards, 5
use, 6

Test
  instructions, 243
  special jigs, 263
Time domain reflectometers, 150
  metallic, 151
    textbook signatures, 152
  optical, 152
Tools, 7
Units, international system of, 280
Wires, 218
  identification, 220
Wrapped joints
  good, 208
  inspection, 169, 207
  reworked, 210
  stripping test, 211